변재택

케네디에서 큰 꿈을

만들기를 바라며, 이 책을

증정드립니다.

2023년 9월

오상익 드림

Overseas Energy Investment of Korea and Japan

Seong-ik Oh

Overseas Energy Investment of Korea and Japan

How did Two East Asian Resources-Rare Industrial Giants Respond to Energy Security Challenges

palgrave
macmillan

Seong-ik Oh
Ministry of Land, Infrastructure
and Transport
Sejong Special Self-Governing City
Korea (Republic of)

ISBN 978-981-99-0284-2 ISBN 978-981-99-0285-9 (eBook)
https://doi.org/10.1007/978-981-99-0285-9

This Palgrave Macmillan imprint is published by the registered company Springer Nature Singapore Pte Ltd.
The registered company address is: 152 Beach Road, #21-01/04 Gateway East, Singapore 189721, Singapore

For the late Judge Choon-ho Park,

I have seen farther by standing on the shoulder of this giant.

Acknowledgements

I have to begin by expressing my heartfelt gratitude to Dr. Kent E. Calder. As my Ph.D. advisor at SAIS, Johns Hopkins University, his intellectual stimulus and ceaseless motivation are precious during my academic journey. As one renowned expert on energy security and East Asian politics, Dr. Calder helped me to chart my path in the research and to reach ideas, many of which are reflected in this book. Not only did he broaden my academic horizons, he generously provided emotional support throughout my time in Washington D.C. and after.

Also, I would like to thank my professors who guided my academic journey. The Ph.D. dissertation committee members, Professors Charles F. Doran, Mitchell A. Orenstein and Jae J. Suh, offered me insightful advice and support. Dr. Keun-wook Paik deserves special mention for providing a steady flow of knowledge and insight into the energy industry through our conversation in person and over phone calls from London. Professors at Seoul National University, Fletcher School, Harvard Kennedy School, and SAIS, Professors Francis Fukuyama, Andrew C. Hess, Sung-yoon Lee, Meghan L. O'Sullivan, K. Mi Tak, Stephen M. Walt, the late Cei one Kim, and the late Alan M. Wachman, empowered me in this academic expedition.

The candid comments and suggestions by policymakers, scholars, researchers, and industry insiders in South Korea and Japan were invaluable in understanding the politics and the energy industry in each country. In particular, the face-to-face interviews with seniors, such as Mr. Won-chol O, the Second Senior Economic Advisor to President Park Chung-hee, were great opportunities for me to uncover the history of South Korea and Japan's overseas oil development. Their unstinting provision of information and materials helped me move forward without exhaustion. The support for my research is also given by institutes in Korea and Japan, such as the Korea Energy and Economy Institute (KEEI) and the Institute of Energy Economy, Japan (IEEJ). Furthermore, I took great pleasure in enjoying the vast knowledge accumulated in libraries, such as the National Library of Korea, the Presidential Archives of Korea, the National Diet Library of Japan, and the Library of Congress.

This book would not have seen the light of day without Dr. Se-hyun Ahn's encouragement that my Ph.D. dissertation will be a timely addition to the energy policy discussion under the current geopolitical situation. Editors at Palgrave Macmillan, including Mr. Nandakini Lahiri and Mr. Jacob Dreyer, have been indispensable partners in the effort of bringing this book to completion, and I appreciate their efforts.

Finally, my deepest thanks go to my family members in Korea, including my late grandfather Chan-young Oh, my late grandmother You-saeng Kang, my father Seung-jong Oh, my mother Kyung-soon Koh, my father-in-law Man-joon Yoon, and my mother-in-law Sei-hwa Choi. Most of all, I would like to thank my wife, Vhang-eun Yoon, and my son, Tae-keun Oh, who have shared the most treasured chapters of my life. While writing this book, Vhang-eun carefully read my manuscripts and gave me useful comments and constructive criticism. Her consistent encouragement and support, in good times and in bad, helped me to reach my destination of publishing this book.

Contents

About the Author

Seong-ik Oh received his bachelor's degree from Seoul National University and master's degrees from Harvard Kennedy School and Fletcher School, Tufts University, respectively. Also, he earned a Ph.D. from Johns Hopkins University's School of Advanced International Studies (SAIS). Continuing much of his career in public service, from the Korean Ministry of Maritime Affairs and Fisheries, the Ministry of Land, Infrastructure and Transport, to the Presidential Committee on Regional Development, he is currently a director general at the Ministry of Land, Infrastructure and Transport. He also teaches as adjunct professor at the Graduate School of Public Administration, Yonsei University.

Abbreviations and Terms

ANRE	Japan's Agency for Natural Resources and Energy
AOC	Arabian Oil Company
BAI	Korea's Board of Audit and Inspection
BOE	Barrels of Oil Equivalence
CCS	Chief Cabinet Secretary under Japan's Prime Minister
CEFP	Council on Economic and Fiscal Policy
CGP	Clean Government Party
Chaebol	Korea's big business
Cheongwadae	The Office of the South Korean President until May 2022
DCMC	Daehan Coal Mining Corporation
DPJ	Democratic Party of Japan
DRP	Democratic Republican Party of South Korea
DSP	Democratic Socialist Party of Japan
DUP	Democratic Unification Party of South Korea
E&P	Exploration and Production
EMRD	Energy & Mineral Resources Development Association of Korea
EPB	Korea's Economic Planning Board
FKI	Federation of Korean Industries
GNP	Grand National Party in South Korea
Himotsuki oil	Japanese refiners' tied oil purchase from Western oil majors as a return for a long-term loan at a favorable interest rate
Hinomaru oil	The oil developed by Japanese companies abroad
HOT	Head of Terms
JCP	Japan Communist Party

Jishukaihatsu oil	Another name for *Hinomaru* oil
JNOC	Japan National Oil Corporation, the successor to the JPDC
JOGMEC	Japan Oil, Gas and Metals National Corporation
JPDA	Japan Petroleum Development Association
JPDC	Japan Petroleum Development Corporation
JSP	Japan Socialist Party
Kantei	Japan's Office of the Prime Minister
KCIA	Korea Central Intelligence Agency
Keidanren	Japan Federation of Economic Organizations
Keiretsu	Japan's big business
KNOC	Korea National Oil Corporation
KOGAS	Korea Gas Corporation
KOSCO	Korea Oil Storage Corporation
KPA	Korea Petroleum Association
LDP	Liberal Democratic Party in Japan
MCI	Japan's Ministry of Commerce and Industry
MER	Korea's Ministry of Energy and Resources
METI	Japan's Ministry of Economy, Trade, and Industry (the successor to MITI)
MITI	Japan's Ministry of International Trade, and Industry (the successor to MCI)
MKE	Korea's Ministry of Knowledge Economy
MLIT	Japan's Ministry of Land, Infrastructure, Transport and Tourism
MOCIE	Ministry of Commerce, Industry and Energy (the predecessor to MKE)
MOF	Ministry of Finance
MOSF	Korea's Ministry of Strategy and Finance
MOTIE	Korea's Ministry of Trade, Industry and Energy (the successor to MKE)
MTI	Korea's Ministry of Trade and Industry (the predecessor to MOTIE)
NABO	National Assembly Budget Office of Korea
NDP	New Democratic Party in Korea
OPDA	Overseas Petroleum Development Association in Korea
PAJ	Petroleum Association of Japan
PARC	Policy Affairs Research Council in the LDP
PEDCO	Korea National Petroleum Development Corporation, former entity of KNOC
Self-sufficiency ratio	Originally, the ratio between the <u>developed & imported</u> by national companies (1) and total oil import (2). This has changed to the ratio between the oil that is <u>developed</u> by national companies (1') and total oil import (2). This is pronounced as *Jishukaihatsuritsu* in Japanese and *chachukaepalyul* in Korean.

SOEs	State-Owned Enterprises
SPR	Strategic Petroleum Reserve
Taegeuk oil	The oil developed by Korean companies abroad (newly invented for this research)

List of Figures

Chapter 1

Chapter 5

Chapter 7

Appendices

List of Tables

Chapter 5

Chapter 6

Appendices

1

Introduction

1 Puzzle for Analysis

This study begins with the empirical observation that Japan, when pursuing overseas oil resources, is private sector-oriented, while South Korea (hereinafter referred to as Korea) is relatively public sector-oriented. For oil development in foreign countries, Japan's overseas oil and gas development is led by the private sector with state's subsidization and research and development (R&D). Korea, on the other hand, has heavily invested in direct production of oil and gas through state-owned oil companies (SOEs).

As shown in Fig. 1, Japan relies entirely upon private firms to produce self-developed oil and gas. Unlike Japan, approximately 54% of Korea's self-developed oil and gas was contributed by the Korea National Oil Corporation (KNOC) and Korea Gas Corporation (KOGAS) in 2011. Also, in 2010, 89.9% of Korea's investment in overseas oil and gas development was led by these two SOEs.[1] In short, while Japan's overseas oil and gas development

[1] KNOC's homepage, the material of Korea's Energy and Mineral Resources Development Association, and Ministry of Knowledge Economy, "This Year Investment for Overseas Oil and Gas Development Amounts to $7.8 billion," *Press Release*, Jan 24, 2011. As of the end of 2010, the oil and gas developed by Korean companies amounted to 340 thousand BOE (barrels of oil equivalent)/day, of which KNOC produced 180 thousand BOE/day and KOGAS produced 16 thousand BOE/day. The investment by state-owned companies, KNOC and KOGAS, was $5.43 billion in the total investment of $6.04billion in 2010.

S.-i. Oh, *Overseas Energy Investment of Korea and Japan*,
https://doi.org/10.1007/978-981-99-0285-9_1

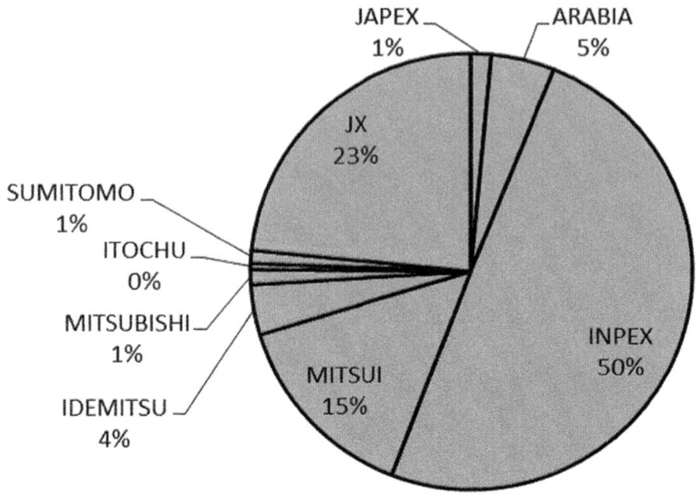

Fig. 1 Overseas oil and gas production by Japanese companies in 2011 (*Sources* Japan Petroleum Development Association, "Statistics: Production of Oil and Natural Gas" *Sekiyu Kaihatsu Jihō [Petroleum Development Magazine]*, May 2011, 68–71: Japan Petroleum Development Association, "Statistics: Production of Oil and Natural Gas" *Sekiyu Kaihatsu Jihō [Petroleum Development Magazine]*, May, 2012, 84–87. Legend: INPEX: INPEX Corporation, Mitsui: MOECO, Idemitsu: Idemitsu Oil, Mitsubishi: Mitsubishi Oil Exploration, Itochu: Itochu Corp, Sumitomo: Sumitomo Corp, JX: JX Nippon Oil and Gas Exploration Corporation, JAPEX: Japan Petroleum Exploration, ARABIA: Arabian Oil Company)

is operated by private firms, Korea's overseas oil and gas venture is mainly implemented by SOEs (Fig. 2).[2]

Since the idea to establish an SOE for overseas oil and gas development was raised in both countries, this study aims to determine why the state's efforts to deal with energy security by means of an SOE produced different outcomes in Japan and Korea.[3] Both Japan and Korea have attempted to establish a state-owned oil company for the purpose of securing oil supplies from overseas.[4] Unlike Korea, though, Japan did not establish SOEs for

[2] As mentioned in subsequent sections, this study defines a state-owned oil company as an oil company with very specific company law and where the state's ownership is more than half of the company's shares. In this study, national oil company means an oil company with a certain nationality, which is different from state-owned oil company. For example, Exxon Mobil is national oil company of the United States, but is not a state-owned oil company.

[3] This study follows Theda Skocpol's definition of a state. According to his research, a state is "a set of organizations through which collectivities of officials may be able to formulate and implement distinctive strategies and policies." Theda Skocpol, "Bringing the State Back In: Strategies of Analysis in Current Research," in Peter B. Evans et al., *Bringing the State Back In* (Cambridge, Cambridgeshire; New York: Cambridge University Press, 1985), 20–21.

[4] Both in Japan and Korea, ENI, which used to be a state-owned oil company in Italy, has been mentioned as a successful case among non-Western oil majors.

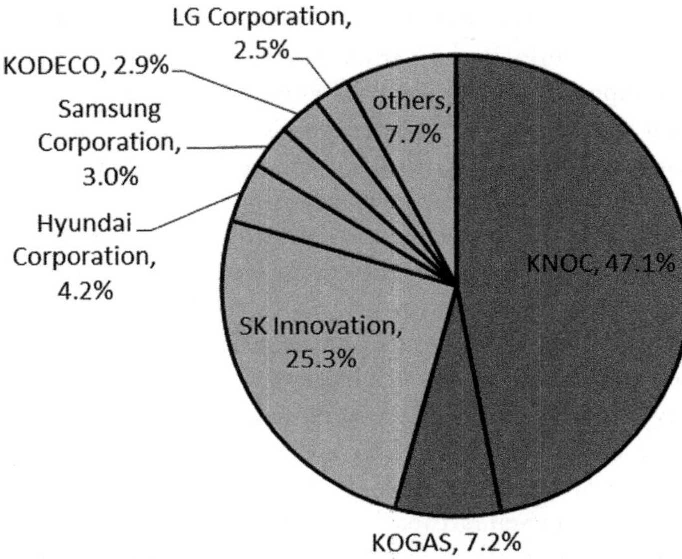

Fig. 2 Overseas oil and gas production by Korean companies in 2011 (*Source* Material of Energy and Mineral Resources Development Association of Korea [EMRD])

overseas oil development, depending instead on private firms. While both Japan and Korea are viewed as "strong states" among industrialized nations and lack domestic resource deposits, the role of the state in Japan has been restrained to financial support for private sector's overseas oil development, but Korea established the KNOC, which is a wholly owned SOE, and the KOGAS, whose majority share is owned by the public sector, including the central government, local governments, and other SOEs.[5] Furthermore, since their founding, the role of these SOEs has been strengthened overall. This phenomenon raises a question worthwhile of study: When two strong states wanted to establish SOEs for overseas oil development, why could one strong state establish and strengthen SOEs, while the other could not? In more general words, why have these two strong states had different results in terms of state capacity in using SOEs for overseas oil development?

This study seeks to understand the variations of the "developmental state" model by examining historical cases of overseas oil development policy.[6] The

[5] Strong states are states in which state leaders are able to utilize the agencies of the state to make people in the society do what they want them to do. Migdal exemplifies Japan, South Korea, North Korea, China, Cuba, Israel, Taiwan and Vietnam as strong states. See Joel S. Migdal, *Strong Societies and Weak States: State-Society Relations and State Capabilities in the Third World* (Princeton, NJ: Princeton University Press, 1988), xiii, xiv.

[6] Chalmers Johnson forms a thoughtful concept of the developmental state to differentiate Japan from traditional capitalist countries. He writes: "In states that were late to industrialize, the state

research focuses on the state's attempt to establish and strengthen state-owned oil companies to secure oil and natural gas supplies, as these attempts lie at the center of important theoretical debates on the state capacity of Japan and Korea. In that vein, this inquiry also examines why the Korean state was more successful than the Japanese state in translating the state's will to establish or strengthen state-owned oil companies into concrete policy outcomes.

This inquiry into state capacity would create an opportunity for academia to reconsider the persuasiveness of the developmental state argument, emphasizing the state's active role in economic activity. While Japanese bureaucrats, such as officials at the Ministry of International Trade and Industry (MITI) and its recent reincarnation, the Ministry of Economy, Trade and Industry (METI), have formulated and pursued national objectives in a strategic and developmental fashion, we have observed cases of incapacity in the postwar era including Japan's delayed response to the tsunami disaster in March 2011.[7] In Korea, the state's capacity in economic development has been visible even in the aftermath of the Asian Economic Crisis of 1997. Previous literature provides some insight into how and when state industrial strategists were been able to achieve their designs.

2 Prevailing Explanations

Previous studies of state capacity examined the financial resources and instruments that states can use to solve particular problems. These studies are expanded upon here to include the relational perspective to investigate the state's power over domestic actors. The standard statist argument holds that the state has not only the ability to impose these preferences upon the public in the face of societal resistance but also its own interests and policy preferences. This argument emphasizes that state capacity is proportional to the state's autonomy vis-à-vis the private sector.

itself led the industrialization drive, that is, it took on developmental functions. These two differing orientations toward private economic activities, the regulatory orientation and the developmental orientation, produced two different kinds of business-government relationships. The United States is a good example of a state in which the regulatory orientation predominates, whereas Japan is a good example of a state in which the developmental orientation predominates." See Chalmers A. Johnson, *MITI and the Japanese Miracle: The Growth of Industrial Policy, 1925–1975* (Stanford, CA: Stanford University Press, 1982), 19. Atul Kohli calls it a cohesive-capitalist state. See Atul Kohli, *State-Directed Development: Political Power and Industrialization in the Global Periphery* (Cambridge, UK; New York: Cambridge University Press, 2004), 10.

[7] Following the tsunami disaster that ensued after the Tōhoku earthquake on March 11, 2011, it has taken more than 100 days for Japan to enact a special law to revive Fukushima Prefecture and eleven months to launch the Agency of Reconstruction.

Within this state-centric approach, Chalmers Johnson is well-known for developing the notion that Japan is a "developmental state" and that its government is the central agent of economic transformation in *MITI and the Japanese Miracle*.[8] Johnson's state-oriented analysis carefully documents the role of MITI in transforming Japan's industrial structure. The state dominance described in his book has been cited as a major piece of evidence to support prevailing notions of Japan as a "strong state." Johnson's thesis that applies the state autonomy argument has also been emphasized by John Waterbury and Ezra N. Suleiman. Waterbury emphasizes that a "change team," a small number of decision-makers insulated from pressure, can formulate and implement key policies.[9] Like Johnson, Suleiman illustrates how the French administrative state plays an important role in policymaking.[10]

Scholars such as Jones and Sakong and Alice Amsden have explained a similar thesis to Korea, explaining the significant role of the Korean state in economic development.[11] In particular, Amsden, in *Asia's Next Giant*, illustrates how the state in South Korea facilitated the process of technological progress and industrial transformation.

However, the state-centric argument cannot explain the mystery of why two strong states, Japan and Korea, experienced different outcomes in overseas oil development in projecting the state's will to realize SOE in their economy. The statist argument would lead us to believe that both states of Japan and Korea would have been able to impose their will to establish and strengthen SOEs for overseas oil development. In other words, the state-centric argument cannot answer why the Japanese government, the central agent of economic transformation in Johnson's terms, could not make its will realized even in the 1960s, the era in which Japan was a developmental state. This gap in the state-centric argument can be attributed to the fact that statists underestimate how political parties, the legislature, and interest group politics tie into economic policy formation. In the case of Johnson, while the role of the legislature is reduced to "safety valve function," which provides

---

[8] Johnson, *MITI and the Japanese Miracle: The Growth of Industrial Policy, 1925–1975*.

[9] John Waterbury, *Exposed to Innumerable Delusions: Public Enterprise and State Power in Egypt, India, Mexico, and Turkey*, Political Economy of Institutions and Decisions (Cambridge, UK; New York, NY, USA: Cambridge University Press, 1993).

[10] Ezra N. Suleiman, *Bureaucrats and Policy Making: A Comparative Overview* (New York: Holmes & Meier, 1984).

[11] Leroy P. Jones and Il SaKong, *Government, Business, and Entrepreneurship in Economic Development: The Korean Case*, Studies in the Modernization of the Republic of Korea, 1945–1975 (Cambridge, MA: Council on East Asian Studies, Harvard University: distributed by Harvard University Press, 1980); Alice H. Amsden, *Asia's Next Giant: South Korea and Late Industrialization* (New York: Oxford University Press, 1989).

bureaucrats with enough space to take initiative and operate effectively, the dominance of MITI over private-interest groups has been overly emphasized. Pluralists, such as Karel van Wolferen, assert that the Japanese government apparatus is fragmented and therefore Japanese ministries could not always realize their own preferences.[12]

One possible answer to the puzzle might be found by examining the relationship between the state and business in economic development. In that regard, it is useful to review some challenges to the state autonomy argument. Peter Evans argues that if state embeddedness means effective communication between the state and society, state autonomy does not necessarily correlate with state capacity. By examining the information technology (IT) industry of newly industrializing countries, he argues that the state must be embedded in the market in order for it to be effective since markets work only if they are connected in other forms of social relations. Taking the state's involvement for granted, he argues "what kind" of state involvement depends on the state structure, the context, and implementation.[13]

However, Evans' expectations for overseas oil development would be different from what actually happened in the early twenty-first century in Korea, in which the state strengthened the demiurge role by heavily investing in its state-owned oil companies. Following his argument, Korea would have played a husbandry role by taking over risk in high-risk oil exploration and development in foreign countries for the benefit of the private sector.[14] This historical fact does not fit his idea that state's role of producer cannot be easily realized without knowing that the business has a limited capital, which is different from the situation in Korea's private sector in the early twenty-first century.[15]

The second approach is that state capacity cannot be materialized or expanded without considering the capacity of the private sector. This literature was initiated by scholars of the new generation in Japan Studies, such as Kent Calder, Richard Samuels, and Daniel Okimoto who counter Johnson's state-centric thesis of the developmental state. Calder analyzes Japan's financial policy, a significant intersection of state and market. Criticizing Johnson's

[12] Karel van Wolferen, *The Enigma of Japanese Power: People and Politics in a Stateless Nation*, 1st ed. (New York: A.A. Knopf, 1989).

[13] Peter B. Evans, *Embedded Autonomy: States and Industrial Transformation*, Princeton Paperbacks (Princeton, NJ: Princeton University Press, 1995), 11. For the meaning of the terms on the role of state, see ibid., 13–14, 77–81.

[14] Evans describes husbandry role as "cajoling and assisting" the private firms to meet global challenges, which ranges from simple signaling to some measures like establishing public organizations to take over risky business. Ibid., 14.

[15] Ibid., 13.

failure in precisely observing Japan's political economy of the postwar period, Calder's detailed historical research is compelling because it shows that strategic moves by the private sector, rather than by the government's administrative guidance, were the main drivers of economic development in Japan. Based on more accurate and affluent research outcomes, he counters Jonson's state-centric argument in Japan Studies by labeling Japan as "corporate-led strategic capitalism."[16]

Introducing the concept of "reciprocal consent," Samuels makes a similar argument that policy outcome is the result of negotiation between state and business. He argues that in Japan's energy industry, the Japanese state's intended commercial role has been impeded as a result of the state's negotiations with the private sector.[17] However, the prediction implied from his argument could not explain the phenomena in the twenty-first century. Japanese government could not reach its goal of establishing a state-owned oil company, even despite the fact that when *Keidanren* is not as powerful as it was in the 1960s; the Korean government, on the other hand, could strengthen existing state-owned oil companies, KNOC and KOGAS, even when Korea's *chaebol* becomes a more powerful actor in Korean economy.

Emphasizing the broader social foundation of MITI's industrial policy, Okimoto, in *Between MITI and the Market*, also asserts that big business, industrial associations, and mixed public–private entities improve the effectiveness of Japan's industrial policy. However, he considers the process of negotiation among political parties and interest groups has little impact on industrial policy formation and does not research such phenomena in detail. While this literature on the role of the private sector expands the horizon of analysis beyond the state-centric argument, this approach is not satisfactory in explaining the divergence in the realization and strengthening of SOEs in Japan and Korea.

The gap in the revisionist approach comes by seeing political economy from the perspective of the private sector. In particular, in Japan studies, while *Keidanren* in the 1960s was exceptional enough to be counted as a veto player, it needs to be clear what pressure groups are and are not veto players in the decision-making system. Considering the importance of legitimacy in policymaking, especially in Japan and Korea, this weakness is all the more noticeable.[18] The legislatures in these two "developmental states" have

---

[16] Kent E. Calder, *Strategic Capitalism: Private Business and Public Purpose in Japanese Industrial Finance* (Princeton, NJ: Princeton University Press, 1993).

[17] Richard J. Samuels, *The Business of the Japanese State: Energy Markets in Comparative and Historical Perspective*, Cornell Studies in Political Economy (Ithaca: Cornell University Press, 1987).

[18] For Max Weber, the modern state is characterized primarily by appeals to rational-legal authority, whereby legitimacy is derived from the belief that a certain group has been placed in power in a legal

gradually gained more power in policymaking through the power to check the executive by its legislative power and budget control, and thus the policymaking process is associated with politics more closely than before.[19] This requires a fuller and more satisfactory account of the executive branch, law makers, bureaucrats, and recognized interest groups who have direct access to the decision-making process and represent interests of societal actors. For this reason, it is necessary to consider the effect of political structure on policy outcomes.

While the literature on political structure has mainly focused on the effect of different constitutional structures on regime stability, accountability, and responsiveness,[20] a few scholars shed light on how and when institutions affect policy outcomes. For example, Shepsle argues that institutional structure significantly affects the existence of equilibrium in the social choice process, and the structure, together with the distribution of preferences, decides what features will be included in the equilibrium state of the collective choice processes.[21] Shugart and Wattenberg also examine how economic policy is affected by the transition of constitutional reform to a mixed electoral system in the 1990s in a few countries, including Britain and Canada.[22] In the literature of political structure and policy outcomes, most research on democratic institutions has been on the institutional checks among legislative, executive, and judicial powers. Here, veto player theory is the recent theoretical work that is worth noting.

---

manner, and that their actions are justifiable according to a specific code of written laws. Richard Swedberg and Ola Agevall, *The Max Weber Dictionary: Key Words and Central Concepts* (Stanford, CA: Stanford Social Sciences, 2005), 148.

[19] Kim Gwang-rim, current member of the National Assembly and former vice minister of Finance and Economy, mentioned that ¾ of power in economic policymaking is distributed to the legislature and ¼ to the government in Korea. See Kyung-ho Seo, "Dominant Power in Economic Policy-Making Transferred from Government to Legislature," *Korea JoongAng Daily*, May 28, 2012.

[20] Examples include: Arend Lijphart, *Parliamentary Versus Presidential Government*, Oxford Readings in Politics and Government (Oxford; New York: Oxford University Press, 1992); Matthew Soberg Shugart, "Leaders, Rank and File, and Constituents: Electoral Reform in Colombia and Venezuela," *Electoral Studies* 11, no. 1 (1992); Juan J. Linz and Arturo Valenzuela, *The Failure of Presidential Democracy* (Baltimore: Johns Hopkins University Press, 1994); Giovanni Sartori, *Comparative Constitutional Engineering: An Inquiry into Structures, Incentives, and Outcomes* (New York: New York University Press, 1994).

[21] Kenneth A. Shepsle, "Institutional Arrangements and Equilibrium in Multidimensional Voting Models," *American Journal of Political Science* 23, no. 1 (1979).

[22] Matthew Soberg Shugart and Martin P. Wattenberg, *Mixed-Member Electoral Systems: The Best of Both Worlds?* Comparative European Politics (Oxford, UK New York: Oxford University Press, 2001).

3 Veto Player Theory and Hypothesis

Veto player theory provides an intuitively strong and theoretically relevant explanation of how political structure strengthens or restrains state capacity, the ability of a state to formulate and implement policies. This strength is particularly noticeable considering that political analysis has been overwhelmed by a simple dichotomy between presidentialism and parliamentarism, an approach that provides only case-by-case empirical conclusions. Veto player theory then becomes a platform from which analyze political structure integrating diverse institutions, such as the government, legislature, and parties.

According to George Tsebelis, veto players are "individual or collective actors whose agreement is necessary for a change of the status quo."[23] He mainly divides veto players into two categories: institutional veto players specified by the constitution, and partisan veto players specified by the political system. However, Tsebelis is cautious about the existence of other potential veto players such as influential ministers or powerful interest groups, at least within their respective policy areas of concern.[24] Each veto player has its own policy preference. The level of policy changeability depends on the winset size of the status quo, or the set of policies that can provide a substitute for the existing one. The size of the winset is affected mainly by the number of veto players or the diversity of veto players. The larger the winset is, the higher possibility of change from a given status quo.[25] In order to compare the size of the winsets, which significantly affect the level of decisiveness in decision-making, the number of "effective veto players" needs to be considered by analyzing the number of veto players who have distinct preferences, which is called the "absorption rule."

While Tsebelis established a useful framework with which analyze the political system's capacity to solve problems when they arise, he leaves a few critical points on how to operationalize the theory unanswered. In that regard, this research attempts to bridge the gaps between existing veto player

[23] George Tsebelis, *Veto Players: How Political Institutions Work* (Princeton, NJ: Princeton University Press, 2002), 19. There are earlier literature applying veto player theory: Ellen M. Immergut, *Health Politics: Interests and Institutions in Western Europe*, Cambridge Studies in Comparative Politics (Cambridge, UK; New York, NY, USA: Cambridge University Press, 1992); Evelyne Huber, Charles Ragin, and John D. Stephens, "Social Democracy, Christian Democracy, Constitutional Structure, and the Welfare State," *American Journal of Sociology* 99, no. 3 (1993). In the case of Immergut, the concept of "veto gate" is utilized to analyze policy change.

[24] Tsebelis, *Veto Players: How Political Institutions Work*, 81. And George Tsebelis, "Decision Making in Political Systems: Veto Players in Presidentialism, Parliamentarism, Multicameralism and Multipartyism," *British Journal of Political Science* 25, no. 3 (1995): 306.

[25] Tsebelis, *Veto Players: How Political Institutions Work*, 19–21.

theory and its applicability to case studies. In spite of the theory's merits in comparative politics, insufficient details in operationalizing the theory have caused inconsistent application of the theory among scholars. Even worse, Tsebelis' application of the theory to Japan concludes that Japan is more subject to change, which does not reflect the reality of Japan's policymaking. This implausible prediction is the result of the application of the theory that is not sophisticated enough; it sees that Japan has only one veto player, the LDP.[26]

This analysis requires us to refine Tsebelis' model in an effort to more clearly predict the direction of policy change. One main issue is the identification of veto players. We need to look at both vertical and horizontal dimensions in the decision-making system by examining the scope of group of individuals that can be regarded as collective veto players (e.g., Congress, a chamber, or committees of chamber) and whether governmental ministries in the executive can be counted as veto players in the vertical dimension, as well as whether certain influential actors, such as interest groups, can be included as veto players in the horizontal dimension.[27]

In the vertical dimension, the subunits of the political organization will be counted as veto players if they have effective veto power in the policymaking process. For example, committees in one legislative chamber and a certain relevant ministry can be distinguished from the whole chamber or the executive. Japan and Korea both adopt the committee system of the United States, where the committee has substantial deliberation authority. Also, agenda power has been explicitly delegated to ministries through specific laws, and in East Asian developmental states, where governmental agencies have played a significant role in policymaking, this approach makes more sense.[28] If a party has multiple important councils that can block decision-making procedures, those respective councils can also be counted as veto players.

In the horizontal dimension, one ambiguity in Tsebelis' model is whether interest groups are veto players because they don't have formal veto power but exert significant influence in contemporary policymaking.[29] In Japan studies,

---

[26] See Tsebelis, "Decision Making in Political Systems: Veto Players in Presidentialism, Parliamentarism, Multicameralism and Multipartyism," 317.

[27] This issue is mentioned as "problem of identification" in Steffen Ganghof, "Promises and Pitfalls of Veto Player Analysis," *Swiss Political Science Review* 9, no. 2 (2003): 3.

[28] Laver and Shepsle support this stance by arguing that ministers have exclusive jurisdiction over their respective policy areas, even in a coalition government, where constraint is imposed by the partners of the government coalition. See Michael Laver and Kenneth A. Shepsle, "Coalitions and Cabinet Government," *The American Political Science Review* 84, no. 3 (1990): 873–90.

[29] The only thing Tsebelis mentions is that "powerful interest groups" can be veto players. This identification is deficient, though, and causes some authors to differentiate interest groups from veto players for a systematic analysis. One example is Witold J. Henisz and Bennet A. Zelner, "Interest

it is unclear how interest group politics connect to economic policymaking,[30] and this problem needs to be resolved by applying veto player theory. In other words, how can the institutional structure of the policy process offer a certain pressure group veto power even if the group's consent is not formally asked when a bill becomes a law? In order to clarify which interest groups are veto players, I borrow the concept of the insider/outsider, whose main differentiation is the recognition by the government.

According to Grant, some insider groups are members of high-level policy committees and are usually asked by the government, thereby having influence over the government's policymaking.[31] In Japan, offering representation with *shingikai*, the ministries openly agree that certain interest groups are core actors whose involvement in policymaking is legitimate and whose preferences must be taken into consideration.[32] In return for this honored access, insider groups provide information to the government and work together in policy implementation, thus creating a lower compliance cost in policymaking.[33] On the other hand, outsider groups are either unable or unwilling to be embedded in a consultative relationship with government officials. This study approximates that veto player interest groups are insider groups according to Grant's terms. The proposition of an insider is the recognition by the government, rather than the interest group's own assessment of its role and that of insider contacts. The reason is that the government's recognition provides political legitimacy for interest groups in policymaking and the interest groups' own assessment tends to be exaggerated. However, this bilateral relationship needs to be fine-tuned by considering its effective role in policymaking.

While veto player theory itself does not display policy direction, the theory has been extended beyond notions of policy decisiveness and resoluteness to explain many specific policy outcomes. If we identify the preferences of veto players, the status quo, and the location of the agenda setter, we can predict the results of policymaking efforts. This is the utility of the

---

Groups, Veto Points, and Electricity Infrastructure Deployment," *International Organization* 60, no. 1 (2006).

30 Calder, *Strategic Capitalism: Private Business and Public Purpose in Japanese Industrial Finance*, 10–11; Robert M. Uriu, *Troubled Industries: Confronting Economic Change in Japan*, Cornell Studies in Political Economy (Ithaca, NY: Cornell University Press, 1996), 252.

31 Wyn. Grant, *Pressure Groups, Politics and Democracy in Britain* (New York: Philip Allan, 1989).

32 Frank J. Schwartz, *Advice and Consent: The Politics of Consultation in Japan* (Cambridge, UK; New York: Cambridge University Press, 1998), 284.

33 Buchannan and Tullock assert that ideal institutional design needs to be based on a minimization of the sum of compliance costs and transaction costs. See James M. Buchanan and Gordon Tullock, *The Calculus of Consent, Logical Foundations of Constitutional Democracy* (Ann Arbor: University of Michigan Press, 1962).

theory. In particular, the link between the number of veto players and the economic policy outcomes has been researched in a few economic policy arenas: capital account liberalization by Kastner and Rector, trade liberalization by O'Reilley, Henisz, and Mansfield, and electric utilities infrastructure project by Henisz and Zelner.[34] Application of the theory is expanded to the correlation between the number of veto players and crisis responses by MacIntyre and even to the length of civil war by David Cunningham.[35]

Furthermore, the application of the theory is not limited to the legislative process in policymaking.[36] In Japan, important administrative guidance of ministries also needs to be consulted with the ruling party, and thus administrative measures, which evade the review process in the legislature, can also be analyzed through the framework of veto player theory.[37] If we recognize that the theory can be applied from the perspective of partisan veto players as well as from institutional veto players, the application of the theory in analyzing policy change is significantly relevant.

However, there are limitations that need to be noted. In particular, the theory cannot travel "politics in hard times" well. During crisis, unusual flexibility of policymaking has been observed. From the perspective of veto player theory, this can be explained in that veto power of veto players tends to be halted, and this situation makes decisive policy changes possible.[38] Therefore, the theory has limits in predicting this possible oscillation of policy change during crisis based on the number of veto players.

Utilizing veto player theory, this study hypothesizes that if there are more effective veto players, the state's capacity to formulate and implement policies is likely to be lower. Applying the number of effective veto players is more feasible than using the nominal number of veto players, considering the

---

34 Scott L. Kastner and Chad Rector, "International Regimes, Domestic Veto-Players, and Capital Controls Policy Stability," *International Studies Quarterly* 47, no. 1 (2003); Robert F O'Reilley, "Veto Points, Veto Players, and International Trade Policy," *Comparative Political Studies* 38, no. 6 (2005); Witold J. Henisz and Edward D. Mansfield, "Votes and Vetoes: The Political Determinants of Commercial Openness," *International Studies Quarterly* 50, no. 1 (2006); Henisz and Zelner, "Interest Groups, Veto Points, and Electricity Infrastructure Deployment."

35 Andrew J. MacIntyre, *The Power of Institutions: Political Architecture and Governance*, Cornell Studies in Political Economy (Ithaca: Cornell University Press, 2003); David E. Cunningham, "Veto Players and Civil War Duration," *American Journal of Political Science* 50, no. 4 (2006).

36 Some scholars have different opinions from the author on the relevance of veto player theory beyond the legislative process in comparing Japan and Korea. See Heon-joo Jung, "Reassessing Reform Outcomes in South Korea and Japan a Decade After Crisis: The Paradox of Political Leadership" (Thesis Ph.D.), University of Pennsylvania, 2008), 264–65.

37 Interview with a METI official, conducted on March 4, 2012.

38 In evaluating the applicability of veto player theory, the definition of crisis is stricter than normally thought. Calder shows a few indicators to distinguish a crisis from a non-crisis. See Kent E. Calder, *Crisis and Compensation: Public Policy and Political Stability in Japan, 1949–1986* (Princeton, NJ: Princeton University Press, 1988), 71–72.

possibility of a divided government. The establishment and strengthening of SOEs cannot be easily understood in terms of models where state capacity is determined only by the level of autonomy or only by the government's relationship with the private sector. On the contrary, the number of effective veto players is crucial in explaining the realization of these policy changes. Veto players have different preferences, and the number of veto players affects the winset or the alternative policy outcome. This results in decisiveness and resoluteness of policy change from the status quo. In this vein, if Japan, the nation with more veto players, cannot achieve the intended establishment or strengthening of an SOE, and if Korea, the nation with fewer veto players, is able to realize the momentous change from a situation where there is no SOE or one with moderate power, veto player theory can explain the puzzle this study examines.

In this study, state capacity, the dependent variable, is measured by the realization of SOEs for overseas oil development. There is no fixed definition of an SOE; therefore it is important to define SOE for the purpose of this research. This study uses a plausible definition in terms of ownership and legitimacy. One attribute of an SOE is whether the government owns more than half the shares of the company, either directly or indirectly. The majority provides the state with the authority to control the company for the sake of national interests at the cost of profit maximization.[39] The other component of an SOE is a specific establishment law that confers legitimacy and requires the company to be accountable to the legislature or relevant ministries. Sometimes, the state is able to evade the legislature by founding quasi-SOEs to achieve the state's intention. However, the quasi-SOE is vulnerable to accusations from the public and may have vague structures of accountability. For these reasons, this study defines a state-owned oil company as one in which the state owns over half of its shares and that has very specific company law.

4 Methodology

4.1 Case Study

The application of veto player theory requires knowledge of the details of the political process of decision-making. Veto player analysis takes into account

---

[39] Valerie Marcel studies state-owned oil companies in oil producing countries with the following definition: A company with majority state ownership that is active in the exploration and production of oil and natural gas. See Valérie Marcel and John V. Mitchell, *Oil Titans: National Oil Companies in the Middle East* (London, Baltimore: Chatham House; Brookings Institution Press, 2006), 5.

the positions and preferences of each actor in the decision-making process, so the accuracy of analysis increases as more precise policy preferences are introduced in the data. Therefore, this study explores the cases of Japan and Korea by examining how the veto players in overseas oil development interact sequentially based on their preferences. Specifically, the case study focuses on how, in each country, the state's efforts to establish state-owned oil companies for overseas oil production is constrained or stimulated in the political process of law making and budget allocation.

For the purpose of this research, four historic incidents (or events) in Japan and Korea's oil development history were selected, respectively: (1) the 1960s and the early twenty-first century in Japan and (2) the 1970s and the early twenty-first century in Korea. In these periods, both Japan and Korea faced a critical need to manage their respective energy issues, and as a result, both states looked into establishing (or strengthening) a state-owned oil company. The 1960s and 1970s in Japan and Korea, respectively, were developmental phases for each nation, in which states are perceived to wield strong power for economic development. The twenty-first century is relevant for both countries because it was during this time that politics became more diverse and democratized than before and economies reached industrialized levels. Testing the hypothesis in these various time periods enhances the persuasiveness of veto player theory in explaining the deviation of the state's involvement in overseas oil development in both Japan and Korea.

Yet, it is important to note that each case generated totally different outcomes: The pursuit of a state-owned oil company ended in vain in Japan, while Korea established and strengthened its SOE. This outcome has set and reinforced the different paths for oil development policy in Japan and Korea—in terms of overseas oil development, Japan is more private sector-oriented while Korea is more public sector-oriented.

4.2 Case of Realization of SOEs for Overseas Oil Development

Energy security is one area in which governments frequently attempt to intervene due to their strategic concerns. This is especially true for industrialized nations, such as Japan and Korea, since their dependence on foreign oil and natural gas creates economic security concerns. In addition, the state has an advantage over businesses in conducting foreign relations with the governments of oil producing countries and their state-owned oil companies. This gives the state another reason to remain deeply involved in securing an energy supply. As Samuels points out, the energy industry has been an "early target for state intervention,"[40] and this is still true in the contemporary world.

Overseas oil development has been introduced in Japan and Korea to ease each nation's difficult positions and enhance predictability and stability. States utilize overseas oil development for energy security for the following reasons: In times of emergency, host countries are expected to prioritize the supply of oil and natural gas to investing countries, while at other times, national companies gain capacity to explore and develop resources by themselves.[41] Japan and Korea, both of which suffered from a lack of energy resources during the oil shocks of the 1970s, have pursued oil ventures in foreign countries.

This study focuses on the establishment and strengthening of an SOE for overseas oil development to sharpen general ideas on the relationship between political structure and policy outcome. Both countries have attempted to establish and strengthen state-owned oil companies in an effort to achieve energy security since the government can support public firms more easily than private ones and manage it for the public purpose of enhancing oil self-sufficiency ratio. At the same time, founding and reinforcing an SOE is an explicitly political economic agenda, because there tend to be conflicts of interest between the state and business. In particular, in all cases the issue is subject to review by the legislature to prevent taxes from being misused and the public sector from over-expanding, and this becomes significant in determining state capacity in democratic countries.

4.3 Comparison of Japan and Korea

Japan and Korea are compared in this study due to their societal similarities, which is considered sufficient from a methodological perspective. In other words, the situation of these two countries provides researchers with conditions that are close to *ceteris paribus*, where John Stuart Mill's method of difference can be applied in the inquiry of social science.[42] Both

[40] Samuels, *The Business of the Japanese State: Energy Markets in Comparative and Historical Perspective*, 18.

[41] The specific arrangements between the host country and investing country would depend on various factors, such as contract, the capacity of the host country and investing country, and the circumstances at that moment.

[42] Gary King, Robert O. Keohane, and Sidney Verba, *Designing Social Inquiry: Scientific Inference in Qualitative Research* (Princeton, NJ: Princeton University Press, 1994), 168. John Stuart Mill expresses the Method of Difference as follows: "If an instance in which the phenomenon under investigation occurs, and an instance in which it does not occur, have every circumstance in common save one, that one occurring only in the former; the circumstances in which alone the two instances differ, is the effect, or an indispensable part of the cause, of the phenomenon." John Stuart Mill, *A System of*

nations are highly homogeneous with few distinctive ethnic or class divisions to polarize society. Japan and Korea achieved economic development following the devastation of World War II and the Korean War, respectively. In their economic development, the state has traditionally loomed large in economic life, and many scholars, such as Chalmers Johnson, ascribe the successful economic development of both countries to government intervention.[43] Strong bureaucracies, as seen in the case of MITI in Japan or the Economic Planning Board (EPB) in Korea, played substantial roles in the economic development of each country. Moreover, the practice of reemploying retired bureaucrats has existed under the guise of *amakudari* (descent from heaven) in Japan and *nakhasan* (parachute) in Korea.[44] Also, big businesses in each country (Japan's *keiretsu* and Korea's *chaebol*) have formed the core of the private sector in both economies.

A few shared characteristics control the possible key variables that can affect policy outcomes. The importance of an informal network in the daily lives of citizens in Japan and Korea has been considered one big difference between those two nations and Western industrialized nations and a significant independent variable.[45] The informal network that is established through school ties and family ties has been associated with formal networks and has been strengthened. The close relationship with the United States, including a strong military alliance, has provided both Japan and Korea with regional stability, which is one significant condition for their economic development (Table 1).

Furthermore, Japan and Korea are equally deficient in natural resources, including petroleum as well as their dependence on petroleum imports to meet energy demands. Japan is the second-largest net importer of oil, and Korea is the fifth largest, as of 2009. Also, these countries have made efforts to lessen their heavy dependence on crude oil by importing liquefied natural

---

Logic, Ratiocinative and Inductive: Being a Connected View of the Principles of Evidence and the Methods of Scientific Investigation, 8th ed. (New York: Harper & Brothers, 1874), 280.

43 Scholars, such as Atul Kohli, emphasize the lineage between Japanese and Korean developmental states. See Atul Kohli, "Where Do High Growth Political Economies Come From? The Japanese Lineage of Korea's 'Developmental State'," *World Development* 22, no. 9 (1994). While agreeing on the effect of Japanese developmental state to Korean one, Eun-mee Kim argues the effect of Japanese colonial legacy is not direct, but indirect and circumstantial. See Eun Mee Kim, *Big Business, Strong State: Collusion and Conflict in South Korean Development, 1960–1990*, Suny Series in Korean Studies (Albany: State University of New York Press, 1997), 40–41.

44 For details on this similarity, see Kent E. Calder, "Elites in an Equalizing Role: Ex-Bureaucrats as Coordinators and Intermediaries in the Japanese Government-Business Relationship," *Comparative Politics* 21, no. 4 (1989): 394–97.

45 Informal networks have been emphasized in a number of works on Japan studies. One recent example is Jennifer Ann Amyx, *Japan's Financial Crisis: Institutional Rigidity and Reluctant Change* (Princeton: Princeton University Press, 2004).

Table 1 The similar energy situation in Japan and Korea as of 2009

	Type of resources	Japan	Korea
Proven reserves	Petroleum	0.04 billion barrel	0
	Natural Gas	738 bcf	0
Production	Crude Oil	5.33 thousand b/d	0
	Natural Gas	181.10 bcf	22 bcf
Imports	Petroleum	4,261.12 thousand b/d	2,142.17 thousand b/d
	Natural Gas	3,280.62 bcf	1,154.48 bcf
Dependence on the Middle East in 2011	Crude Oil	Approximately 90%	Approximately 87%

Sources U.S. Energy Information Administration, "Countries: Japan," http://www.eia.gov/countries/country-data.cfm?fips=JA&trk=p1#ng; U.S. Energy Information Administration, "Countries: Korea, South," http://www.eia.gov/countries/cab.cfm?fips=KS
Note Dependence on the Middle East is based on the e-mail interview with a METI official and a senior researcher at Korea Energy Economics Institution (KEEI)

gas (LNG). As of 2009, Japan is the largest LNG importer and Korea is the second-largest. These two neighboring countries in East Asia feel equally vulnerable in their need for oil due to their absolute dependence on the Middle East for oil and the presence of China, which is the second-largest consumer of oil behind the United States in the world and the second-largest net oil importer as of 2009. Japan and Korea are uncomfortably aware of their vulnerability, and this situation has forced both nations to actively involve themselves in overseas oil and natural gas development to enhance energy security (Table 2).[46]

However, one significant difference between these two nations is their respective structures of political institutions. This is the point of departure that is important for understanding politics and policymaking comparatively. Japan has a parliamentary system with the bicameral Diet, while Korea has a presidential system with the unicameral National Assembly. While Japan's decision-making system has been described as consensual, Korea's has been described as centralized.[47]

[46] For more on energy security concerns facing Japan and Korea, see Kent E. Calder, "Japan's Energy Angst and the Caspian Great Game," *NBR Analysis* 12, no. 1 (2001); Kent E. Calder, *Korea's Energy Insecurities: Comparative and Regional Perspectives* (Washington, DC: Korea Economic Institute of America, 2005).

[47] See, for example, Byung Chul Koh, *Japan's Administrative Elite* (Berkeley: University of California Press, 1989), 265–66; Susan M. Collins and Won-am Park, "External Debt and Macroeconomic Performance in South Korea," in Jeffrey Sachs, *Developing Country Debt and the World Economy*, A National Bureau of Economic Research Project Report (Chicago: University of Chicago Press, 1989), 131, 135.

Table 2 The difference in political institutional settings in Japan and Korea

	Japan	Korea
Political system	Parliamentarism with Constitutional Monarchy	Presidentialism
Head of government	Prime Minister	President
Selection of head of government	Appointed by the Diet (Normally Chairman of major party)	Direct presidential election every five years
Who can propose a bill	Both the Diet members and the ministries	Both the National Assembly members and the ministries
Legislature	Bicameral (the House of Councillors and the House of Representatives)	Unicameral (the National Assembly)
General election	Mixed general election (the House of Councillors: 146 members from multi-member constituency and 96 by PR (proportional representation) system; the House of Representatives: 300 from single member constituency and 180 by PR system)	Mixed general election (245 in single member constituency and 54 by PR system)
Legislative session	Ordinary session for approximately six months once a year, and extra sessions are common	Ordinary session for no more than 100 days once a year, but extra sessions are common

These distinct differences in state structure can also be conceptualized in terms of the number of veto players in the political system. As seen in the previous part of this chapter, another noticeable difference is found in how the two states pursue their own energy security through overseas oil development. In terms of Peter Evans, the Japanese state has played the role of the "midwife" and "husband" rather than "demiurge" in overseas oil development and vice versa for Korea.[48]

[48] This does not necessarily mean the Korean state always chose the role of the demiurge. Evans explains the role of the Korean state as a midwife in the development of the IT industry. This can be attributed to the different characteristics of specific industries. While the oil industry entails massive investment and political consideration rather than innovation, the IT industry requires more innovation. Thus, the characteristic of economic activity defines the spectrum of the role that can appropriately be played by the state.

5 Significance of This Study

This study will add new perspectives to previous research on developmental states, the majority of which has focused on similarities rather than differences among East Asian countries. This study will also differentiate between the two developmental states, Japan and Korea, by showing how state capacity is affected by political structure, especially in terms of the number of veto players. This comprehensive analysis with veto player theory will open new perspectives beyond the controversy of "who governs whom."

From this structural examination, a universally applicable thesis can be derived. One-country case studies explain the political economy of Japan and Korea with independent variables that look unique in each country, such as culture and leadership style. However, multi-country comparisons can reveal meaningful factors that explain differences in policy outcomes. This comparative analysis of Japan and Korea will provide a more systemic explanation by excluding the cultural factor, which has been overly emphasized as a significant independent variable, especially in Japan Studies. In addition, this research will show how the features of political institutions determine the outcome of energy policy.

The analysis of political structure will suggest practical implications for both Japan and Korea. For Japan, the effect of the constitutional structure implicitly reveals one significant reason for "gradualism" even after a historical regime change took place in the August 2009 general election. Also, appropriate examination of why Japan's policy has seen little comparative change will bring to light other solutions for the problems Japan faces beyond often-mentioned ones, such as strong leadership or new political culture. For Korea, this research will show how the power of an "imperial president" can be lessened from the perspective of the number of veto players.[49] While a few politicians and commentators mention the need to change the constitutional structure from a presidential system to a parliamentary one, this study implies that parliamentarism is not the only possible solution to keep strong executives in check.

From a methodological perspective, this research emphasizes primary material, such as proceedings of legislative bodies, more so than most previous

[49] Originally, "imperial presidency" was used to describe the modern presidency of the United States by the US historian Arthur M., Schlesinger Jr. The term was in response to the idea that the US Presidency was out of control and that the presidency had exceeded its constitutional limits. In Korea, the term is used in response to a similar concern. See Arthur M. Schlesinger, *The Imperial Presidency* (Boston: Houghton Mifflin, 1973).

literature in this field.[50] The proceedings of the Japanese Diet and Korean National Assembly have not been studied well in energy security issues. One possible reason for this might be the perception that the legislature is only a rubber stamp. The powerful influence of the strong bureaucracy, emphasized by Chalmers Johnson, might be another explanation. However, the proceedings of legislature interpellation inform us of reliable facts and prevalent ideas at the time. Also, the interviews with important policymakers in the discussion of the state-owned oil company, such as O. Won-chol, a former Senior Economic Secretary to President Park Chung-hee, uncovers hidden, but significant historic events. Furthermore, the examining of historic documents in Korea's Presidential Archive and contemporary news articles, such as *Asahi Shinbun* in the 1960s and *Maeil Economic Daily* in the 1970s, will fill the blanks left by policymakers and interest group archives.

Furthermore, this study will offer not only facts and analysis on the energy security of Japan and Korea, which has not been extensively researched, but it will also provide information that is useful to predict the energy security policies of other developing countries. Assuming the contraction of a global energy supply caused by the geopolitical situation, developing countries' policy formulation then becomes an important issue. The experience of Japan and Korea and the analysis of what has contributed to their different approaches in particular institutional settings may provide clues to predict what path developing countries may choose for energy security based on their initial sociopolitical conditions.

6 Structure of the Book

In this study, the introductory chapter is followed by five chapters including the conclusion. Chapter 2, "The Establishment of the JPDC in the 1960s," illustrates how the postwar politics of Japan obstructed the government's attempt to establish a state-owned oil company. The interaction among key decision-makers, such as MITI, the Ministry of Finance (MOF), two councils of Liberal Democratic Party (LDP), the prime minister, two chambers of the Diet, and interest group, produced a resolute outcome from the status quo with no SOEs for overseas oil development.

[50] Samuels utilizes history books of interest groups, secondary sources mainly written by Japanese commentators and interviewers. While his research is a seminal one in explaining the structure of Japan's energy industry, he rarely uses primary sources. See Samuels, *The Business of the Japanese State: Energy Markets in Comparative and Historical Perspective*.

Chapter 3, "The Establishment of PEDCO in the 1970s," shows a stark contrast to the case in the previous chapter. This chapter explains how a small number of veto players, such as the president, the Ministry of Trade and Industry (MTI), the EPB, and one chamber of the National Assembly, produced a decisive policy outcome. The 1970s is an important era for development by autocratic regimes, and the chapter also shows how the regime decreased the number of veto players in a constitutional setting.

Chapter 4, "The Attempt to Establish a Japanese State-Owned Oil Company in the Early Twenty-First Century," reemphasizes the importance of the number of veto players in the policy process by discussing METI's failed attempt to create an SOE for overseas oil development under Prime Minister Koizumi, who was known a strong leader in postwar Japan politics. This case illustrates how the presence of multiple veto players narrows the range of change in an effort to abolish a public corporation and make a new SOE. In particular, this case shows how Koizumi and METI, who had opposing preferences on the establishment of a state-owned oil company, reached a compromise by producing a resolute outcome. Also, this chapter shows how the Japanese state evaded difficulties by establishing a quasi-SOE through the rarely used market measure called the golden share.

Chapter 5, "A Business-Friendly President and the Strengthening of State-owned Oil Company," deals with a case in which the presence of a small number of veto players results in the strengthening of the role of an SOE. This case shows that weak private sector capacity is not a condition in which SOEs can be established or strengthened. Indirectly, through the case of Korea, the chapter also shows that democratization through the revision of the constitution does not necessarily lead to an increased number of veto players.

Chapter 6, "Conclusion," summarizes key arguments and illustrates the theoretical and practical implications of this research. A few questions that are worth examining in the future will also be suggested.

Chapter 7, "Epilogue," explains the story of another decade after the year 2010 because previous chapters have examined the evolution of structure in the overseas oil venture of Japan and Korea until the first decade of the twenty-first century. The changes in overseas energy development policy taken place in Japan and Korea show the other aspect of the relationship between decision-making structure and policy consistency.

Now, we begin to explore the puzzle this study raises with a detailed examination of how political structure affects overseas oil development policy through the establishment and strengthening of SOEs. This theoretically guided case study, mainly through veto player theory, starts with the case of Japan in the 1960s when the Japanese state wanted to establish state-owned oil company.

2

The Establishment of the JPDC in the 1960s

For someone who is familiar with statist argument, such as Johnson's developmental state, it is hard to understand why the Japanese state in the 1960s could not make an SOE for overseas oil development. Normally, developmental state argument tells us that when the state decided to enact certain policies, these policies are accepted and implemented. This was not the case in Japan, however, and this chapter explains this unanticipated outcome through a veto player framework.

The case of Japan in the 1960s supports the main argument of this book: A state-owned oil company could not be established when many actors can block the initiative. In this chapter, I analyze the case qualitatively to further our understanding of the dynamics of the many veto players in Japan's policy-making. This chapter shows how the interaction among veto players affected Japan's attempt to realize a state-owned oil company.

In the 1960s, the Japanese state tried to establish a state-owned oil company. MITI's original idea was to establish a state-owned enterprise (SOE) to promote Japan's acquisition of equity oil, such as the oil developed in the Khafji oil field by the Japanese Arabian Oil Company (AOC). The proposal was expected to expand to a vertically integrated state-owned oil company like Italy's ENI and France's TOTAL. However, MITI's attempt to make an SOE for overseas oil production between 1961 and 1967 failed, and the final outcome was the establishment of the Japan Petroleum Development Corporation (JPDC), which supports the private sector's overseas oil ventures financially.

S.-i. Oh, *Overseas Energy Investment of Korea and Japan*,
https://doi.org/10.1007/978-981-99-0285-9_2

In this case study, the outcome is mainly attributed to the interaction among a large number of veto players. The existence of many effective veto players narrowed the size of the winset, which all veto players can agree with. The veto players include not only the MITI, the MOF, the Policy Affairs Research Council (PARC), and the Executive Council in the LDP, but also interest groups, such as Japan Federation of Economic Organizations (*Keidanren*), a then-powerful interest group in Japan's political economy.

1 Japan's Political Institutions in the 1960s

1.1 Institutional Veto Players: The Prime Minster, Ministries, and the Diet

In the 1960s in Japan, seven institutional veto players were part of discussion on the establishment of state-owned oil companies: the prime minister; the MITI; the MOF; the Upper House and the Lower House; and the Standing Committee of International Trade and Industry in each of the chambers. Based on the constitution, the prime minister and the Diet were also veto players in Japan's policymaking. The MITI and the MOF were also veto players. In the discussion, the former was agenda setter and the latter was regulatory ministry.

Prime Minister

In Japan, the prime minister has not always been a strong and assertive leader, but he has always been a primary veto player.[1] The prime minister is the leader of the ruling party and the head of the Cabinet, selected indirectly by representatives in the Lower House of the Diet, not by the general public. When the LDP had a majority in the Lower House, the party president became the prime minister at the same time. As the head of the Cabinet, he or she shares the right to submit bills to the Diet with the Diet member. While most prime ministers have been inclined to be more reactive than assertive, Ikeda Hayato and Sato Eisaku, who had experienced civil service careers, have been described as technocratic leaders who were involved in

[1] Kenji Hayao, *The Japanese Prime Minister and Public Policy*, Pitt Series in Policy and Institutional Studies (Pittsburgh: University of Pittsburgh Press, 1993), 14. Some scholars, like Wolferen, argue that the Japanese prime minister has less power than the head of government of any Western country. See Karel G. van Wolferen, "The Japan Problem," *Foreign Affairs* 65, no. 2 (1986): 290.

problem solving rather than in setting an agenda, and tended to emphasize continuity, stability, and predictability.[2] In budgeting, the role of the prime minister is significant in process because he or she has to maintain support for the finance minister and control over the party. The policies of the prime minister have been important in determining the smoothness of the budget process, but still much less active in budgeting than leaders in other countries.[3]

In the 1960s, the prime minister was assisted by a small inner staff organization in the *Kantei*, the Prime Minister's Office, and the Cabinet Secretariat.[4] The total number of staff in the *Kantei* was eight: five private secretaries, the chief secretary, and two deputy cabinet secretaries. Also, in reality, the bureaucracy penetrated the prime minister's staff in the *Kantei* more than the prime minister penetrated the bureaucracy.

The strongest point of contrast to Korea's *Chŏngwadae*, the Office of the President in the 1970s, was the absence of any real staff for formulating and analyzing policy. Therefore, the prime minister had to depend on the bureaucracy for information and expertise unless he maintained a strong network of personal advisors.[5] This means that the preference of the prime minister is located in the intersection of that of the ministry concerned and that of the MOF in counting effective veto players in veto player framework. This is in contrast to Japan's political situation during the Koizumi Administration, where the prime minister clashed with METI.

Ministerial Veto Players: The MITI and the MOF

In the establishment of SOEs for overseas oil development in the 1960s, MITI and the MOF were veto players in government decision-making. In Japan, a representative of a developmental state, the ministries were known to

[2] Hayao, *The Japanese Prime Minister and Public Policy*, 15–16, 27. In the general election of January 1949, Ikeda Hayato, who had just retired as Vice Minister of Finance, and Sato Eisaku, recently retired as Vice-Minister of Transportation, were brought into the Diet by Yoshida. See Johnson, *MITI and the Japanese Miracle: The Growth of Industrial Policy, 1925–1975*, 45.

[3] John Creighton Campbell, *Contemporary Japanese Budget Politics* (Berkeley: University of California Press, 1977), 165.

[4] The Management and Coordination Agency was established in 1984 as an external organ of the PMO for efficient and responsive government operations. In the Cabinet Secretariat, the Cabinet Office for External Affairs and the Cabinet Security Office were newly established and the Cabinet Research Office and the Cabinet Public Relations Office were expanded to the Cabinet Information Research Office and the Cabinet Office of the Director General of Public Relations, respectively, in the 1986 reorganization. Hayao, *The Japanese Prime Minister and Public Policy*, 168–79.

[5] Ibid., 181–83. For example, Prime Minister Sato Eisaku had his research group deliberate the issue of the status of US military bases upon the return of Okinawa to Japanese sovereignty.

exert independent power in policymaking in the respective jurisdiction. On the contrary, the Cabinet has not been the collective decision-making organization or core of government authority, but it endorses what is put before it with meetings.[6] In Japan's policymaking process, MITI, Japan's strategist, and the MOF, Japan's regulator, tend to have different sets of preferences.[7]

(a) **Ministry of International Trade and Industry (MITI)**

In 1949, MITI was established by reorganizing the Ministry of Commerce and Industry (MCI) that existed from 1925 to 1949. In the early period, MITI consisted of eight bureaus and four agencies: the Minister's Secretariat; the Trade Bureau; the Trade Promotion Bureau; the Enterprise Trade Bureau; the Textile Trade Bureau; the General Merchandise Trade Bureau; the Machinery Trade Bureau; the Chemical Trade Bureau; the Iron and Steel Trade Bureau; the Resources Agency; the Small and Medium Enterprise Agency; the Industrial Technology Agency, and the Patent Office. Also, four regional coal bureaus; and eight regional trade and industry bureaus were established as exterior subdivisions.[8]

Thanks to broad jurisdiction over most key sectors of Japan's industry, MITI was able to take a general equilibrium approach to industrial policy among many industries. No industry or interest group could exercise dominant influence over MITI or the substance of its industrial policy. MITI's broad jurisdiction originated from the merging of the MCI into the Ministry of Munitions, the centralized wartime organization.[9] The powers over foreign exchange, foreign direct investment, and technology licensing authorized by the occupying Allied Forces were also added to MITI's purview, and the ruling LDP permitted MITI unusual leeway to oversee the operation of the country's manufacturing and service sectors. MITI's rational organization into vertical and horizontal units made it possible to aggregate diverse interests through internal checks and balances. [10]

Until 1973, when the Agency for Natural Resources and Energy (ANRE) was established, the Mining and Coal Bureau took charge of oil policymaking and implementation. Under the Petroleum Business Law enacted in 1962,

---

[6] Aurelia George Mulgan, "Japan's 'Un-Westminster' System: Impediments to Reform in a Crisis Economy," *Government and Opposition* 38, no. 1 (2003): 84.

[7] Calder, *Strategic Capitalism: Private Business and Public Purpose in Japanese Industrial Finance.*

[8] Ministry of Economy Trade and Industry, "History of METI," http://www.meti.go.jp/english/aboutmeti/data/ahistory2009.html.

[9] Johnson, *MITI and the Japanese Miracle: The Growth of Industrial Policy, 1925–1975*, 168–70.

[10] Daniel I. Okimoto, *Between MITI and the Market: Japanese Industrial Policy for High Technology* (Stanford, Calif.: Stanford University Press, 1989), 112–19.

MITI's petroleum administration was given extensive powers over the oil industry, including refining and sales permits, oversights on import, production, and pricing of oil. The practice of *amakudari*, where retired government officials went to public and private corporations, often took place in oil companies, associations, and energy-related public corporations.[11]

(b) **Ministry of Finance (MOF)**

In the 1950s and 1960s, the MOF held fairly formidable powers for a national treasury.[12] Its jurisdiction, which located the MOF at the top of the inter-ministerial hierarchy, ranged as follows: budget; taxes; customs and tariffs; finance; securities; banking; and international finance.[13] While the function of the MOF could not be as strong as the French Trésor, which overwhelmingly concentrated power over the entire French financial system, Japan's MOF could enjoy a high level of power consistently throughout the high-growth period. The MOF even enjoyed the power to discharge the governor of the Bank of Japan, Japan's central bank.[14]

Moreover, the MOF held especially great power and high status through a monopolistic role in the budget process.[15] The Budget Bureau (*Shukeikyoku*) is the core bureau in the MOF, and the bureau director is always a leading candidate for the vice minister's post. Budget examiners (*shukeikan*) tend to enjoy higher level of authority toward politicians and other officials than would be expected from their ranks.[16] At the micro level, five criteria for budget allocation have been considered: logic, framework, balance, timing, and accurate calculations.[17] Along with this, macroeconomic issues such as economic growth, revenue estimation, taxes and bonds, and fiscal policy are also considered.[18] Macro-budgeting is the main area in which the finance minister makes key decisions. In terms of public exposure and reputation,

---

[11] Martha Ann Caldwell, "Petroleum Politics in Japan: State and Industry in a Changing Policy Context" (The University of Wisconsin – Madison, 1981), 15–16.

[12] Calder, *Strategic Capitalism: Private Business and Public Purpose in Japanese Industrial Finance*, 95.

[13] Campbell, *Contemporary Japanese Budget Politics*, 43–44.

[14] Calder, *Strategic Capitalism: Private Business and Public Purpose in Japanese Industrial Finance*, 95–96.

[15] Campbell, *Contemporary Japanese Budget Politics*, 43–44.

[16] Ibid., 49–51. In 1967's Standing Committee on Commerce and Industry, where typically the Bureau Chief is the lowest level of official participants, the Division Director-Level Examiner of the MOF appeared to show his opinion on the establishment of the JPDC. See Japan House of Representative, Standing Committee on Commerce and Industry, *26th Committee Meeting*, Interpellation, July 4, 1967.

[17] Campbell, *Contemporary Japanese Budget Politics*, 62.

[18] Ibid., 71–90.

the finance minister is the highest post after the prime minister, and is often considered as a prerequisite for being prime minister as seen in the cases of Ishibashi, Ikeda, Sato, and Tanaka. While the finance minister projects his or her economic beliefs and preferences, he or she also serves in a conciliatory role between the views of the MOF and pressures coming from other ministries, the party and big business, especially when the "prospect of major loss or unwanted change which threatens the established order" is demonstrated.[19]

Generally, the MOF is wary of making new public corporations due to the anticipated administrative costs and the potential conflicts with guiding principles of the MOF. However, in cases where short-term returns for the MOF could be expected, establishment of public corporations was allowed. New organizations are often viewed as inexpensive measures to lessen strong pressures from line ministries and the party, because the first year of the new organization does not require a large budget. In addition, severing the part of the new organization for the general account disguises the growth of government spending. Furthermore, potential positions for retired MOF officials made the establishment not impossible. Normally, the line ministries request many new bodies in spite of the MOF's conservative principles, and several requests are granted frequently at the very end of revival negotiations.[20]

Legislative Institution: Japan's Bicameral Diet and Committees

Japan's 1947 Constitution stipulates a bicameral legislature, so the two chambers are basically institutional veto players who must make decisions collectively. While the dominance by a single party distorted the political structure, the bicameral Diet, which was stipulated as "highest organ of state power and sole law-making authority of the State," was one necessary part of policymaking in Japan.[21] The committee system, adopted from the US Congress, and the consensus norm, through which opposition parties had veto power in the legislature, increased the number of veto players.

As the highest organ of state power, the House of Representatives (the Lower House of the Diet, or *Shūgiin*) has the power to select the prime minister, and the prime minister and the Cabinet must maintain the support of the Lower House in order to remain in office. The Lower House also has the power to control the budget and ratify treaties.

[19] Ibid., 152–55. Calder, *Crisis and Compensation: Public Policy and Political Stability in Japan, 1949–1986*, 71.

[20] Campbell, *Contemporary Japanese Budget Politics*, 208–09.

[21] Article 42 in *The Constitution of Japan*, 1st.

Table 1 House of Representatives election result in the 1960s

Election date	Nov 20, 1960	Nov 21, 1963	Jan 29, 1967
LDP	296 (63.4%)	283 (60.7%)	277 (57.0%)
JSP	145 (31.0%)	144 (30.8%)	140 (28.8%)
DSP	17 (3.7%)	23 (4.9%)	30 (6.2%)
CGP	–	–	25 (5.1%)
JCP and others	9 (1.9%)	17 (3.6%)	14 (2.9%)
Total seats	467	467	486

Source J. A. A. Stockwin, *Governing Japan: Divided Politics in a Resurgent Economy*, 4th ed. (Malden, Mass.: Blackwell Pub., 2008), 275
Note The debate on state-owned oil company took place from 1961 until 1967, so this table excludes the election outcome of December 27, 1969

Although the House of Representatives is more powerful, the House of Councillors (*Sangiin*) is not powerless. Unless a bill is passed with a two-thirds majority in the Lower House, all bills must pass not only in the Lower House but also in the Upper House. As seen in Table 1, not only has the LDP never had a two-thirds majority,[22] but also many LDP members in the Upper House try to maintain some independence from the party, and deliberations here often provide opportunities for opposition parties to delay government bills they oppose.[23]

This situation has made passing bills in the Upper House rather difficult, even after they were passed in the Lower House. The Lower House and the Upper House have an ability to "resist, change or retard" the executive branch's legislative proposals as veto players.[24]

Both in the Upper House and the Lower House, there are standing committees and special committees that investigate the government. The committee has the power to kill any bills referred to it. More time has been spent in committees than in the Plenary Session, and ministers bring senior officials of the ministry to the committee to answer difficult or technical questions.[25] The standing committees in each chamber are normally paired with the relevant ministries, and thus the MITI's initiative to establish a state-owned oil company was discussed in the Standing Committee on International Trade and Industry in the 1960s.

---

[22] Hayao, *The Japanese Prime Minister and Public Policy*, 33–36.

[23] Jung, "Reassessing Reform Outcomes in South Korea and Japan a Decade after Crisis: The Paradox of Political Leadership," 79–80.

[24] Mike Mochizuki, "Managing and Influencing the Japanese Legislative Process: The Role of Parties and the National Diet" (Thesis [Ph.D.], Harvard University, 1982, 1982), 45. This ability is called "viscosity."

[25] J. A. A. Stockwin, *Governing Japan: Divided Politics in a Resurgent Economy*, 4th ed. (Malden, Mass.: Blackwell Pub., 2008), 164–66.

The norms of consensus-based negotiation were developed after 1960 when Prime Minister Ikeda tried hard to mollify the apprehensive political atmosphere of the post-1960 United States-Japan Mutual Security Treaty crisis period with his robust efforts to establish some practical rapport with the opposition.[26] Thus, the low number of bills that failed resulted from the LDP's tendency to negotiate with the opposition parties rather than the LDP's majority status in the Diet until 1993. This was reinforced by general public norms that opposition parties needed to be included in the legislative process. Thus, the LDP could not fully utilize its majority status in the legislation against the opposition.[27] In addition to the presence of the opposition parties, the Diet's role as a mediator between the state and society forced the LDP, the ruling party, to accommodate those interests that could not be ignored, and required the state to change course in response to serious problems.[28] This consensus-based decision-making norm made the preference of the chambers not easily absorbed to the preference of the ruling party, and the norm differentiated the characteristic of Japan's parliamentary system from other Western ones.

Even during the time that the LDP held a majority, the Diet did not approve bills sponsored by the LDP and the Cabinet automatically. One relevant example is the ratification of the Korea–Japan Continental Shelf Joint Development Agreement. After signing in January 1974, the bill for the agreement was submitted on May 17th to the Diet for ratification, but was abrogated with the end of the regular Diet session on June 3rd. During the Diet sessions of 1975 and 1976, opposition by the Asia–Africa study group in the LDP and the political turmoil caused by the Lockheed scandal posed difficulties for the approval of the agreement. In April 1977, following the general election, Prime Minister Fukuda made a drastic maneuver to ratify the agreement. During the night of April 27, 1977, the LDP members, together with the Democratic Socialist Party members, delegated and passed the bill in the foreign affairs committee in the Lower House with a vote of 15 in favor and two opposed. Considering the composition of the Upper House, where only 10 of 20 foreign affairs committee members belonged to the LDP and just under half of their counterparts in the Upper House were LDP members, ratification in the Lower House was deemed more viable. Taking advantage of

---

[26] Calder, *Crisis and Compensation: Public Policy and Political Stability in Japan, 1949–1986*, 96. Jung, "Reassessing Reform Outcomes in South Korea and Japan a Decade after Crisis: The Paradox of Political Leadership," 423–29.

[27] Hayao, *The Japanese Prime Minister and Public Policy*, 133.

[28] Johnson, *MITI and the Japanese Miracle: The Growth of Industrial Policy, 1925–1975*, 50.

Diet procedural rules, the LDP extended the length of the regular session after passing the Plenary Session of the Lower House, and finally, the agreement was ratified on June 8, 1977.[29]

1.2 Partisan Veto Players: The LDP and Opposition Parties

The obvious contrast in decision-making structure between Japan and Korea is shown in partisan veto players in Japan's politics in the 1960s. The councils of the ruling party, the LDP, played significant roles as veto players in policymaking. In addition, a consensus-based decision-making norm in the Diet provides the opposition parties with opportunities to be veto players. Also, this power of the opposition parties made the chambers of the Diet become an independent veto player in Japan's policymaking; otherwise the Diet would have been regarded as rubber stamp of the ruling party's will.

The Ruling Party in the 1960s: Liberal Democratic Party (LDP)

The LDP maintained a majority in the Lower House from 1955 until 1993, and during this time, the two organs of the LDP played an important role in policymaking. In the LDP, while the party conference, the Assembly of Members of the Houses of the Diet, and the Executive Council were formally constituted as decision-making organs, every important policy decision of the party was formally deliberated by the PARC (or *Seimu chōsakai*) and decided by the Executive Council (*Sōmukai*). [30] While this policy review was based on the LDP Constitution, the practice had been consolidated in the Japanese policymaking process.

In the LDP, often referred to as a "coalition of sub-parties" more so than a unified party,[31] intra-factional balance was institutionalized in these two councils through a formal policymaking process. The factional split had

[29] When the House of Councillors fails to take final action within thirty (30) days after the receipt of the budget passed by the House of Representatives, excluding recesses, the decision of the House of Representatives becomes the decision of the Diet. This applies also to the Diet approval required for the conclusion of treaties. See Articles 60 and 61 in *The Constitution of Japan.*

[30] Nathaniel B. Thayer, *How the Conservatives Rule Japan*, Studies of the East Asian Institute, Columbia University (Princeton, NJ: Princeton University Press, 1969), 207–67. Haruhiro Fukui and University of California Berkeley. Center for Japanese and Korean Studies., *Party in Power: The Japanese Liberal-Democrats and Policy-Making*, Publications of the Center for Japanese and Korean Studies (Berkeley: University of California Press, 1970), 82–83.

[31] Giovanni Sartori, *Parties and Party Systems: A Framework for Analysis* (Cambridge [Eng.]: Cambridge University Press, 1976), 90.

been reinforced by Japan's multi-member electoral system where three or five members were elected in each electoral district until Hosokawa's electoral reform in 1994.[32] For Japanese politicians at that time, this unusual Japanese electoral system, where multiple candidates of the same party frequently ran against one another for election from any given district, had been the single most important source of personal political vulnerability, especially in the LDP.[33] This system also reinforced the viability of factions in the LDP, and thus lowered the coherence of the party. Each faction's provision of money, posts, and earmarks was important in electoral competitions.[34] Factionalism seriously impedes the mobility of Japanese policymaking in non-crisis periods, while making the ruling party more receptive to societal demands once forces have created a powerful need for policy change on the part of Japan's conservative party.[35]

This factionalism was reflected in the LDP's decision-making system. As Campbell states, the impact of factionalism on budgeting worked through the pattern of appointments to various veto player positions, such as the prime minister, the PARC chairman and the MOF minister.[36] Considering the fact that factional splits were formally filtered in these two councils in policy-making, the number of partisan veto players can be counted as two rather than one. Both the PARC and the Executive Council had independent power to block the draft laws and budgets, respectively, and the Executive Council did not readily provide its consent to the PARC.[37]

The PARC has divisions (*bukai*) that correspond to government agencies, and therefore, the various divisions under PARC are the most significant party representatives among the ministry, the LDP and interest groups. It was in the divisions of PARC that the party's policy was basically formed, and the division's recommendations were reviewed in the Policy Deliberation Commission, which has 15 or 20 members. Basically, the PARC supports the ministry's policies and serves as its representative in the larger legislative body. At the same time, the PARC divisions hear interest group representatives, whose opinions are relayed to the ministry. Members of the PARC

---

32 For example, the three candidates with the most votes would be elected in a three-seat district.

33 Calder, *Crisis and Compensation: Public Policy and Political Stability in Japan, 1949–1986*, 62–63.

34 Gary W. Cox, Frances McCall Rosenbluth, and Michael F. Thies, "Electoral Reform and the Fate of Factions: The Case of Japan's Liberal Democratic Party," *British Journal of Political Science* 29, no. 1 (1999): 35–37.

35 Calder, *Crisis and Compensation: Public Policy and Political Stability in Japan, 1949–1986*, 194–95.

36 Campbell, *Contemporary Japanese Budget Politics*, 134–37. For more research on the policy role of the LDP factions, see Mathew D. McCubbins and Michael F. Thies, "As a Matter of Factions: The Budgetary Implications of Shifting Factional Control in Japan's LDP," *Legislative Studies Quarterly* 22, no. 3 (1997).

37 Thayer, *How the Conservatives Rule Japan*, 261.

divisions joined on a voluntary basis through the formation of informal tribes (*zoku).* Membership of the PARC's Policy Deliberation Commission, which reviewed the decisions made at divisions, was distributed based on membership in a certain faction.[38] It is noteworthy to mention that while Article 41 of the LDP Constitution stipulated that an industry person can be included in the PARC, this was not realized.[39] Therefore, the power of interest group had been limited in the LDP's decision-making.

Every policy recommendation initiated by the PARC finally reached the Executive Council with the goal of making it into an official party decision.[40] The prior approval by the LDP Executive Council was put in place in 1962, when Akagi Munenori, the then-LDP Executive Council Chair, tendered a written request to Ohirsa Masayoshi, the then-Chief Cabinet Secretary, for the government to have Cabinet-drafted bills referred to the Executive Council in advance of Cabinet decisions on bills.[41] The Executive Council was the commanding decision-making body for most issues (also playing a role as a legislative chamber), but it was too unwieldy to take part in negotiations with the MOF. Therefore, the council's decisions were entrusted to the party executive for implementation.[42]

Opposition Parties

While opposition parties had never gained a majority in either chamber until 1993, this does not mean that Japanese conservative policymakers had consistently dominated either the definition of issues or the ultimate shape of policy outputs in Japan. Several distinctive policies, such as children's allowances, originated with opposition parties. [43] The unanimity norm on Diet procedures and the division of opinion in the LDP also permit opposition parties to use the parliamentary process either to veto or to keep

[38] Fukui and University of California Berkeley. Center for Japanese and Korean Studies., *Party in Power: The Japanese Liberal-Democrats and Policy-Making*, 83–89. Stockwin states that *Zoku*, or LDP policy tribes, began to be noticed in the 1970s. See Stockwin, *Governing Japan: Divided Politics in a Resurgent Economy*, 143.

[39] Thayer, *How the Conservatives Rule Japan*, 229–30.

[40] Fukui and University of California Berkeley. Center for Japanese and Korean Studies., *Party in Power: The Japanese Liberal-Democrats and Policy-Making*, 89.

[41] Aurelia George Mulgan and Australian National University. Asia Pacific School of Economics and Management, *Japan's Failed Revolution: Koizumi and the Politics of Economic Reform* (Canberra: Asia Pacific Press, Asia Pacific School of Economics and Management, 2002), 130.

[42] Campbell, *Contemporary Japanese Budget Politics*, 145.

[43] Calder, *Crisis and Compensation: Public Policy and Political Stability in Japan, 1949–1986*, 473.

the government's policy initiatives in check.[44] The opposition parties influenced the legislation with several measures phase by phase: (1) Direct contact with bureaucrat, and participation in deliberation committee (*shingikai*) in the pre-legislation phase, (2) interpellation, supplementary resolution, and delaying tactics during parliamentary session, and (3) utilizing interest groups to voice.[45] During the 1960s, opposition members of the Diet were from four opposition parties: The Japan Socialist Party (JSP), the Democratic Socialist Party (DSP), the Clean Government Party (CGP), and the Japan Communist Party (JCP). In the Diet's consensus-based decision-making system, four opposition parties meant four additional veto players in the policymaking process.

(a) **Japan Socialist Party (JSP)**

For many years including the 1960s, the JSP had been the largest and most significant party of opposition to Japan's conservative government.[46] The JSP was founded in 1945, split in 1951, and reunited in 1955. During the early 1960s, the JSP adopted a highly creative and pragmatic strategy, as seen in Eda Saburo's 1962 vision for Japan. The party also set the national agenda for small business, as well as regional and welfare policy. While the JSP gained more than 140 seats, approximately 30% of total seats in the Lower House before the 1969 general election, its weak organization and lack of appeal beyond industrial workers made it difficult to gain more political ground.[47]

(b) **Democratic Socialist Party (DSP)**

The DSP was founded in 1960 by the labor leader Nishio Suehiro, in protest against the Marxist influence within the JSP during the dispute over revising the Japan–US Security Treaty. While the DSP modeled itself after the British Labor Party to be a "people's party," it never developed into a major party. Its support in the Dōmei labor union federation consisted of a relatively narrow base, concentrated especially in the Nagoya and Kansai

---

[44] Jung, "Reassessing Reform Outcomes in South Korea and Japan a Decade after Crisis: The Paradox of Political Leadership," 432–33.

[45] Ibid., 322–31.

[46] Stockwin, *Governing Japan: Divided Politics in a Resurgent Economy*, 198.

[47] J. A. A. Stockwin, "The Japanese Socialist Party under New Leadership," *Asian Survey* 6, no. 4 (1966): 188; Junnosuke Masumi, *Contemporary Politics in Japan* (Berkeley: University of California Press, 1995), 464.

areas. Moreover, by the 1970s the party had developed rather hawkish views on defense, bringing it into sharp confrontation with the JSP.[48]

(c) **Clean Government Party (CGP)**

The CGP is the only significant party to be founded by a religious group, the Nichren Buddhist sect, in the postwar era. Formed in 1964, the party appeared in the 1967's general election for the first time and gained remarkable success in general elections. During the 1960s, the party drew support from the less educated, less well-off and less organized sections of big city populations, and took on a generally moderate conservative policy orientation.[49]

(d) **Japan Communist Party (JCP)**

After being legalized in 1945 by the US Occupation, the JCP attempted to ride the tide of the labor union movement. However, the party had minimal electoral significance until the late 1960s when it resurged in the general elections.[50]

1.3 Interest Group Veto Players in the 1960s

Japan Federation of Economic Organizations (*Keidanren*)

The Federation of Economic Organizations (*Keidanren*) was an all-inclusive, unwieldy association that functioned as the supreme coordinating body of big business. Established in 1946, it maintained close contact with all sectors of the business community to adjust and harmonize conflicting views and interests of the many businesses and industries it represented. *Keidanren*'s membership included not only more than 100 major national trade associations and over 750 large corporations, but also public corporations, such as the Japan Development Bank, Japan Export–Import Bank, and the Overseas Economic Cooperation Fund. *Keidanren*'s 12 standing committees covered every aspect of the national economy including energy policy, and these committees had constant contact with members of the Diet, the relevant

[48] Stockwin, *Governing Japan: Divided Politics in a Resurgent Economy*, 199.
[49] Ibid., 198.
[50] Ibid., 198–99.

ministries and the subcommittees under the LDP PARC.[51] *Keidanren* came to be viewed as an "omnipotent power" that could actually affect the government, rather than merely a pressure group.[52] In the case of the Special Measures Law attempted by Sahashi Shigeru of MITI, Ishizaka Taizō, Chair of *Keidanren*, stood against the draft bill, resulting in the failure of the legislation.[53]

In energy policymaking, *Keidanren* played an informal leading role. When MITI raised a proposal to reduce the country's heavy oil consumption, *Keidanren* provided a forum, called *zadankai*, in which various stakeholder industries could discuss energy policy issues. *Keidanren* also emphasized adjustment of conflicting interests in the short term and the development of nuclear power in the long term.[54] The *Keidanren* energy committee's leading role in energy policymaking continued during the debacle on Petroleum Industry Law and the discussion in the government and the Diet concerning a state-owned company.

Petroleum Association of Japan (PAJ or *Sekiyu Renmei*)

The PAJ was formed to defend the industry's interest against oil user groups. The industry-specific nature raises an expectation that the PAJ be a more viable and active veto player in Japan's policymaking process. However, this expectation could not be met. Its political power was not powerful enough to halt other interest groups' encroaching attempts at involving downstream oil business. In addition, MITI hesitated to recognize its role in energy policymaking. Instead, MITI facilitated the creation of the Japan Petroleum Development Association (JPDA), a competing pressure group against the PAJ. The PAJ's direct participation in *shingikai* took place in 1965, which is later than that of *Keidanren* in 1962. Therefore, it could not play a decisive role in fuel policymaking.

The PAJ, the interest group representing Japanese downstream producers, was established on November 1, 1955, in response to potential revisions of three energy laws that year. Those revisions included unfavorable changes against Japan's downstream production, such as an increase in tariffs on oil imports and the restraint of heavy oil use for rationalization and subsidization

---

[51] Chitoshi Yanaga, *Big Business in Japanese Politics*, Yale Studies in Political Science 22 (New Haven,: Yale University Press, 1968), 42–44.

[52] Masumi, *Contemporary Politics in Japan*, 298.

[53] Calder, *Strategic Capitalism: Private Business and Public Purpose in Japanese Industrial Finance*, 65.

[54] Caldwell, "Petroleum Politics in Japan: State and Industry in a Changing Policy Context," 73–74.

of Japan's declining coal mining industry. In order to counter these unfavorable changes in energy policy, oil refiners, and wholesalers created the PAJ.[55]

As seen in the oil industry's failure to restrain the Fishery Industries Association (*zengyōren*) from gaining foreign exchange allocation for its own independent crude oil import in January1955's Diet, the PAJ was not strong enough politically at its beginning.[56] Also, the PAJ did not have easy access to the government's policymaking process. MITI's *Enerugi Kondankai* in 1961 and the Energy subcommittee under Industrial Structure Investigative Committee in May 1962 both excluded the PAJ when discussing a comprehensive energy policy. While a partial representation might be possible thanks to the *shingikai* membership of Mr. Nanbu (the then-Toa Nenryo Vice President and former Vice Chairman of PAJ), it was not until 1965 when the Energy Investigative Committee was established, that the PAJ joined the energy-related committee.[57]

In addition, there were splits among the 16 member companies based on the size and affiliation. Domestically capitalized firms, such as Idemitsu and Daikyō, and foreign affiliates, such as Nippon Oil and Shōwa, were consistently divided on issues of security and supply. At the same time, however, sizeable companies such as Nippon Oil and Idemitsu had opinions that were different from those of Shōwa and Daikyō.[58] These fault lines were further solidified due to the unanimous decision-making system in the PAJ Articles of Association.[59] The only issues upon which association members regularly agreed were consumption and import taxes. Although the association was publicly annoyed by the heavy-handedness of MITI and its *Kondankai* (Roundtable), these internal disunions left PAJ unable to formally repudiate the *Kondankai* report of December 1961. Its members banded together

[55] Petroelum Association of Japan, *Sengo Sekiyu Sangyōsi [Postwar History of Oil Industry]* (Tokyo: Japan Petroleum Association, 1985), 133–34.

[56] Caldwell, "Petroleum Politics in Japan: State and Industry in a Changing Policy Context," 78–80.

[57] Suzuki Shigeru, *Nihon No Enerugi Kaihatsu Seisaku [Japan's Energy Development Policy]* (Tokyo: Mineruba shobō, 1985), 154; Samuels, *The Business of the Japanese State: Energy Markets in Comparative and Historical Perspective*, 198. Caldwell, "Petroleum Politics in Japan: State and Industry in a Changing Policy Context," 102–04.

[58] Samuels, *The Business of the Japanese State: Energy Markets in Comparative and Historical Perspective*, 199.

[59] (Article 25) An executive board meeting is possible if two-thirds of the executive members that are selected by member companies are present. The decision of the board meeting is made through unanimous agreement of present members. However, the executive members selected from outside member companies are excluded from the decision-making. See Petroelum Association of Japan, *Sengo Sekiyu Sangyōsi [Postwar History of Oil Industry]*, 137.

to oppose a state-owned oil company but remained undecided about the proposal for the Petroleum Industry Law.[60]

Japan Petroleum Development Association (JPDA)

The Japan Petroleum Development Association (JPDA) was established on November 20, 1961, at the beginning of MITI's initiation of the Petroleum Industry Law and state-owned oil company for overseas oil development. While it included only four member companies (Teikoku Oil, Japan Petroleum Exploration [JAPEX], the AOC, and North Sumatra Offshore Development Corporation [NOSODECO], its political influence was considerable. Four hundred high-ranking figures, including the then-MITI Minister Sato Eisaku, attended the launch party on December 18, 1961.[61] In energy policymaking, the JPDA could be represented because Mr. Fukuda Hideo, the then-JAPEX Vice President, was a member in the Comprehensive Energy Committee in 1962. Also, because the JPDA was MITI's creation, it gained unofficial recognition by the government for policymaking.

2 The Debate on Petroleum Industry Law and a State-Owned Oil Company

In 1961, the Petroleum Industry Law was considered as a measure to maintain order in Japan's oil market against the backdrop of trade liberalization. One main issue of the bill up for consideration was the proposal to found a state-owned oil company to deal with the overseas oil developed by Japanese companies, such as the AOC. The proposal was excluded from MITI's draft bill during preliminary consultation with interest groups, but the proposal was revived by the LDP Executive Council and discussed in the Diet. The Diet, in 1962, produced an unbinding resolution to recommend the establishment of a state-owned oil company.

---

[60] Samuels, *The Business of the Japanese State: Energy Markets in Comparative and Historical Perspective*, 199.

[61] Japan Petroleum Development Association, *Sekiyu Kogyō Renmei Nijunen No Ayumi[20 Year-Footprints of Japan Petroleum Development Association]* (Tokyo: Japan Petroleum Development Association, 1982), 21–22.

2.1 Background: Trade Liberalization and the Arabian Oil Company

The idea of founding a Japanese state-owned oil company for overseas oil development was raised against the backdrop of trade liberalization associated with the liberalization of foreign currency exchange and the success of oil ventures conducted by the AOC. The MITI considered the idea, and it was officially proposed by the Energy Roundtable Meeting (*Enerugi Kondankai*) chaired by Prof. Arisawa Hiromi at the beginning of the debate on the Petroleum Industry Law in the early 1960s.[62]

Trade Liberalization

Facing pressure from Western industrialized countries in the late 1950s, the Japanese government, in September 1960, decided to liberalize crude oil imports from October 1962, forcing policymakers to consider how to maintain order in the domestic oil market.[63] The liberalization was implemented by removing MITI's foreign exchange allocation function, which was based on the Foreign Exchange and Foreign Trade Control Law. This function had been one of the most powerful measures for MITI's industrial policy and figured predominantly in implementing industrial policy prescriptions for sectors like petroleum refining, where foreign currency was a continuous requirement.[64] The liberalization of trade and foreign exchange was reminiscent of the "invasion of black ships (*kurofune*)," which forced Japan to end its isolationist policy in the nineteenth century, but also required new economic order because the government was set to lose an important measure of control over the domestic market.

[62] Arisawa Hiromi was famous for making so-called priority-production policy (*Keisha seisan hōshiki*) as Yoshida's adviser. During the Yoshida Administration, this policy concentrated on breaking industrial bottlenecks centering with coal by strategically focusing investment capital in the steel, coal, shipping, and fertilizer industries. To achieve this, financial policy was integrated with materials policy to increase production. It is important to note that this policy acted as a shrewd compromise between all competing economic ideologies. It managed to revive the economy without challenging private companies; at the same time, it enlarged the allocative and distributive functions of the state by raising wages and channeling funds to existing management. This strategy has been evaluated as broadly effective. See Calder, *Strategic Capitalism: Private Business and Public Purpose in Japanese Industrial Finance*, 31; Laura Elizabeth Hein, *Fueling Growth: The Energy Revolution and Economic Policy in Postwar Japan*, Harvard East Asian Monographs (Cambridge, Mass.: Council on East Asian Studies Distributed by Harvard University Press, 1990), 118–28.

[63] Caldwell, "Petroleum Politics in Japan: State and Industry in a Changing Policy Context," 92.

[64] Calder, *Strategic Capitalism: Private Business and Public Purpose in Japanese Industrial Finance*, 34.

Facing this situation, businesses and the government expressed different preferences for how to shape the new economic order. *Keidanren* requested a self-controlling way (*Jishu chōsei hōshiki*) as a new industry system in the open economy including the oil market. The PAJ representing oil refiners argued that the government's direct control through an individual law targeting a specific industry is not desirable, and that the Fair Trade Law needed to be loosened in order to make preemptive cartels possible.[65]

Nevertheless, the government pursued an interventionist approach. Trade liberalization generated "MITI's neurosis," a bureaucratic crisis over what role MITI would play in the future. MITI thus prepared a new fundamental law called Law of Special Measures for the Promotion of Designated Industries (Special Measure Law or *Tokutei sangyō shinkō rinji sochi hō*), which pursues the principle of the public–private cooperation system (*Kanmin kyōchō hōshiki*).[66] For Sahashi Shigeru, an influential mainstream MITI official in the early 1960s, this was a "third approach" that was neither a laissez-faire nor controlled economy.[67]

In line with this approach, a majority of MITI officials leaned toward continuous governmental intervention in the oil market. They considered that the possible problems resulting from the liberalization of crude oil imports were caused by the special characteristics of oil in the international market that could not have been effectively dealt with through existing laws, such as the Fair Trade Law and Foreign Capital Law. Taking these concerns into consideration, MITI considered enacting legal regulations and establishing a policy company, which was expected to support economic policies and to guide the framework of industry to benefit the public.[68] These preferences were expressed through the Petroleum Industry Law made by MITI and the report by MITI's *Enerugi Kondankai*, which was composed mostly of scholars and did not include industry insiders. Also, the Japan Petroleum Development Association (*sekiyu kogyo renmei*) that represents oil developers was allegedly the idea of Takase Mitsuya, the then-MITI Petroleum Division Chief.[69]

---

[65] Petroelum Association of Japan, *Sengo Sekiyu Sangyōsi [Postwar History of Oil Industry]*, 147.

[66] Johnson, "MITI," in Robert A. Scalapino and Social Science Research Council (U.S.), *The Foreign Policy of Modern Japan* (Berkeley: University of California Press, 1977), 248–52.

[67] Quoted Comment of Sahashi Shigeru in Masumi, *Contemporary Politics in Japan*, 299.

[68] Petroelum Association of Japan, *Sengo Sekiyu Sangyōsi [Postwar History of Oil Industry]*, 152.

[69] Samuels, *The Business of the Japanese State: Energy Markets in Comparative and Historical Perspective*, 317.

Japan's Overseas Oil Ventures

The oil developed by Japanese oil companies, called *Hinomaru* oil or *Jishukaihatsu* oil, could not find a market beyond Japan's domestic oil market.[70] However, there was less room to purchase *Hinomaru* oil than the produced, and therefore, establishing a state-owned oil company was regarded as a solution to the problem.

As oil demand increased in Japan, the interest in overseas oil development also increased, and Japanese businesses attempted to initiate foreign oil ventures. This early phase of post-World War II oil ventures saw the establishment of the AOC and the North Sumatra Oil Development Corporation (NOSODECO). Unlike the AOC, the NOSODECO was not a big problem in sales of its developed oil. The NOSODECO was established in 1960 for a project in which Japan offered 18,845 million yen of credit to Indonesia in exchange for 5.6 million kl of oil.[71] Owing to the then-Prime Minister Kishi's support, the yen credit was financed by a loan from Japan's Export–Import Bank, which at the time was supported by a loan from the World Bank. This led to complaints from the bank that its funds were not intended to be used for North Sumatran oil development.[72] The investors in the NOSODECO involved 52 companies, including refiners. Also, because the oil quality was good and the quantity was relatively small, the oil from the NOSODECO could be accepted by Japan's refiners.[73]

However, the situation created by the AOC was quite problematic. The AOC, initiated by Yamashita Taro and established in February 1958 for the oil exploration in the Neutral Zone between Saudi Arabia and Kuwait, was Japan's first foreign oil venture. Although forty Japanese companies (mostly steel firms, public utilities, and general trading companies that wanted their own captive supply of crude oil) invested in this private company,[74] no Japanese refinery companies affiliated with Western oil majors showed

[70] *Hinomaru* oil is another name for Japan's equity oil, which means the oil developed by Japanese companies or the oil reserves in which Japanese companies own shares. Originally, *Hinomaru* means the national flag of Japan.

[71] Comment by Mr. Morozumi, then-MITI mining bureau chief, at Japan House of Councillors, Standing Committee on Commerce and Industry, *19th Committee Meeting*, Interpellation, July 20, 1967.

[72] "Kishi, Sato and Ikeda Decided to Included NOSODECO in Japan EXIM Bank Loan," *Asahi Shinbun*, Mar. 11, 1960.

[73] Japan Petroleum Development Association, *Sekiyu Kogyō Renmei Nijunen No Ayumi[20 Year-Footprints of Japan Petroleum Development Association]*, 6–7.

[74] Johnson places the AOC in the category of broadly defined national policy companies, which belong to a gray area between enterprises made by government initiatives and purely private companies. See Chalmers A. Johnson, *Japan's Public Policy Companies*, AEI-Hoover Policy Studies (Washington: American Enterprise Institute for Public Policy Research, 1978), 57–58.

interest in the project initiated by Yamashita Taro.[75] While it was considered a "star of hope" at the outset of operations, the AOC's sizeable oil production became a headache due to a lacking sales network and over-supply of oil in the world market. This was yet another reason to consider establishing a state-owned oil company.

Nationalism played a prominent role in development of the AOC, the first *Hinomaru* oil project in the post-World War II era. Patriotism was the reason that Mr. Ishizaka Taizo, who had been chairman of *Keidanren* from 1956 until 1968, was persuaded into serving as the AOC's chairman. Heeding the advice of Dr. Rudolf Heberlein, Chairman of Swiss Air, who said that overseas oil exploration could be the "greatest thing for Japan,"[76] Mr. Ishizaka accepted Mr. Taro's request and served as Chairman of the AOC from February 1958 until February 1971.[77] On June 11, 1957, the Kishi Cabinet decided to support the AOC via the Japan Export–Import Bank.[78] The Cabinet declared that the acquisition and right to develop in Saudi Arabia would be extremely effective given the increase in oil demand and dependence on imports. Former Prime Minister Yoshida Shigeru also agreed to the idea of overseas oil exploration, commenting that Japan would not be judged as a "first-class nation" until it had its own petroleum resources.[79] Even the JSP supported this "national oil project."[80]

---

[75] Japan National Oil Corporation, *Sekiyukōdan Nijūnensi [20-Year History of Japan National Oil Corporation]* (Tokyo: Japan National Oil Corporation, 1987), 5–6; Shōji Tarō, *Yamashita Tarō to Hinomaru Genyu [Mr. Yamashita Taro and Hinomaru Crude Oil]* (Tokyo: Enerugi Horamu, 2007), 37–38. One Japanese author explains the reason for refiners' absence among investors in the AOC. He argues that Yamashita Taro did not ask for help from Japanese refiners affiliated with Western oil majors. See Yamaoka Junichiro, *Tanaka Kakuei Fujirareta Shigen Senryaku: Sekiyu, Uran, Sosite Americatono Toui [Tanaka Kakuei's Sealed Resource Strategy: Oil, Uranium and Battle with the U.S.]* (Tokyo: Soshisha, 2009), 89.

[76] Arabian Oil Company, *The 35-Year History of the Arabian Oil Company 1958–1993: A Bridge between the Arabian Gulf and Japan in War and Peace* (Tokyo: Arabian Oil Company, 1995), 39–40. For Mr. Ishizaka's life, see Harold R. Kerbo and John A. McKinstry, *Who Rules Japan?: The Inner Circles of Economic and Political Power* (Westport, Conn.: Praeger, 1995), 69.

[77] Arabian Oil Company, *The 35-Year History of the Arabian Oil Company 1958–1993: A Bridge between the Arabian Gulf and Japan in War and Peace*, 238–46.

[78] Japan National Oil Corporation, *Sekiyukōdan Nijūnensi [20-Year History of Japan National Oil Corporation]*, 6–7. See also Samuels, *The Business of the Japanese State: Energy Markets in Comparative and Historical Perspective*, 194–95. Yamashita Taro's personal network with Prime Minister Kishi and Kobayashi Ataru, President of Development Bank from 1951 until 1957, was also effectively utilized for financing the AOC. See Johnson, *Japan's Public Policy Companies*, 42.

[79] Arabian Oil Company, *The 35-Year History of the Arabian Oil Company 1958–1993: A Bridge between the Arabian Gulf and Japan in War and Peace*, 41.

[80] Caldwell, "Petroleum Politics in Japan: State and Industry in a Changing Policy Context," 86–87. However, it should be noted that not every opinion leader supported the project. Mr. Matsunaga, a tycoon of the power generation, was largely silent on the matter. Prof. Wakimura, then Dean of the Faculty of Economics at the University of Tokyo and a respected leading national authority on the petroleum industry, strongly opposed the private oil venture, arguing that purchasing oil

The AOC found a giant oil well in the Neutral Zone in January 1960 and began producing and shipping oil back to Japan in April 1961. As a result, annual oil production increased from 1.5 million kl in 1961 to 6.5 million kl in 1962 and was expected to reach 10 million kl in 1963.[81] However, this initial success quickly turned into a headache for policymakers due to a change in the global oil market at the time. There was a worldwide excess supply of crude oil, including the Soviet Union's sales of oil at a low price. In the domestic market, many Japanese oil refiners had long-term oil purchase contracts with Western oil majors, which accounted for more than 80% of crude oil demand in Japan.[82] In addition, the Japanese refinery sector argued that the oil produced by the AOC was relatively expensive and that the high sulfur content could cause pollution. The AOC's oil thus became a political issue that needed to be resolved.

2.2 Debate on the Petroleum Industry Law and a State-Owned Oil Company

Inception of the Proposal in Petroleum Industry Law

When MITI initiated enactment of the Petroleum Industry Law in 1961, the proposal to create a state-owned oil company was considered with the support of *Enerugi Kondankai* (Energy Roundtable meeting) chaired by Prof. Arisawa. While considerations of political expediency meant the proposal was excluded from MITI's draft bill submitted to the LDP for review before the 1962 regular Diet session, the proposal was revived during subsequent review by other veto players.

In November 1961, MITI announced its willingness to submit the Petroleum Industry Law draft during the regular Diet session, aiming to protect Japanese oil markets after the oil import liberalization of October 1962.[83] The original draft kept some contents of previous Petroleum Industry Law that had existed during wartime. Companies were required to submit

from Western oil majors is more economically efficient than is undertaking risky oil exploration. See Arabian Oil Company, *The 35-Year History of the Arabian Oil Company 1958–1993: A Bridge between the Arabian Gulf and Japan in War and Peace*, 41–42.

[81] Comment by Mr. Kawade, then-MITI Mining Bureau Chief, at Japan House of Representative, Standing Committee on Commerce and Industry, *28th Committee Meeting*, Interpellation, April 11, 1962.

[82] Arabian Oil Company, *The 35-Year History of the Arabian Oil Company 1958–1993: A Bridge between the Arabian Gulf and Japan in War and Peace*, 74–75.

[83] "Petroleum Industry Law Will Be Submitted to Regular Diet Session," *Asahi Shinbun*, Nov. 25, 1961.

production plans and import plans, MITI retained its power to recommend sales price changes, and MITI could cancel business permission.[84] On December 14, 1961, seven of the eight members of the *Kondankai* expressed support for the original draft bill, which included MITI's power to establish stability in the oil business.[85] In the oil policy report of December 22, 1961, along with concerns about the monopolization of the oil market by a small number of companies, the *Kondankai* also expressed the need to accommodate domestic oil sales for Japan's oil development companies, such as the AOC. The interim report of the Europe Petroleum Study Group on the same day asserted the need to enact the law.[86] The report also stated that European countries had ensured a cheap and stable energy supply by controlling a certain portion of the energy supply independently of the Western oil majors, and that Japan should follow this model by securing approximately 30% of its petroleum product supply under national control.[87] Furthermore, a majority of *Kondankai* asserted a need to establish a half-public half-private style state-owned company with investment from the government and big oil demanders like refiners, utility companies, and steel companies.[88] As concrete examples, the *Kondankai* recommended two options: One was to establish a full-fledged state-owned oil company to deal with oil development, refining, and sales. The other was to found a state-owned investor from which oil refiners would have bought special crude oil (*tokushu genyu*).[89]

[84] Nippon Oil, *Nihonsekiyu Hyakunenshi [100-Year History of Nippon Oil]* (Tokyo: Nippon Oil, 1988), 591. MITI excluded the articles stipulating strong control, such as the government's oil import and companies' obligation to stockpile oil.

[85] "Petroleum Industry Law Needed after Liberalization," *Asahi Shinbun*, Dec. 15, 1961. The members were Hiromi Arisawa (Dean of Hosei University), and seven others. One minority member was Prof. Watanabe from the University of Tokyo. He was later excluded from the energy subcommittee under the Industrial Structure Research Council.

[86] Nippon Oil, *Nihonsekiyu Hyakunenshi [100-Year History of Nippon Oil]*, 590–91. This study tour group was initiated by Prof. Arisawa, an influential academic figure in Japan. The study tour group visited the United Kingdom, France, Italy, and West Germany, and published a book, *Enerugi Seisaku no Shin Tenkai* [*New Development of Energy Policy*].

[87] This 30% became a policy target for Japan's oil self-sufficiency for a certain period. See Suzuki Shigeru, *Nihon No Enerugi Kaihatsu Seisaku [Japan's Energy Development Policy]*, 160–61.

[88] Ministry of International Trade and Industry, *Tsūshōsangyōsesakusi [the History of International Trade and Industry Policy]*, 17 vols., vol. 13 (Tokyo: Ministry of International Trade and Industry, 1991), 507–09. Prof. Arisawa seemed to be impressed by ENI, the Italian state-owned oil company. His strong interest in ENI can be inferred from the final report of Research Mission for European Oil Policy. See Arisawa Hiromi, *Enerugi Seisaku No Shindankai: Ōshū No Enerugi Kakumei [New Stage of Energy Policy: European Energy Revolution]* (Tokyo: Daiyamondosha, 1963).

[89] Comment by then-MITI minister Sato Eisaku at *28th Committee Meeting*.

MITI was committed to some form of national oil corporation, and the ministry supplemented the *Kondankai* recommendations with four additional proposals: (1) A policy company dealing with the purchase of *Hinomaru* oil and refining, sales, and investment; (2) an investment company for refiners that would purchase *Hinomaru* oil; (3) a petroleum stabilization fund for subsidizing domestic oil development and refiners; and (4) a special oil wholesaler to buy oil developed by the AOC and NOSODECO. MITI's preference as expressed in their 1962 report was that "the establishment of a public policy company is no less important or fundamental vis-à-vis a stable supply structure than the Petroleum Industry Law."[90]

As mentioned already, the situation of Japan's oil industry justified the idea of a state-owned company. On the one hand, Japan's upstream oil industry was very weak, and only four Japanese companies had explored oil: Teikoku and JAPEX in the domestic area and the AOC and the NOSODECO in the Middle East and Indonesia, respectively. On the other hand, Japan's downstream industry was relatively big, but Western oil majors supplied more than 80% of oil to Japan. Many *Keiretsu*-related refiners were affiliated with Western oil majors, and were thus required to use oil provided by those companies.[91] In addition, the companies that were not affiliated with Western oil producers also had to import the foreign companies' oil in return for cheap loans. For Western oil majors, Japan was a fast-growing market in which they could sell overproduced oil. This was a negative for the AOC, Japan's first successful overseas oil venture, because the AOC was not prepared to sell its oil in the world market and expected a certain market share in its domestic market. However, with the constrained supply of oil in Japan, the AOC's expansion in *Hinomaru* oil production from 6 million kl in 1961 to 10 million kl in 1962 only increased the concern regarding sales of the oil, and its high sulfur ratio lowered the oil's attractiveness. In this context, the establishment of a state-owned oil company became an important issue in Japan.

---

[90] Samuels, *The Business of the Japanese State: Energy Markets in Comparative and Historical Perspective*, 199–200.

[91] This tie-up took the form of joint ventures, and this partnership weakened political opposition within the occupation authorities that prevented Japan from rehabilitating its petroleum industry, which was considered war-related industry. Between 1949 and 1952, Japanese major companies partnered with foreign oil companies. Mitsubishi Oil partnered with Tidewater Oil, Tōa Nenryō with Standard Vacuum (February, 1949), Kōa Oil with Caltex Oil Products (July, 1949), Maruzen Oil with Union (August, 1949), Nihon Sekiyu with Caltex (1951), and Showa Oil with Shell (December, 1952). See Llewelyn Hughes, "The Limits of Liberalism?: Long-Run Petroleum Prices and Government Intervention in Petroleum Markets in Japan, France, and the United States" (Thesis [Ph.D.], Massachusetts Institute of Technology, 2009), 122.

Therefore, in the process of preliminary consultation between MITI and businesses for drafting a final bill, one large concern was whether the bill would include a state-owned wholesaler, which would have evolved into a typical state-owned oil company. In particular, the PAJ (*Sekiyu Renmei*), representing refiners, and the JPDA (*Sekiyu Kōgyō Renmei*), representing oil developers, made efforts to change the draft bill on their own behalf. Japanese refiners also worried that a state-owned wholesaler would force them to buy *Hinomaru* oil and that the government would intervene in downstream business more directly. Japanese oil developers supported the establishment of a state-owned oil wholesaler who would buy special crude oil that they developed in domestic areas or overseas. This state-owned wholesaler could have been expanded to a vertically integrated state-owned oil company like Italy's ENI. As mentioned during the NHK 7 O'Clock News of April 9, 1962, it would have been inevitable to merge the AOC and the NOSODECO to form the exploration and development (E&P) arm of the state-owned company.[92] This would have generated a significant problem if Japan's state-owned wholesaler bought only the oil produced by a few specific Japanese private companies.[93] Also, in a favorable political situation, it is possible for a state-owned wholesaler to expand its function to E&P, as seen in the case of KOGAS in the early twenty-first century in Korea.

During preliminary consultations, the idea of a state-owned oil wholesaler was included a few times over the course of nine revisions of MITI's draft bill, but the idea was ultimately excluded from the final draft due to strong opposition from Japan's refiners.[94] In a situation where the private sector strongly opposed MITI's intervention in the oil market through the law, the inclusion of a state-owned wholesaler was not viable. The refiners such as Idemitsu and Caltex, together with the Association of Electricity Providers, the Coal Association and the Steel Association, opposed MITI's control of the refinery

[92] The news stated that the government considered establishing a state-owned oil company, whose core companies would be the AOC and the NOSODECO. Although denied by Sato Eisaku, the then-MITI Minister, this news shows that the state-owned oil wholesaler would have been the stepping-stone to a vertically integrated state-owned oil company, including not only refinery and sales, but also exploration and development. See a comment by Kato Seiji (Japan Socialist Party) in *28th Committee Meeting*.

[93] According to Minister Sato Eisaku, the United States and the United Kingdom sent messages to the Japanese government through diplomatic channels asking not to discriminate against foreign companies. A comment by the then-MITI Minister Sato Eisaku in Japan House of Representative, Standing Committee on Commerce and Industry, *21st Committee Meeting*, Interpellation, Mar. 23, 1962.

[94] Petroelum Association of Japan, *Sengo Sekiyu Sangyōsi [Postwar History of Oil Industry]*, 159.

business and suggested a five-year temporary law.[95] Businesses were suspicious of MITI's attempts to intervene in the economy, and even Western oil majors were watching the debate in Japan.[96]

Aggregating the opinions of all sectors, *Keidanren* proposed a maximum five-year sunset law and recommended that the regulations should be the minimum required to stabilize the market. However, *Keidanren* also showed an equivocal position on overseas oil development. At the energy subcommittee meeting under its Energy Countermeasure Committee of February 17, 1962—in which not only oil demanders like Nippon Oil, Idemitsu Oil, Mitsubishi Oil, and Tokyo Electric, but also oil developers like Teikoku Oil, the Arabian Oil, and JAPEX participated—the opinion of *Keidanren* was drafted as follows.

> Measures should be taken to consume the oil developed by the AOC in the domestic market. While this should follow the discussion in the business, the government needs to support [the sales of the AOC's oil] as much as possible.[97]

Considering the fact that the *Keidanren* is an all-inclusive economic organization and that the then-chairman of the organization worked as chairman of the AOC, it is likely that the *Keidanren* energy committee could not fall completely on one side of either oil developers or oil refiners. While *Keidanren* strongly opposed MITI's over-intervention in the market, its stance on the proposal of a state-owned oil company was a little ambiguous. This position was confirmed at *Keidanren*'s Energy Countermeasure Committee of February 22, 1962, and its final position was submitted to the government and the Diet as follows.

---

[95] "Proposal of Businesses for Petroleum Industry Law, Identical among Oil, Coal and Electricity," *Asahi Shinbun*, Feb. 15, 1962; "Petroleum Industry Law is against Liberlaization. Steel Industry Decided to Oppose," *Asahi Shinbun*, Feb. 16, 1962.

[96] "US Oil Capitals Watch Petroleum Industry Law Bill with Strong Interest," *Asahi Shinbun*, Feb. 17, 1962.

[97] "Approximately 5-Year Temporary Law, No Power to Licence Facility, Keidanren's Attitude on Petroleum Industry Law," *Asahi Shinbun*, Feb. 17, 1962. Samuels misunderstands that Keidanren showed "clear and determined opposition," on national oil corporation, but *Keidaren* was cautious about the issue. It seems that Samuels neglected the weakness of *Keidaren*, which was caused by the nature of an all-inclusive and nation-wide economic organization. This may be attributed to Samuels' attempt to define Japan's state-market relation in his framework of "reciprocal consent."See Samuels, *The Business of the Japanese State: Energy Markets in Comparative and Historical Perspective*, 201–02; "Approximately 5-Year Temporary Law, No Power to Licence Facility, Keidanren's Attitude on Petroleum Industry Law."

> On the matter of the special oil developed by the AOC, the government should make an effective policy to secure its domestic sales, and the private sector will consult in order to cooperate for the secured sales.[98]

With the goal of getting the law passed in mind, the then-MITI Minister Sato Eisaku tried to persuade the private sector by revising the draft to remove the article that would have given MITI power to recommend standard prices and add an article stipulating that the law could be abolished or weakened according to the situation.[99] With the final meeting on February 27, 1962 between MITI officials including Minister Sato Eisaku, and the representatives of *Keidanren*, such as Vice Chairmen Sato Kiichi and Uemura Kōgorō, MITI finished the preliminary consultations and finalized its draft bill.[100] Facing oppositions from the private sector, MITI's final draft bill of the Petroleum Industry Law omitted plans for a state-owned oil company.

From the perspective of Sato and MITI in general, the absence of a proposal for a state-owned oil company in the draft bill was politically expedient. In order to maintain MITI's power to intervene in the oil market, they believed that the legal regulation replacing the foreign exchange allocation system should be enacted in this 1962 Diet regular session before the scheduled liberalization of crude oil import in October 1962.[101] It seems that his alleged personal connection with the AOC would have obstructed passage of the law during the legislation process if MITI had maintained the proposal for a state-owned wholesaler. Sato Ryotaro, the first son of Sato Eisaku, was known to work at the AOC at that time, and Mr. Kano, who had worked together as Deputy Chief Cabinet Secretary when Mr. Sato was Chief Cabinet Secretary, was an executive member at the AOC. Former Prime Minister Kishi, Mr. Sato's biological brother, supported Japan's overseas oil

98 "Necessary Minimum Regulation, Keidanren Response Committee Finalized Opinion," *Asahi Shinbun*, Feb. 22, 1962.

99 "Big Demanders Show Disapproval to the Color of Control in Petroleum Industry Law Bill," *Asahi Shinbun*, Feb. 2, 1962.

100 "Representative of Keidanren Proposed to MITI Minister on Petroleum Industry Law," *Asahi Shinbun*, Feb. 28, 1962. Here, Samuels slightly overstates the impact of the meeting between MITI Minister Sato and representatives of *Keidanren*, which was held at Kaiun Club on February 27, 1962. Though Samuels describes the event as if Sato Eisaku fulfilled what *Keidanren* representative requested, the reality is different. Sato had already given way in some measures, such as recommending standard prices. In the meeting, Sato stated "the currently considered draft bill weakened the color of control, so there are few things for *Keidanren* to worry about." While Samuels cites Ōi to argue that there was no major difference between the industry's position and final legislation, *Keidanren*'s proposal of a five-year temporary law was in fact not accepted in the Diet. See Samuels, *The Business of the Japanese State: Energy Markets in Comparative and Historical Perspective*, 202.

101 Comment by then-MITI Minister Sato Eisaku at *28th Committee Meeting*. Mr. Sato mentions that the bill is the best option from a political viewpoint.

development and was suspected of corrupt relations with the AOC.[102] Thus, inclusion of the proposal would have risked both MITI's control and Mr. Sato's political life.

Revival of the Proposal During the Review of the LDP

During the review process in the LDP, the fate of the state-owned oil company changed. On March 2, 1962, the Division of Commerce and Industry under the LDP PARC approved MITI's draft of Petroleum Industry Law.[103] This meant that the PARC was not enthusiastic about the idea of a public oil company, but this situation was altered in the Executive Council, another important partisan veto player in Japan's policymaking. When a vice ministers' meeting on March 5th decided to submit MITI's draft bill for Petroleum Industry Law to the Diet on the condition that the LDP Executive Council would approve it,[104] the Executive Council revived the proposal to establish a state-owned oil company.[105] Considering the Executive Council seldom overruled a policy recommended by the PARC with the exception of budget draft bills, this change was unusual.[106] At that time, the Chair of the Executive Council was Mr. Akagi Munenori of the Kishi faction,[107] who may have gone along with Kishi in supporting Japan's overseas oil development.[108] With the prompt implementation of a comprehensive energy plan, the Executive Council meeting of March 6, 1962 raised one more condition for approval of the draft bill: MITI must consider establishing a special public corporation to buy and sell domestically produced oil as well as overseas oil

---

[102] Comment by Yoshida Hosei, upper house member from Japan Socialist Party at Japan House of Councillors, Standing Committee on Commerce and Industry, *28th Committee Meeting*, Interpellation, April 27, 1962.

[103] "Division of Commerce and Industry Approved Petroleum Industry Law," *Asahi Shinbun*, Mar. 2, 1962.

[104] Ministers were so busy that every draft bill was thoroughly examined in the Vice Minister meeting. See footnote of Fukui and University of California Berkeley. Center for Japanese and Korean Studies, *Party in Power: The Japanese Liberal-Democrats and Policy-Making*, 89–90.

[105] "Petroleum Industry Law and Others Are Informally Decided to Be Submitted to Coming Diet Session," *Asahi Shinbun*, Mar. 6, 1962.

[106] Fukui and University of California Berkeley. Center for Japanese and Korean Studies, *Party in Power: The Japanese Liberal-Democrats and Policy-Making*, 91.

[107] Masumi, *Contemporary Politics in Japan*, 446. And Liberal Democratic Party of Japan, "Liberal Democratic Party - Terms in Office of Top Party Leaders," http://www.jimin.jp/english/the_president/leaders/104392.html.

[108] The Kishi faction was formally dissolved on October 31, 1962, and 25 of its 49 members formed Kōyū kurabu under the joint direction of Mr. Akagi Munenori and Mr. Kawashima Shōjirō. See Fukui and University of California Berkeley. Center for Japanese and Korean Studies, *Party in Power: The Japanese Liberal-Democrats and Policy-Making*, 110.

developed by Japanese companies. In addition, there were other minority opinions: Some maintained that because the law strengthened bureaucratic control, it should include a time limit; meanwhile, others argued the law should be stronger.[109] The LDP Executive Council had given a new life to the proposal of a Japanese state-owned oil company.

Support for State-Owned Oil Company in the Cabinet Meeting

In the Cabinet meeting of March 7th, the MITI minister mentioned that the ministry would prepare for the establishment of a state-owned oil company and that it would submit the proposal to the Diet one year later. People became interested in whether the company would only be a state-owned wholesaler that would buy *Hinomaru* oil or be a state-owned oil company that also had its own refinery.[110]

Discussion in the Diet and Attached Resolution Supporting SOE

In the 1962 Diet regular session, the bill was passed with a resolution that the government needed to establish a state-owned oil wholesaler. The need to establish a policy company was raised mainly by opposition party members, and Nakagawa Shunji, a member of the LDP in the Lower House, also strongly supported that idea. Representatives of various interest groups had opportunities to express their opinions both in the Lower and the Upper House. While the proposal to establish a state-owned oil company did not materialize in the 1962 regular session and only produced the attached resolution, the interpellation and the hearing explicitly revealed the situation of Japan's oil industry.

The role of opposition parties made the Diet play the role of real veto player in policymaking. The opposition members pointed out that a state-owned company that would purchase special oil, such as the oil developed by the AOC, had to be formed before the liberalization of crude oil imports expected in October of that year. MITI Minister Sato Eisaku was reluctant to incorporate their suggestions into the bill on the spot, and whenever Diet members argued the need to establish a state-owned wholesaler for *Hinomaru*

[109] "Liberal Democratic Party Approved the Draft Bill Conditionally," *Asahi Shinbun*, Mar. 6, 1962.

[110] "Comprehensive Energy Plan Including Wholesaler Will Be Examined," *Asahi Shinbun*, Mar. 7, 1962. Though MITI's draft bill passed the LDP councils, there was still a change of the draft. Due to the strong opposition of businesses, the article that would give MITI the power to enter business sites for inspection was removed. "In the Draft of Petroelum Industry Law, the Power to Enter for Inspection Removed," *Asahi Shinbun*, Mar. 7, 1962.

oil based on *Kondankai*'s interim report of December 1961, he argued that MITI would examine the proposal with care.[111] This gap between MITI's preferences and MITI's behavior could be attributed to the inevitable choice they had to make in order to prevent the bill from being blocked in this legislation process. To pass the draft bill in the Diet during this session, which was the only opportunity for MITI to change the status quo of no legal regulation, Minister Sato needed to strike a balance in the spectrum between the free market and strong governmental intervention in dealing with this delicate issue. Also, the interest of Western oil majors could not be neglected. The level of governmental intervention could affect their profit-making, and, therefore, Western oil majors revealed their opinions via diplomatic channels. Interest groups' positions shown at the hearings of March 29th and April 25th in the Lower House and the Upper House, respectively, were quite contrasting, especially between the PAJ, representing refiners, and the JPDA, representing oil developers. Mr. Mimura, Chairman of the JPDA, supported the Petroleum Industry Law and also argued for the need to establish a state-owned oil wholesaler for *Hinomaru* oil and oil stockpiling. Mr. Itō, Chairman of the Central Committee of National Petroleum Development Labor Union, also assisted Mr. Mimura. On the other hand, Mr. Nanbu, Vice-Chair of the PAJ, hesitated to accept the AOC's oil due to its high sulfur content.[112]

Varied and contrasting views of the AOC might have further burdened MITI. On the one hand, the AOC was admired as Japan's first success in overseas oil development, having improved the country's balance of payments. On the other hand, the AOC's contract terms in which Japan provided Saudi Arabia and Kuwait with 57% of all profits plus infrastructure investments and rental fees was considered an indignity for some, primarily because the AOC offered better terms for the host countries than the Western majors would. The rumors that the AOC was only lobbying for political benefit might be unjustly used to undermine the credibility of the argument for oil developers.[113] The negative rumors included MITI Minister Sato's alleged personal connection with the AOC. For the MITI minister, the current draft law was "politically best."[114]

---

[111] Japan House of Representative, Standing Committee on Commerce and Industry, *23rd Committee Meeting*, Interpellation, Mar. 28, 1962.

[112] Japan House of Representative, Standing Committee on Commerce and Industry, *24th Committee Meeting*, Interpellation, Mar. 29, 1962.

[113] Comment by Itagawa Shōgo, Diet member of Japan Socialist Party, at Japan House of Representative, Standing Committee on Commerce and Industry, *27th Committee Meeting*, Interpellation, Apr. 10, 1962; Samuels, *The Business of the Japanese State: Energy Markets in Comparative and Historical Perspective*, 195.

[114] A comment by Sato Eisaku in Japan House of Representatives, Standing Committee of Commerce and Industry *28th Committee Meeting*.

The final outcome in the Diet made some differences from the perspective of the private sector. Both the Lower and the Upper house asked MITI to establish a system to receive the developed oil by Japanese companies by attaching a supplementary resolution (*futai ketsugi*).[115] In the Lower House, the majority LDP and the opposition JSP and DSP unanimously suggested such a resolution on April 11, 1962. In the Upper House, Mr. Nakata, a Diet member of the JSP, raised a similar suggestion, and it was also passed unanimously on April 27, 1962.

Revision of the draft law during the Diet session showed that the decisions by Japan's veto players in the public sector were not always aligned with the position of interest groups including *Keidanren*. With the suggestion of Tanaka Takeo and Itō Ushirō from the JSP and the DSP respectively, the article on reviewing the bill was removed in the Lower House.[116] This can be attributed to the opposition parties' concern for Japanese-owned capital. The suggestion also received support from the LDP. The removed article was originally added due to a request from *zaikai*, which had argued that Petroleum Industry Law should be temporary. Now, the draft bill on a state-owned wholesaler was to be submitted during the following year's Diet session.

3 New Initiatives for a State-Owned Oil Company

This section describes how MITI, MOF, and *Keidanren* play as veto players in Japan's policymaking. Whenever MITI proposed ideas on the state-owned company, the MOF, the financial regulator, had rejected MITI's proposal. This situation of stalemate continued until 1967, when MITI proposed the moderate idea to found a special financier for the private sector's overseas oil development. The atmosphere could be changed with the help of *Keidanren*. MITI's proposal was accepted by the MOF, and the Diet passed the Japan Petroleum Development Corporation (JPDC) Law in 1967. The JPDC was thus born, not as a state-owned oil company, but as a national upstream funding agency for private oil developers.

---

[115] The supplementary resolution is an explicit way for opposition parties to influence the bureaucracy. It provides the ministries with guidelines for the implementation of legislation. See Jung, "Reassessing Reform Outcomes in South Korea and Japan a Decade after Crisis: The Paradox of Political Leadership," 325.

[116] Article 4 in the supplementary provisions in the draft bill stated that seeing the oil situation and economic situation domestically and internationally, the review will be conducted and the law will be loosened or abolished. *28th Committee Meeting*. Comment by Itō Ushirō, the Lower House member from the DSP.

3.1 MITI's Initiative and MOF's Obstruction

MITI began to advance its proposal at a time when the Petroleum Industry Law did not guarantee an accommodating environment for special oil, especially the oil developed by the Japanese-owned AOC. In June 1962, MITI asked the Industrial Structure Delegation Committee's Energy subcommittee, chaired by Prof. Arisawa, to review the needs and to detail a proposal for state-owned oil wholesalers. This had been opposed by the *zaikai* with a concern that the establishment of a policy company strongly implied "controlling" (*tōseisaku*). The impact on foreign-affiliated refiners was expected to be enormous and the required budget size was also a problem. Therefore, the proposal carried with it the danger of arousing significant controversy.[117]

On August 4, 1962, when refiners and oil developers had not yet reached agreement on the purchase of *Hinomaru* oil, MITI decided to push the proposal for a special oil wholesaler to buy oil developed by the AOC and NOSODECO, one of four options that had previously been discussed.[118] However, soon after, a compromise was reached between the AOC and refiners, lowering the persuasiveness of founding a state-owned oil wholesaler. MITI also missed its opportunity to reflect the proposal in the budget draft for 1963.[119]

During discussion of the state-owned wholesaler proposal in the oil subcommittee in 1963, MITI proposed the idea of a petroleum stabilization fund. The function was intended to be an investment for refiners, loans for oil stockpiling, and loans for oil development. Interest groups repeated their own positions. The JPDA asked to set up the system to buy *Hinomaru* oil and to establish a special policy corporation for oil policy. The PAJ, though, opposed any attempt to found a state-owned oil wholesaler. In this discussion, *Keidanren* suggested strengthening the state's subsidization by introducing a so-called success repayable loan, in which the borrower is required to return the funds only when the project is successful.[120] While the oil subcommittee suggested a state-owned oil company,[121] the Finance

117 "MITI Asked Industrial Structure Delegation Committee to Review Crude Oil Policy Company," *Asahi Shinbun*, June 5, 1962.

118 Ibid.

119 Japan Petroleum Development Association, *Sekiyu Kogyō Renmei Nijunen No Ayumi[20 Year-Footprints of Japan Petroleum Development Association]*, 33.

120 Ibid., 35–37.

121 Ibid., 39.

Ministry's decision to prohibit ministries from establishing public companies obstructed fruition of any related proposal.[122] The MOF's rejection was announced in early December 1963 at the beginning of the budget modification process with ministries and the LDP. MITI omitted the proposal and the MOF distributed 200 million yen for oil exploration of JAPEX instead.[123]

In the subsequent two years, MITI's effort did not produce anything. In 1964, MITI proposed the Overseas Oil Development Loan Business Unit, and its budget, 300 million yen of public investment and 550 million yen of public loans, was included in MITI's budget request for 1965.[124] However, the idea did not attract support. In 1965, MITI Minister Miki proposed to establish the Crude Oil Corporation (*genyu kōdan*), which would have purchased and sold *Hinomaru* oil and supported oil development. For this proposal, the PAJ opposed and the JPDA supported again.[125] The Oil & Gas Resources Development Delegation Committee replied to MITI Minister Miki that Japan needed a core entity to deal with crude oil development, refining, and sales in order to increase its self-sufficiency ratio from less than 15 to 30%.[126] Also, in September 1964, the Overseas Energy Research Group chaired by Prof. Arisawa suggested an "integrated oil company" that would receive support from the government.[127] In spite of support from the *shingikai*, the idea could not be materialized due to MOF's decision not to permit any public corporation.[128] MITI had to hold the proposal, and instead asked for 4.1 billion yen to support overseas oil development projects.[129] The final amount distributed by the MOF totaled two billion yen.[130]

---

122 "Ministry of Finance Decided Not to Permit Any Public Companies," *Asahi Shinbun*, Dec. 3, 1963.

123 Japan Petroleum Development Association, *Sekiyu Kogyō Renmei Nijunen No Ayumi[20 Year-Footprints of Japan Petroleum Development Association]*, 41.

124 Comment by Mr. Kanai, then-MITI accounting division chief, at Japan House of Representative, Standing Committee on Commerce and Industry, *2nd Committee Meeting*, Interpellation, Dec. 17, 1963.

125 "PAJ Did Not Support MITI's Crude Oil Corporation at the Meeting," *Asahi Shinbun*, July 28, 1965.

126 "Oil Shingikai Supported Crude Oil Corporation for Overseas Resouces Development," *Asahi Shinbun*, Aug. 21, 1965.

127 Comment by Itagawa Shōgo, Diet member of Japan Socialist Party, at Japan House of Representative, Standing Committee on Commerce and Industry, *37th Committee Meeting*, Interpellation, May 18, 1965.

128 Japan Petroleum Development Association, *Sekiyu Kogyō Renmei Nijunen No Ayumi[20 Year-Footprints of Japan Petroleum Development Association]*, 55–56.

129 "MOF Will Not Permit Public Corporation," *Asahi Shinbun*, Jan. 10, 1966.

130 Japan Petroleum Development Association, *Sekiyu Kogyō Renmei Nijunen No Ayumi[20 Year-Footprints of Japan Petroleum Development Association]*, 56.

3.2 MITI's Changed Proposal and MOF's Reluctant Acceptance

In the mid-1960s, the conditions for oil policymaking became more favorable to MITI, due to the increasingly amicable stand of veto players. The Japanese private sector began to consider oil business as a new lucrative business opportunity, and thus interest groups including *Keidanren* moved in that direction. Japan's domestic political scandal in 1966 added momentum to overcome MOF's consistent obstruction of a public corporation.

In July 1966, MITI proposed a moderate idea of the JPDC that would support the private sector financially. Under this plan, the JPDC would implement loans, equity investment, loan guarantees, exploration machine lending, and basic research for oil and natural gas development. The first year's budget of 1.5 million yen would have been financed by the Industrial Investment Account.[131] Officially, this proposal was based on the reply of Comprehensive Energy Research Committee in February 1967. The committee was composed of 20 regular members and 16 temporary members, and Mr. Uemura Kōgorō, *Keidanren* Vice Chair, and Prof. Arisawa, a supporter of state-owned oil company, were involved in the committee as the committee chair and the oil subcommittee chair, respectively. The committee's first final report was submitted on February 22, 1967, just before a political compromise was reached among veto players, and it stated that Japan needs to develop equity oil to meet up to 30% of domestic oil demand by 1985. For the autonomous oil supply amounting to approximately 140 million kl, the exploration cost was estimated by MITI to be 300 billion yen for the explorations of 45 wells and the development cost was 800 billion yen.[132]

This moderate idea was initiated by Mr. Morozumi Yoshihiko, who succeeded Ōjimi Yoshihisa as MITI Mining Bureau Chief in March 1966. While Mr. Morozumi had been one main initiator of the Special Measure Law with Mr. Sahashi in the early 1960s, now his policy orientation at MITI was internationalist rather than nationalist. Also, Sahashi Shigeru retired from the vice-ministership on April 25, 1966.

For the private sector, it was natural to support MITI's proposal this time. After listening to Mr. Morozumi's explanation, the *Keidanren* energy committee announced its official support for the idea of the JPDC on October 17, 1966, and showed its intention to sway the MOF, which

131 Ibid., 56–58. "Japan Petroleum Development Corporation Will Be Established," *Asahi Shinbun*, July 29, 1966.

132 *26th Committee Meeting*.

continued to oppose the plan. The MOF's funding was directed only to JAPEX, a half-public half-private oil development company, but the business environment and financial situation made Japan's businesses more involved in oil development. Japan's general trading companies became more interested in oil business ranging from investment in oil refineries and oil development to sales of petroleum products.[133] Also, the business world, including Teikoku Oil and Alaska Pulp, established Alaska Petroleum Development Company in September 1966, and Yawata Steel founded the Kyushu Oil Development for the exploration of Southeast Kalimantan area of Indonesia in November 1966.[134] Still, the PAJ criticized the report of Comprehensive Energy Research Committee that autonomous oil development would negatively affect the profit of international oil majors.[135] However, Japanese refiners who had opposed the Crude Oil Corporation now did not obstruct this proposal firmly, because it meant the JPDC would not force them to buy *Hinomaru* oil.[136] The proposal that the JPDC would support the Japanese oil developers' overseas activities was possible because the AOC's oil was smoothly sold to Japanese refiners at that time.[137]

Facing MITI's even milder proposal, the MOF's stance still did not change at first. The MOF had considered that JAPEX could be used for overseas oil development instead of founding a new public corporation.[138] In addition, the government's administrative reform effort provided the MOF with a good reason to oppose the proposal. According to the reform plan, no new public company could receive a budget without authorization by the Ministry of Administration. Also, existing institutions needed to be scrapped if ministries wanted to build new public corporations.[139]

In the reshuffle of posts in the LDP and the Cabinet in December 1966, the change in leadership of main veto players created a rather favorable environment for MITI. The then-Prime Minister Sato was a former MITI minister, and Nishimura Naomi, Chair of the LDP PARC, was

---

[133] "Big General Trading House is Eager for Oil Business," *Asahi Shinbun*, June 12, 1966.

[134] Japan National Oil Corporation, *Sekiyukōdan Nijūnensi [20-Year History of Japan National Oil Corporation]*, 9.

[135] "Sekiyu Renmei Criticized the Reply of Energy Committee," *Asahi Shinbun*, Feb. 25, 1967.

[136] Japan Petroleum Development Association, *Sekiyu Kogyō Renmei Nijunen No Ayumi[20 Year-Footprints of Japan Petroleum Development Association]*, 59.

[137] Comment by Mr. Morozumi, then MITI mining bureau chief, at Japan House of Representative, Standing Committee on Commerce and Industry, *18th Committee Meeting*, Interpellation, June 9, 1967.

[138] "Ministry of Finance Will Repress Establishement of New Public Corporation," *Asahi Shinbun*, Aug. 30, 1966.

[139] "Government and Ruling Party Review Administrative Reform from Zero-Base," *Asahi Shinbun*, Aug. 30, 1966.

from the Sato faction. Shiina Etsusaburō was Chair of LDP Executive Council and had worked in MITI for many years, eventually serving as Minister from December 8, 1960 until July 18, 1961. More importantly, the Minister of Finance had changed from Mr. Fukuda, who favored conservative fiscal policy, to Mr. Mizuta Mikio, who had been Finance Minister under Ikeda and supported Ikeda's Income Doubling Plan through an expansionist approach.[140]

Under enormous pressure from ministries and interest groups, the MOF changed its stance little by little. In the previous year, the MOF announced that there would be "no" permission for new public corporation, but for the 1967 budget, MOF's position was that there would be no new public corporation "in principle." In the final party-government negotiation on February 28, 1967, the MOF minister accepted five new corporations, including the JPDC, among 25 proposals from the ministries.[141] During this political decision-making process among the veto players, the review of Ministry of Administration on 108 special corporations (*Tokushu hōjin*) was not given much consideration. For the Ministry of Administration to save face, the JPDC would be an entity created by restructuring existing JAPEX, and thus launching the JPDC itself did not increase the total number of public corporations.[142]

On the other hand, the "Black Mist" political scandal of 1966 created conditions which demanded that the LDP respond to the general populace and to pressure groups.[143] In August 1966, LDP diet member Tanaka Shōji was arrested on charges of extortion and fraud. The Transportation Minister, Defense Agency Director, and Agriculture Minister were accused of misuse of public authority, weakening the relative political strength of the LDP. Approval ratings of the Sato cabinet in public opinion polls dropped to 25%, and an anti-Sato movement arose, which was tied to a call for both party disciplinary action and the LDP presidential election. Prime Minister Sato received less support than previously estimated in the LDP election of December 1, 1966. The replacement of Tanaka Kakuei, the then-LDP

140 Campbell, *Contemporary Japanese Budget Politics*, 154.

141 "Will Ministry of Finance under Pressure Accept New Public Corporation," *Asahi Shinbun*, Dec. 19, 1966. "Budget, Unskillful Finish," *Asahi Shinbun*, Feb. 28, 1967. On the detailed procedure of revival negotiation, see Campbell, *Contemporary Japanese Budget Politics*, 172–99.

142 Quoted Points of the Draft Law at Japan House of Representative, Plenary Session, *14th Plenary Meeting*, Interpellation, May 18, 1967. Like JAPEX, the Nuclear Fuel Corporation and the Guidance Center for Small and Medium Company were restricted and incorporated into new public corporations.

143 "Budget, Unskillful Finish." The progress of the incident partly fits Calder's definition of a political crisis. See Calder, *Crisis and Compensation: Public Policy and Political Stability in Japan, 1949–1986*, 71–72.

Secretary General, drew attention in the Cabinet reshuffle following Sato's reelection. The LDP also received a concerted attack by the media and public, and the opposition parties were poised to boycott December's extraordinary session by being absent in the interpellations and deliberations on the supplementary budget. In the general election of January 1967, following dissolution of the Lower House in December 1966, the LDP maintained the majority, but their share of seats decreased from 283 to 277, in spite of the fact that the Lower House seats had increased by 19.[144]

While the JPDC launch was accepted, MITI could not achieve full financial support for the management of the JPDC due to the compromise reached between ministers, bypassing working-level consultation. According to Mr. Iwase, an MOF examiner (*shukeikan*), the decision to found the JPDC was not made at the working level. The autonomous oil supply target of 30% and MITI's cost estimation were not well communicated with the MOF, either.[145] Therefore, this reluctant permission at the ministerial level only resulted in a budget reduction of four billion yen for the fiscal year of 1967, and there was no long-term financing plan. The MOF minister's understanding was that the demands on public spending for the JPDC were evaluated as being not burdensome if the governmental financial support would be provided with strict standards.[146]

3.3 Finalized as a State Financier, Not a State-Owned Oil Company in the Diet

In the Diet session delegating the JPDC draft law from May 18 until July 21, 1967, the main issues were the need to share oil policy target among ministries, especially the MOF and MITI, and the need to make a concrete financing plan. How JAPEX would be merged into and separated from the JPDC in three years was also a question, and in line with that, whether the JPDC would be able to become a state-owned oil developer was also uncertain. While this issue was mainly raised by an opposition member, the Lower House and the Upper House, two collective veto players in Japan's decision-making process, could not make any change from the original JPDC

[144] Masumi, *Contemporary Politics in Japan*, 92–94.

[145] *26th Committee Meeting*.

[146] *14th Plenary Meeting*. Comment by then-MOF minister Mizuta Mikio.

proposal. Due to increased concerns about energy security created by the Six-Day War that coincided with the Diet session, the JPDC draft law itself easily passed.[147]

While oil had not been as politically attractive an issue as steel or fertilizer, the year of 1967 was different due to Japan's energy situation and increased concern for its energy supply following the third Arab–Israeli War, which took place in the middle of the Diet session. In 1967, Japan was the foremost global oil importer and the third-largest oil consumer. Therefore, the oil price hike caused by the Six-Day War and the subsequent shortage of oil tankers worried Japan's Diet members. At that time, more than 90% of oil was imported from the Middle East and Near East Asia, and Japan had only 21 days of crude oil reserves and 25 days of petroleum products. This deepened concerns among policymakers, even in terms of national security.[148]

Therefore, there were a broad range of questions on energy security about oil stockpiling, diversification of oil import sources, and autonomous oil supply. These issues were related to the possible areas in which the JPDC could expand into a public corporation. On July 5 and July 11, 1967, Mr. Itagawa Shōgo and Mr. Nakamura Jūko, Diet members from the JSP, asked to add a development function to the planned functions of the JPDC.[149] For supporters of a state-owned oil company, such as Diet members from the JSP, the temporary merger of JAPEX, a state-owned oil developer that had been established for domestic oil development in 1955, to the JPDC seemed like a good opportunity to make an ENI-style company. While it could not accept the revision of the draft law at the moment, the LDP, the JSP, and the DSP together suggested an attached resolution for that cause—the government needed to make positive efforts to expand the JPDC's tasks to include exploration and development projects, the purchase of *Hinomaru* oil and oil stockpiling.[150] Another potential added function was to support domestic oil and gas exploration. Mr. Okada Toshiharu from the JSP approached this issue from the perspective of the economic development of local areas. However,

---

[147] The Six-Day War began on June 5, 1967 and was over on June 11, 1967, which was during the 1967 Diet session in Japan. During the war, Japan announced its neutral position.

[148] Comment by Mr. Morozumi at *18th Committee Meeting*. Discussion between Mr. Kondo Shinichi, Diet memebr of Japan Socialist Party, and Mr. Morozumi at *19th Committee Meeting*.

[149] Japan House of Representative, Standing Committee on Commerce and Industry, *27th Committee Meeting*, Interpellation, July 5, 1967. Japan House of Representative, Standing Committee on Commerce and Industry, *28th Committee Meeting*, Interpellation, July 11, 1967.

[150] *28th Committee Meeting*. The oil stockpiling and its financial support became the JPDC's task in the fourth revision of the JPDC Law in June 1978, just before the second oil shock. In this revision, the JPDC was also renamed as the JNOC (Japan National Oil Corporation). See Japan National Oil Corporation, *Sekiyukōdan Nijūnensi [20-Year History of Japan National Oil Corporation]*, 471.

Mr. Morozumi, MITI Mining Bureau Chief, persuaded him that subsidies are efficient for domestic oil and gas development that normally has low potential for profit-making.[151] His wish was merely added to the resolution and did not receive more attention later.[152]

The Diet discussion revealed a hidden aspect of power of the MOF in the decision-making process. Mr. Suzuki Hajime, a Diet member of the DSP, raised discussion of a rumor of a behind-the-scenes deal in which the MOF would take the chair of the JPDC on the ground that *amakudari* from the MOF would be helpful to receive a budget for the JPDC.[153] This rumor was raised at the Diet management committee of the Upper House at that time. In 1967, the number of *amakudari* in 108 special public corporations was known to be 398 among a total 757 executive member level posts, and the *amakudari* from the MOF was 71, which was the most among ministries. An unrelated person's appointment looked problematic, as seen in the example of the former Self Defense Force Chief of Staff's appointment to Chair of the Housing Corporation.[154] In this context, who would become Chair of the JPDC was one issue yet to be resolved for MITI.[155]

3.4 JPDC's Activities for Overseas Oil Development

As seen above, the debate on the establishment of a Japanese national oil company in the 1960s ended with the production of the public corporation JPDC (which later became Japan National Oil Corporation, JNOC). Its role was constrained to a non-competitive form of state assistance as a special financial institution rather than an oil developer. In the support scheme of the JPDC, half of the project costs could be financed with the JPDC's success repayable loan, and later the portion of financial support increased to 70% of the total cost.[156] Japan adapted its success repayable loans from the

---

[151] Discussion between Mr. Okada Toshiharu, Diet member from Japan Socialist Party, and Mr. Morozumi at *26th Committee Meeting*.

[152] The financial support for domestic oil and gas development was included in the JPDC's task in the second revision of the JPDC Law in June 1975, after the first oil shock. See Japan National Oil Corporation, *Sekiyukōdan Nijūnensi [20-Year History of Japan National Oil Corporation]*, 471.

[153] Comment by Suzuki Hajime, House of Representative member from Democratic Socialist Party, at *18th Committee Meeting*. Upon Mr. Suzuki's argument, the then-MITI Vice Minister Uno Sosuke denied this on the spot. The then-MITI minister Kano also disowned the allegation on July 4, 1967.

[154] Comments by Kase Kan, Diet member from Japan Socialist Party, at Japan House of Councillors, Standing Committee on Diet Management, *16th Committee Meeting*, Interpellation, June 8, 1967.

[155] The controversy on *amakudari* to the JPDC was resolved by appointing Mr. Mimura Kiichi, former chair of JAPEX, as the first chair of the JPDC.

[156] Japan National Oil Corporation, *Sekiyukōdan Nijūnensi [20-Year History of Japan National Oil Corporation]*, 476–77.

West German model, with the difference that the JPDC's success repayable loan was not project-based but company-based. This was the outcome of the discussion with the MOF, which wanted the company to repay the loan if they were able to, in spite of the failure of the project.[157] Along with a few host countries' practices of permitting oil rights on project basis, this financing scheme made the one-project, one-company system prevalent in Japan's overseas oil development. In 1972, MITI shifted the JPDC's financing source from the Industrial Investment Special Account to the newly established Coal & Oil Special Account (*sekitoku kaikei*).[158]

In this environment, Japanese big businesses joined overseas oil development in the late 1960s. Mitsubishi Corporation acquired the Abu Dhabi onshore oil field with other Mitsubishi Group companies, such as Mitsubishi Heavy Industry, Mitsubishi Mining, Mitsubishi Refining, and Mitsubishi Petrochemical. Mitsubishi Corporation also established a special project company called Middle East Oil in 1968, for which half of the investment was supported by the JPDC.[159] Mitsui & Co. started its oil exploration history with a Thailand offshore oil exploration project that was suggested by Continental Oil (its current name is Conoco Philips), a friend to Mitsui Mining in 1969. Mitsui Group established Mitsui Oil Exploration (MOECO) for this project in the Gulf of Thai with 17 group companies participating including Mitsui & Co. in 1969.[160] Mitsui also joined Japan's first oil E&P in Africa in 1970 with Imperial Oil and Teijin. Sumitomo Corporation invested in the Saba offshore in Malaysia with Sumitomo Steel and Teikoku Oil by purchasing shares of the concession of SNEA, a French oil company known as Elf, in 1969. Marubeni began its oil development business in 1969 by joining Teikoku Oil's Saba-Teiseki in Indonesia.[161] Itochu Corporation began its overseas oil exploration in Indonesia's West Java offshore in 1969, which is Itochu Group's centennial anniversary year.

[157] Ibid., 17. The MOF's intention could not be realized, because the private sector evaded it by establishing project companies. Once a project failed, the company went bankrupt. Therefore, the distinction between company-based and project-based became useless.

[158] Ibid., 34.

[159] Ibid., 73–74.

[160] Ibid., 70. Unlike other Japanese companies, Mitsui & Co., Ltd. started its oil E&P in Thailand, not in Indonesia. This seems to be related to its long history of business in Thailand. Mitsui & Co. had commenced business activities in Thailand since 1906 and Mitsui Thailand was established in 1959. See Mitsui & Co. Thailand Ltd., "Company Profile," http://www.mitsui.co.th/en/company_profile/company_overview.php.

[161] JAPEX originally intended to invest in the project. However, Malaysia, which did not have a good relationship with Indonesia at that time, did not like that, so JAPEX just provided loans. The reason JAPEX got involved in this oil development project in a limited fashion was because it might be seen as too close to the Indonesian government otherwise. See Comment by Mr. Morozumi, then-MITI mining bureau chief, at *19th Committee Meeting*.

Through a production-sharing agreement, Itochu succeeded in finding the Ardjuna oil field.[162] Utility companies also joined the oil ventures. Tokyo Electric and Osaka Electric participated in oil development in Qatar with small refiners. The JPDC supported this project not only through finance but also through technology. Among Japanese oil developers, Teikoku Oil, which had been mainly involved in oil development in Japan's domestic gas development, joined overseas oil exploration. In 1970, it invested in Zaire in West Africa with Gulf Oil, a US oil major, and the project was rather successful.[163]

However, the overall effect on autonomous oil development since the launching of the JPDC was not satisfactory for the Japanese government in terms of its energy security objectives. From 1967 until 1985, the JPDC's total loan and equity investment to the private sector climbed to 1125 billion yen.[164] Contrary to the expectation, while MITI's original target of *Hinomaru* oil production was 30% of Japan's oil demand by 1985, the oil self-sufficiency ratio (*Hinomaru* oil/imported oil) in 1985 was 10.7%. The oil developed by the AOC, which began its oil production prior to the establishment of the JPDC, accounted for 47% of *Hinomaru* oil in 1985.[165]

4 Summary and Conclusion

Japanese overseas oil development policy in the 1960s illustrates that the interaction among many veto players produced a resolute policy outcome, which maintains the non-existence of a state-owned oil company in Japan and creates only a state financier. In policymaking on the establishment of a state-owned oil company in the 1960s, the number of veto players were the following sixteen: the prime minister; MITI; MOF; two LDP councils of the LDP; four opposition parties; the two Diet chambers; the standing committee in each chamber; *Keidanren*; PAJ; and JPDA. This case counts the last three interest groups as veto players mainly because they were included in *shingikai*, the formal channel through which they could affect policy outcome. Applying the absorption rule, the preference of the prime minister is included in the overlapped preference area among other institutional and

[162] Japan National Oil Corporation, *Sekiyukōdan Nijūnensi [20-Year History of Japan National Oil Corporation]*, 68.

[163] Ibid., 82–83. It is interesting to note that the developed oil was imported through a swap with the help of the Gulf because the transportation cost of produced oil in Zaire to Japan was too high.

[164] Ibid., 518–19.

[165] Ibid., 492–93.

partisan veto players, and the preference of the JPDA, the brainchild of MITI, is absorbed to that of MITI. Also, the preference of the standing committee in each chamber is accepted by the chambers, and the preferences of opposition parties are absorbed to that of the Diet. As a result, the number of effective veto players becomes eight.

The resolute outcome in the decision-making process is differed from MITI's original intention, which was to establish a state-owned oil company. The initiative of MITI, the agenda setter, for the founding of a state-owned oil company was obstructed, compromised, and transferred into the establishment of a state financier to assist the private sector's overseas oil development. This outcome confirmed the relevance of veto player theory in that the outcome tends to be more private-regarding than public-regarding when passed through many veto players.[166]

While this case supports the veto player approach, it also portrays an interesting dynamic in Japan Studies. In the decision-making of collective veto players, Japan's consensus-based decision-making provided factions and opposition parties with opportunities to voice their opinions regarding the outcome, and this made veto power of the LDP councils and two chambers of the Diet real. Also, the strength and limit of the private sector in Japan's policymaking was clear. *Keidanren*, in particular, played an important role in the policymaking process related to a state-owned oil company in the 1960s, and thus this politically powerful economic organization could be categorized as veto player. However, it is noteworthy that this prominent organization, which was been perceived omnipotent in the past, also had some weaknesses. Its stance on a state-owned oil company was vague between the PAJ and the JPDA. Some of *Keidanren*'s requests were rejected by MITI and the Diet. This shortcoming of *Keidanren* might be attributed to its nature of all-inclusive national economic organization and its relationship in the complex web of veto players. The internal and external shocks, such as the Black Mist Incident and the Arab–Israeli War, added additional pressure to neutralize strong veto players like the MOF.

This chapter showed the relationship and developments between many veto players involved in creating outcomes. The outcome of Japan's 1960s initiative stands in stark contrast with what came out of Korea's initiative to create a state-owned oil corporation in the 1970s, a case that will be analyzed in the following chapter.

[166] If a policy is responsive to broad popular demand, it can be called public-regarding. If a policy is responsive to narrow and special interests, it can be called private-regarding. For details, see Stephan Haggard and Mathew D. McCubbins, *Presidents, Parliaments, and Policy*, Political Economy of Institutions and Decisions (Cambridge; New York: Cambridge University Press, 2001), 47–51.

3

The Establishment of PEDCO in the 1970s

South Korea in the 1970s shows the flip side of the main argument in this book: A state-owned company is more likely to be established when there are fewer actors who can block the initiative. This result is rather unexpected to some scholars who know that Korea's economic development is modeled after Japan's. In this chapter, I analyze the case qualitatively to further our understanding of the dynamics of fewer veto players in Korea's policymaking and how interactions among these veto players affected Korea's realization of a state-owned oil company.

Korea's Ministry of Trade and Industry (MTI) made efforts to establish a state-owned oil company and succeeded in establishing the Korea Petroleum Development Corporation (PEDCO, the former entity of KNOC) in 1979. While the original idea was to establish a state-owned enterprise (SOE) to deal exclusively with oil stockpiling, the proposal was expanded to a vertically integrated state-owned oil company, with activities ranging from oil E&P to the refining and sales of petroleum products. While the difficult economic situation in the aftermath of the first oil shock in 1973 obstructed the initiative for establishing a state-owned oil company, the MTI and its successor revived the idea due to the domestic oil fever and Japan's delayed ratification of the Korea–Japan Continental Shelf Joint Development Agreement.

This outcome was possible, though, mainly due to a lack of veto players in the centralized presidential system. The instructions of the president were dealt with only by the MTI and the EPB. Unlike Japan, the ruling party did

S.-i. Oh, *Overseas Energy Investment of Korea and Japan*,
https://doi.org/10.1007/978-981-99-0285-9_3

not have the power to review policy, and interest groups were too immature to express their voice. Korea's case in the 1970s proves that a small number of veto players make the realization of decisive policy change more possible.

1 Korea's Political Institutions in the 1970s

1.1 Institutional Veto Players: The President, the MTI and the EPB

Strong President

Korea's constitution under the *Yushin* regime (from October 1972 until October 1979) called for a strong presidential system.[1] Korea has maintained a presidential system since the establishment of Korean government in 1948, with the exception of 1960–1961.[2] During the *Yushin* period, electoral laws were favorable to the president. The president was elected through the Electoral College, called *Tongilchuch'e Kukminhoeŭi*, the chair of which was the president himself.[3] As set forth by the constitution, the president had a six-year term with no term limit.

[1] "*Yushin*" means revitalizing, reforming, restoring, or renovation. In Japanese, the term is pronounced *issin*, used by the Meiji leaders in 1868. President Park explained that *Yushin* was the second revolution, which was subsequent to his May 16th Military Coup in 1961. He defined the ideology of *Yushin* as the process of selecting the path that should be embarked upon in order to achieve national stability and prosperity and facilitate peaceful unification. The medium-range economic goal was to achieve $10 billion in export mark and $1,000 in GDP per capita by the 1980s in order to overwhelm North Korea in terms of national power. While Eun-mee Kim sees the *Yushin* as an attempt to earn public support and political legitimacy with economic delivery, Bruce Cumings states that President Park's reasons should not be dismissed as purely self-serving. In the early 1970s, dramtic changes in US foreign policy, such as Nixon's state visit to China (PRC), the Nixon Doctrine and Nixon's announcement of the withdrawal of a full division of the US troops from Korea, created a sense of crisis in Seoul. See Chung-hee Park, *Korea Reborn: A Model for Development* (Englewood Cliffs, NJ: Prentice-Hall, 1979), 14; Won-chol O, *The Korea Story: President Park Jung-Hee's Leadership and the Korean Industrial Revolution*, 1st. ed. (Seoul, Korea: Wisdom Tree, 2009), 41–42, 237; Kim, *Big Business, Strong State: Collusion and Conflict in South Korean Development, 1960–1990*, 140; and Bruce Cumings, *Korea's Place in the Sun: A Modern History*, 1st ed. (New York: Norton, 1997), 358–59.

[2] The draft of Korea's first constitution, submitted on June 12, 1948, stipulated a parliamentary system with a unicameral National Assembly. However, Rhee Syng-man, the most influential politician at that time, opposed the parliamentary system, stating it was "undemocratic." Therefore, the draft had to be revised to the presidential system. Rhee Syng-man became Korea's first president. See Korea National Assembly Secretariat, *Taehanmin'gukkukhoe Yuksimnyŏnsa [60-Year History of the National Assembly of the Republic of Korea]* (Seoul: Korea National Assembly Secretariat, 2008), 11–12.

[3] President Park Chung-hee, the only presidential candidate, got 2,359 votes among 2,361 votes in the presidential election by *Tongilchuch'e Kukminhoeŭi*. Young Heo, *Hankuk Hŏnbŏmnon [Korean Constitution]* (Seoul: Pakkyŏngsa, 2011), 114.

Under the *Yushin* Constitution, the president could dissolve the National Assembly, in addition to submitting draft bills and vetoing laws passed by the National Assembly.[4] The *Yushin* Constitution also gave the president the right to announce preventive emergency decrees without the consent of the National Assembly in times of natural calamity, economic crisis, and internal turmoil or threats to national security.[5] The president also appointed the members of the Cabinet including the prime minister, which was a delegation institution, rather than a decision-making institution.

President Park Chung-hee soon used his centralized authority to institute a mercantilist economic policy, and under his rule, *Chŏngwadae*, Korea's Office of the President, became a hotbed for neo-mercantilism. Unlike previous presidents who had small numbers of secretaries mainly for protocol and to facilitate ministerial work, President Park strengthened his Secretariat as a policy-planning organization to lead Korea's foray into industrialization. After 1969, President Park afforded his economic team in *Chŏngwadae* an even more active role, and the team directed Korea out of the 1970s cloud of stagflation and oil shocks.[6] Before this period, the President had intervened in the economy indirectly through confidant-led ministries in order to avoid demoralizing the line ministries and to coordinate ministries more easily, but after 1969 he strengthened the policymaking role of the *Chŏngwadae* Secretariat, and it became directly involved in policy formulation and implementation.[7] For example, the task of developing the Heavy and Chemical Industrialization (HCI) policy was assigned to the secretary's office in *Chŏngwadae* without consulting with the EPB or MTI, which were typically the participants in industrial policy making.[8] Emergency Decree No.3, drafted to cope with the economic crisis caused by the first oil shock in 1973, was also assigned not to the relevant ministry, but to the task force team in *Chŏngwadae*.[9] The rationale behind President Park's decision was to avoid

---

[4] Ibid., 753–54.

[5] The power of emergency decrees was adopted from the Constitution of the French Fifth Republic. During the period of secretly drafting the *Yushin* Constitution, President Park sent Mr. Kim Ki-ch'oon, an official of the Ministry of Legal Affairs to Paris for a one-year study on the emergency power in Article 16 of the French Constitution. Youn-sik Choi, "Personalization of Power and Yushin Constitution," *Journal of Korean Political And Diplomatic History* 33, no. 1 (2011): 89.

[6] Cumings, *Korea's Place in the Sun: A Modern History*, 323.

[7] Kim, Pyong-guk, "The Labyrinth of Solitude: Park and the Exercise of Presidential Power," in Pyong-guk Kim and Ezra F. Vogel, *The Park Chung Hee Era: The Transformation of South Korea* (Cambridge, MA: Harvard University Press, 2011), 161.

[8] Kim, *Big Business, Strong State: Collusion and Conflict in South Korean Development, 1960–1990*, 141.

[9] O., *The Korea Story: President Park Jung-Hee's Leadership and the Korean Industrial Revolution*, 307. For the confidential work, only selected government officials and a few economists from the Korea

premature disclosure of these important plans, which could have invited divisive inter-ministerial competition and stalemate.[10] The President himself also played a proactive role in policymaking when he felt the need. His policy control was backed by the Minister-level Chief of Staff, the Vice Minister-level Senior Secretaries, and the aides of the Secretariat who were recruited mainly from the state's bureaucracy and the military sector. They were consistently bestowed with political favors, and in exchange, they returned to the ministries with a renewed sense of mission and loyalty.[11]

Ministerial Veto Players: The MTI/MER and the EPB

When the creation of a state-owned oil company came to the forefront in the 1970s, the MTI typically played the role of agenda setter, and the EPB was a ministerial veto player as the budget controller. However, in December 1977, the work of agenda-setting to found a state-owned oil company was transferred to the newly established Ministry of Energy and Resources (MER), an independent energy ministry that was separate from the MTI.

(a) MTI/MER[12]

The MTI, which was equivalent to Japan's MITI, was established as one of eleven ministries and four bureaus during Korea's inception in 1948. The MTI's jurisdiction included commerce, fisheries, mining, industry, electricity, weights and measures, patents, trade, and even shipping.[13] The government increased its role in economic development in the 1960s and 1970s, and President Park Chung-hee's HCI policy in particular provided the MTI with an opportunity to expand its organizational responsibilities.[14] The number

Development Institute (KDI) were called. See Lee-Jay Cho and Yoon Hyung Kim, *Economic Development in the Republic of Korea: A Policy Perspective* (Honolulu, Hawaii: East–West Center: Distributed by the University of Hawaii Press, 1991), 187.

10 Kim, *Big Business, Strong State: Collusion and Conflict in South Korean Development, 1960–1990*, 142.

11 Kim and Vogel, *The Park Chung Hee Era: The Transformation of South Korea*, 148–49.

12 Literature on Korea's economic development has focused mainly on the role of the Economic Planning Board (EPB), and thus it is difficult to find trustworthy literature like Johnson, *MITI and the Japanese Miracle: The Growth of Industrial Policy, 1925–1975*. Therefore, the author relies on the ministry's history book, laws, and official website for the description of MTI and MER.

13 Article 22 in "Law on Government Organization," ed. National Assembly (National Assembly, 1948). In 1955, the Agency for Ocean Affairs (AOA) was established under the MTI to solely manage ocean-related tasks such as fisheries, shipping, ship building, port construction and management and the coast guard.

14 Kim Byung-kook, "The Leviathan: Economic Bureaucracy," in Kim and Vogel, *The Park Chung Hee Era: The Transformation of South Korea*, 225.

Table 1 MTI's size of organization compared to other ministries as of 1977

Number of unit	Assistant vice minister	Bureau	Division
MTI	5	12	45
Other ministries	2–4	6–9	22–34

Source Comment by Park Noh-Chung, Committee Staff Director, at Korea National Assembly, Standing Committee on Internal Affairs, 20th Committee Meeting, Interpellation, December 5, 1977
Note Other ministries included the Economic Planning Bureau (EPB), the Ministry of Foreign Affairs, the Ministry of Finance, and the Ministry of Agriculture and Fisheries

of divisions increased from 25 in 1948 to 53 in 1973, and the number of assistant vice ministers increased from two to five, whose responsibilities included taking charge of commerce and trade, international cooperation, heavy industry, light industry, and resources.[15] Table 1 shows the extent of the MTI's expansion in 1977 when compared to other ministries. Considering that the AOA, one agency under the MTI, was dismantled and that its functions were redistributed to other ministries in 1961, this organizational expansion illustrates the importance of the MTI in Korea's economic development.

Energy issues were handled mainly by the mining bureau and the energy development bureau in the MTI. However, the former focused on coal production and the latter dealt with only electricity and nuclear power generation. Thus, in order to take on a comprehensive energy policy after the first oil shock took place, the idea to have the Ministry of Resources was raised in the government in 1975 and was officially discussed and realized in 1977.[16]

One of the main functions of the MER was to consolidate scattered energy-related functions. Thus, the MER merged the mining bureau and the energy development bureau from the MTI, the function of resources survey from the Bureau of Science and Technology, and the regulation of heating and inspection from the Agency for Industrial Advancement and the second State Minister.[17] Another aim of this new ministry was to relieve the

[15] Ministry of Knowldge Economy, "History: Ministry of Trade and Industry," http://www.mke.go.kr/intro/intro/history01.jsp.

[16] "Government Considers Establishment of Ministry of Resource," *Kyung-hyang Shinmun*, Sept. 3, 1975. The idea was raised just after the EPB abolished the Resource Bureau in August, 1975. The Bureau was created just after the first oil shock.

[17] The Bureau of Government Legislation, *Organization of Ministry of Energy and Resources* (Seoul, 1977); Korea National Assembly, Standing Committee on Internal Affairs, *20th Committee Meeting*, Interpellation, Dec. 5, 1977. Comment by Choi Won-taek, Minister of General Affairs. The second State Minister was involved in oil exploration in Pohang onshore, based on the President's order.

burden of the MTI by severing energy issues from the MTI.[18] Energy policymakers considered the tasks of the MER difficult to navigate, and, therefore MTI members were reluctant to move to the new ministry and wanted to stay in the now-powerful MTI.[19] As a result, while the MER required 275 officials for 17 divisions, only 117 officials were recruited from the existing ministries.[20] This might be one reason why the MTI preferred an independent energy ministry, not an energy agency under the MTI like in Japan.[21] If it had been an energy agency, the difficult energy issues would have been the MTI minister's task. The expansion of the government through the establishment of a new ministry did meet some criticism in the committee and plenary sessions of the National Assembly, including for example, from Kim, Soo-han, a New Democratic Party (NDP) member. Nevertheless, the bill stipulating the establishment of the new energy ministry passed the Standing Committee on Internal Affairs with a vote of 12 to 5 and the Plenary Session with a vote of 125 to 41.[22]

The newly established MER was not so weak in terms of power. Jang Ye-joon, who was an experienced EPB vice minister, construction minister, and MTI minister, became the MER minister. The ministry seemed to recruit technical expertise. Some of the working-level officials in the MER with a background in engineering were ranked *kigam* (bureau chief level), *kichŏng* (division chief level), *kizwa* (assistant director level) or *kisa* (officer in charge level).[23] The newly born ministry was actively involved in energy issues, such as Minister Jang's resource diplomacy to the United States and Canada, the

---

[18] "Resource Agency Is Considered to Sever Energy Issues from MTI," *Kyung-hyang Shinmun*, Aug. 26, 1977.

[19] *20th Committee Meeting*. Comment by Mr. Park Byung-bae, Assembly member from Democratic Unification Party.

[20] "Actively Implement Overseas Resources Development and Import," *Maeil Economic Daily*, Dec. 21, 1977.

[21] One more idea was to consolidate energy-related functions to the task of the Bureau of Science and Technology, but it was soon abandoned. See "Economic Ministerial Meeting Agreed to Establish the Ministry Taking Full Charge of Resource and Submit Bill to Assembly in Coming September," *Maeil Economic Daily*, Aug. 13, 1977; "Resource Agency Is Considered to Sever Energy Issues from MTI."

[22] *20th Committee Meeting*. "Organizational Inflation Accumulating," *Kyung-hyang Shinmun*, Dec. 14, 1977. The bill was the Law on Government Organization, which is under the jurisdiction of the Bureau of General Affairs. Therefore, the Internal Affairs committee dealing with the Bureau handled the establishment of the MER.

[23] Legislation, "Organization of Ministry of Energy and Resources." Mr. Kim Hui-sool, the Petroleum Bureau Chief managing the establishment of PEDCO, Korea's state-owned oil developer, had an engineering background and began his government career as *kisa*. He was not recruited through the high-level government official exam.

groundwork to establish Korea's state-owned oil company.[24] The MER also set up a *shingikai*-style Energy & Resources Response Committee, which dealt with comprehensive policy, development planning, and international cooperation issues. This committee consisted of seven subcommittees with 24 intellectuals, who were mainly opinion leaders and heads of government-affiliated institutions like the Korea Institute of Energy Conservation (KIEC), rather than industry people.[25] However, the MER was not well organized in terms of internal work boundaries at the outset as seen in the involvement of MER's planning bureau chief in other bureau's work. MER's planning bureau chief in other bureau's work initiated the discussion on U$1 billion Petroleum Loan from Morgan Guarantee Trust for offshore oil development in the Korea–Japan Continental Shelf Joint Development Zone, which was the work of Resources Development Bureau. This unorganized work style in the MER became news.[26]

(b) **Economic Planning Board (EPB)**

Among the ministries, the EPB was the most important economic ministry, playing a role as a veto player in Korea's policymaking.[27] The EPB was introduced as a powerful planning agency on July 22, 1961, even before General Park became president. It was formed by merging the Bureau of Budget from the MOF and the Bureau of Statistics from the Ministry of Internal Affairs. Additionally, the EPB took planning functions from the newly established Ministry of Construction.[28] The EPB was managed by the Deputy Prime Minister, the elevated position for the chief of the national planning agency, in order to coordinate and control conflicts among the various economic ministries.[29]

---

[24] "Energy Minister Jang's Official Trip to Strengthen Resource Diplomacy in American Continent," *Kyung-hyang Shinmun,* April 12, 1978; "Petroleum Development Corporation Launched Earlier," *Maeil Economic Daily,* Jan. 30, 1978.

[25] "Appointment of 24 Members for Energy & Resources Response Committee," *Dong-A Ilbo,* April 18, 1978.

[26] "He Is Good at Speaking English, So He Touched Other's Business," *Kyung-hyang Shinmun,* Feb. 6, 1978.

[27] Kim, *Big Business, Strong State: Collusion and Conflict in South Korean Development, 1960–1990,* 143. Japan's Ministry of International Trade and Industry (MITI) was equivalent to Korea's Ministry of Trade and Industry (MTI). However, while the MITI was the most significant ministry involved in industrial policy-making, the MTI had been considered second-in-command to the EPB.

[28] Economic Planning Board, *Kaepalnyŏndaeŭi Kyŏngjech'ŏngch'aek: Kyŏngjekihwoegwŏn 20nyŏnsa [Economic Policy in the Era of Development: 20-Year History of the Economic Planning Board]* (Seoul: Economic Planning Board, 1982), 6–7.

[29] Jones and SaKong, *Government, Business, and Entrepreneurship in Economic Development: The Korean Case,* 48.

Along with the EPB minister's high position as the only deputy prime minister in the Cabinet,[30] its purview also included the power to allocate budgets to the ministries and the power to approve and supervise loan capital and foreign direct investment (FDI).[31] The power of the EPB was roughly equivalent to that of Japan's ESB (Economic Stabilization Board), which existed during 1946–1952 and was arguably the most dominant and effective economic planning apparatus in Japanese history.[32] However, the real significance of the EPB's planning function was rather political, as it intervened in other ministries' jurisdictions and indoctrinated society with President Park's strident "can-do" (*hamyŏn toenda*) spirit and bold "grow-at-all-costs" strategy by showing economic target to be achieved.[33] However, as with the Korea Central Intelligence Agency (KCIA), the EPB was kept within President Park's tight grip with the support of his presidential secretariat at *Chŏngwadae*.[34]

Legislative Institutional Veto Player: The National Assembly and the Committee

Korea's unicameral National Assembly was another institutional veto player in the 1970s. However, the Assembly and its Standing Committee could not be effective veto players because the ruling party occupying majority seats played as a rubber stamp for the president in the Assembly where decisions were made based on majority rule rather than consensus. This tendency was reinforced by the electoral system's favoring of the ruling party.

(a) Unicameral National Assembly

The constitution that preceded the *Yushin* era, which had been drafted by the KCIA and Rupert Emerson, a political scientist at Harvard in 1963, had already substantially weakened the Korean legislature by transforming the

[30] There had been only one deputy prime minister under the president and prime minister in Korea until 1990, when one more deputy minister in charge of the Ministry of Unification was appointed. See Il SaKong, *Korea in the World Economy* (Washington, DC: Institute for International Economics, 1993), 26.

[31] Kim and Vogel, *The Park Chung Hee Era: The Transformation of South Korea*, 100.

[32] Calder, *Strategic Capitalism: Private Business and Public Purpose in Japanese Industrial Finance*, 56–57.

[33] Kim Byung-kook, "The Leviathan: Economic Bureaucracy," in Kim and Vogel, *The Park Chung Hee Era: The Transformation of South Korea*.

[34] Kim Byung-kook, "The Labyrinth of Solitude," in ibid., 148.

National Assembly from a bicameral to a unicameral system.[35] The *Yushin* Constitution maintained this structure. While the Assembly had the power to dismiss the prime minister and the Cabinet members, in an effort to keep the president's power in check, it was difficult for the Assembly to utilize the power during the *Yushin* era.[36]

These already-weakened parliamentary elements were further offset by strong presidentialism. The constitutional revision was easy for the president, as he only had to ask for the referendums. However, under the *Yushin* Constitution, the Assembly had to obtain the approval of *Tongilchuch'e Kukminhoeŭi*, the set of members who were selected to delegate important issues and elect the Korean president, as well as one-third of the Assembly members, in order to propose the referendum, which made a revision of the constitution from the legislative side almost impossible.[37] Also, the State Affairs Inspection by the National Assembly was abolished and a 150-day limit on the length of the total annual session was newly added to the *Yushin* Constitution.[38]

During the Yushin period, the president also had the advantage of selecting the members of the National Assembly, and thus the assembly could not be an effective veto player. In the multi-electoral system, where two members were selected from each constituency, two-thirds of the National Assembly members were selected in 73 constituencies.

As seen in Table 2, this electoral system created favorable conditions for the ruling party members in every constituency, essentially guaranteeing them victory, because the voters tended to select a ruling party candidate and an opposition party candidate if they could choose two.[39] The Remaining 73 members were selected by *Tongilchuch'e Kukminhoeŭi*, and those 73 selected members were called *Yujŏnghoe* (the abbreviation of *Yushin Jŏnguhoe*).[40]

The ratification of the Japan–South Korea Agreement on the Continental Shelf Joint Development, adopted on January 30, 1974, is a good example

---

35 Heo, *Hankuk Hŏnbŏmnon [Korean Constitution]*, 109.

36 The elements of a parliamentary system were placed in Korea's presidential system due to the swift change from the original constitutional draft of the parliamentary system to the presidential system in 1948.

37 Unlike the previous constitution, the *Yushin* Constitution introduced the indirect presidential election by the Electoral College called *Tongilchuch'e Kukminhoeŭi*. The direct presidential election was revived in the 1987 Constitution.

38 Heo, *Hankuk Hŏnbŏmnon [Korean Constitution]*, 114–15.

39 Ji-yŏn Sim, "Park Chung-hee Chŏngbuhaŭi Chŏngdangkutopunsŏk (2): 1972–1979 [the Analysis of Party Formation During Park Chung-hee Administration (2): 1972–1979]," *Korean Party Studies Review* 1, no. 1 (2002): 198.

40 In the National Assembly election, which took place just after *Yushin*, 73 were from the DRP, 73 from *Yujŏnghoe*, 52 from the NDP, 19 were independent, and 2 were from Democratic Unification Party. Heo, *Hankuk Hŏnbŏmnon [Korean Constitution]*, 115.

Table 2 The party composition of the National Assembly in the 1970s

Party	9th National Assembly (Mar 11, 1973–Mar 11, 1979)	10th National Assembly (Mar 15, 1979–Oct 27, 1980)
Yujŏnghoe	73	77
DRP (Democratic Republican Party, ruling party)	73	68
NDP (New Democratic Party)	52	61
DUP (Democratic Unification Party)	2	3
Independent member	19	22
Total members	219	231
(Total constituencies)	(73 local constituencies)	(77 local constituencies)

Source Korea National Assembly Secretariat, *Taehanmin'gukkukhoe Yuksimnyŏnsa* [60-Year History of the National Assembly of the Republic of Korea], 415, 475
Note The data are based on the result of the general elections; therefore this does not reflect the slight change in numbers after the elections

of the decisiveness of Korea's unicameral legislature.[41] In Korea, the ratification bill was discussed during the regular session in late 1974, and it was approved on December 17, 1974. During the Standing Committee on Foreign Affairs on December 13, 1974, Lee Chŏl-sŭng, a member of the opposition NDP, suggested that it would have been better to deal with the ratification in the agreed extraordinary session of February 1975, rather than on December 13, 1974, in order to wait and see the Japanese Diet's ratification and China's possible protest. This proposal was not fully accepted, but only three days were given for further review of the agreement.[42] On the evening of December 16, 1974, the next delegation session was held only with majority party members, and it took 15 minutes to approve the agreement in the Standing Committee on Foreign Affairs.[43] In that committee meeting, in which the need to ratify the agreement for the national interest was emphasized, only Park Joon-gyu, a majority party member, inquired about the possibility of Korea's independent exploration in the joint development zone in the event that Japan failed to ratify the agreement during

[41] The official title of the agreement is "Agreement between Japan and the Republic of Korea Concerning Joint Development of the Southern Part of Continental Shelf Adjacent to the Two Countries." See Treaty Bureau of Ministry of Foreign Affairs and Trade, ed., *Tongbuga Haeyangbŏmnyŏnggwa Yuenhaeyangbŏphyŏbyakchip [Maritime Law of Northeast Asia and United Nations Convention on Law of the Sea]*, 1 vols., vol. 1 (Seoul: Ilchogak, 2006), 582–653.

[42] Korea National Assembly, Standing Committee on Foreign Affairs, *11th Committee Meeting*, Interpellation, Dec. 13, 1974.

[43] "Reason to Pass Late in the Night," *Kyung-hyang Shinmun*, Dec. 17, 1974.

the regular Diet session that began in December 1974. Hoping Korea's prior ratification would be followed by Japan's approval, Park Joon-gyu affirmed the ruling party's stance that Korea would begin exploration and development in the joint development zone between Korea and Japan if Japan did not approve the agreement.[44] In the Plenary Session, after two suspensions of the session due to the opposition parties' resistance to the procedural schedule, the agreement was rapidly ratified despite some physical obstruction among the members of the Assembly,[45] typical of a unicameral legislature. As the ratification of this agreement shows, Korea's National Assembly had been weakened and was no longer an effective veto player under the *Yushin* Constitution.

1.2 Partisan Veto Players

In the 1970s, the National Assembly was controlled by two ruling entities, the DRP and *Yujŏnghoe*. Both entities were loyal to the president, and this arrangement was different from the coalition setting. Their preferences were included in those of the president, and unlike Japan, the role of Korean ruling entities was not to review the government's policy, but to assist the government's legislation for the president in the 1970s. Therefore, they were two partisan veto players, albeit ineffective. In the 1970s, two entities played a role as the ruling party, and the NDP, in addition to other small parties, existed as opposition parties.

Two Ruling Entities for the President: DRP and *Yujŏnghoe*

Since the general election of November 1963, the first general election after General Park Chung-hee gained power in 1961, the Democratic Republican Party (DRP) had been the dominant party.[46] The DRP was also Korea's first coherent and effectively organized noncommunist political party.[47] In order to support the then-General Park Chung-hee, Kim Chong-p'il, President Park's nephew-in-law, founded a reasonably effective political organization by

44 Korea National Assembly, Standing Committee on Foreign Affairs, *12th Committee Meeting*, Interpellation, Dec. 16, 1974.

45 "90th National Assembly Regular Session Ended," *Maeil Economic Daily*, Dec. 18, 1974.

46 Huntington describes the DRP as the only one party capable of governing. Samuel P. Huntington and Harvard University. Center for International Affairs, *Political Order in Changing Societies* (New Haven: Yale University Press, 1968), 419. Huntington describes DR as the only one party capable of governing.

47 Cumings, *Korea's Place in the Sun: A Modern History*, 353.

receiving approximately 1,200 officers from the military and allegedly using a substantial amount of money from the government. As a former chief of the KCIA, he had opportunities to watch the organizational effectiveness of North Korea's Communist Party, and he followed Leninist principles of organization in forming the DRP. A powerful administrative secretariat was set up at the party's central office in Seoul, and four-man secretariats in each electoral district and an eight-man bureau in each province were created to research local political issues, develop support, create organizations, and select candidates.[48] The party was led by President Park as chairman of the DRP from August 1963 until his death in October 1979.[49] During the 1960s, the party's coherence was further strengthened by Korea's 1963 Constitution, which stipulated that the candidates for the National Assembly elections had to be nominated by his or her party. Under these rules, if a member of the National Assembly left the party, he or she could not maintain their status as a member of the National Assembly.[50]

However, the DRP was totally ignored during the process of preparing the *Yushin* Constitution, and could not participate in the election of members of and the presidential election by *Tongilchuch'e Kukminhoeŭi*. Furthermore, the creation of *Yujŏnghoe* in the constitution diminished the status of the DRP as the only dominant majority party to a majority party in name only.[51] From the perspective of Samuel Huntington, the DRP in the 1960s was a political institution that could channel and accommodate newly socialized citizens in Korea, which was undergoing rapid economic development.[52] The DRP in the 1970s, however, became an "ambassador party," in Sartori's terms, or a party that was peripheral to the sphere of the government and uninvolved with it.[53]

Yujŏnghoe was a political entity created in the *Yushin* Constitution, and thus was a different political entity from the DPJ. *Yujŏnghoe* did not have

---

[48] Huntington and Harvard University. Center for International Affairs, *Political Order in Changing Societies*, 259.

[49] Originally, the Party Constitution stipulated that the term of the party chair is 2 years; however, in October, 1967, the article was changed so that the president could maintain the status as party chair during his term. "The Term of Party Chair Extended. Party Constitution Changed," *Kyung-hyang Shinmun*, Oct. 4, 1967.

[50] Article 36 and 38 in "Taehanmin'guk Hŏnbŏp [the Constitution of Republic of Korea]," ed. The Constitution Delegation Committee (1962). The minimum number of assembly members to get the status of negotiation body in the National Assembly was 20.

[51] Sim, "Park Chung-hee Chŏngbuhaŭi Chŏngdangkutopunsŏk(2): 1972–1979[the Analysis of Party Formation During Park Chung-hee Administration (2): 1972–1979]," 184–85.

[52] Cumings, *Korea's Place in the Sun: A Modern History*, 353–54.

[53] Sim, "Park Chung-hee Chŏngbuhaŭi Chŏngdangkutopunsŏk(2): 1972–1979[the Analysis of Party Formation During Park Chung-hee Administration (2): 1972–1979]," 185–86. Sartori, *Parties and Party Systems: A Framework for Analysis*, 19.

a party organization, but rather a quasi-party composed of the National Assembly members from *Tongilchuch'e Kukminhoeŭi*, and supported President Park with the DRP. *Yujŏnghoe* members were selected not by the general populace, but by the president; therefore they were sensitive only to the president's intentions. President Park could encourage loyalty from politicians in the DPJ and *Yujŏnghoe* through his nomination power, and he made it clear to politicians that their fates were in his hand.[54]

One of the main reasons that he created this additional supporting political entity may have been to make the legislative system more "efficient." As he says in his book, *Korea Reborn*:

> For the National Assembly to operate efficiently, it is most important for the two major political parties represented to cooperate on problems affecting the nation, but to compete fairly on political matters. Political parties in a modern democracy are expected not only to represent the interests of people who support them, but also to help shape the national consensus, and often lead it by helping the public to better understand the issues that confront the nation. If the majority and minority parties could share a common perception of the reality and objectives of the nation, and if such a perception could lead to exchanges of better alternatives, political style in general and the political atmosphere could be made to serve productive purposes.[55]

The benefit of inventing two supportive entities for President Park can be seen in the electoral results and the formation of the Assembly. *Yujŏnghoe* and the DRP together composed an ultra-majority in the assembly, making up two of the three negotiation bodies with the NDP in the National Assembly.[56] The speaker, one of two vice speakers of the assembly, and all chairmen of 13 standing committees were from *Yujŏnghoe* and the DRP.[57] These were quite favorable conditions for the president's policymaking. Indeed, 557 out of 603 administrative bills were passed from March 11,

---

[54] Sim, "Park Chung-hee Chŏngbuhaŭi Chŏngdangkutopunsŏk(2): 1972–1979[the Analysis of Party Formation During Park Chung-hee Administration (2): 1972–1979]," 187–88.

[55] Park, *Korea Reborn: A Model for Development*, 53–54.

[56] A political entity with no fewer than 20 assembly members could be a negotiation body in the Assembly. In September, 1973, 23 independent members organized a negotiation body called the "Independent Committee," but this existed for only 11 months because 6 independent members registered for the NDP in August, 1974. See Article 35 in "Kukhoepŏp [the National Assembly Law]," ed. National Assembly (1972). See also Korea National Assembly Secretariat, *Taehanmin'gukkukhoe Yuksimnyŏnsa [60-Year History of the National Assembly of the Republic of Korea]*, 419.

[57] Korea National Assembly Secretariat, *Taehanmin'gukkukhoe Yuksimnyŏnsa [60-Year History of the National Assembly of the Republic of Korea]*, 74–75, 418.

1973, until October 27, 1980, and the recommendation bills and resolution bills challenging the *Yushin* regime were effectively abrogated.[58]

Opposition Party

In Korean politics, the opposition party could not enjoy the veto power Japanese opposition parties had. A majoritarian model of decision-making diminished the power of the opposition party due to the majority of the ruling parties. The absolute majority of the ruling entities restrained the opposition parties from gaining any post of chairman of any standing committees in the National Assembly. Consequently, as seen in Table 3, law drafts submitted by opposition parties were less likely to be passed compared to those submitted by the ruling party, even if the opposition members outnumbered the ruling party. Further, in a unicameral national assembly, law making and budget procedure could not be effectively blocked as it was in Japan.

Table 3 Weak presence of the opposition parties during the *Yushin* period

	9th National Assembly (Mar 11, 1973–Mar 11, 1979)		10th National Assembly (March 15, 1979–October 27, 1980)	
Party	Ruling parties	Opposition parties and independent members	Ruling parties	Opposition parties and independent members
Seats	146	73	145	86
Chairman of Standing Committees	13	0	13	0
Submission of Draft Bills	33	76	4	1
Passage of Draft Bills	27	13	3	0

Source Choi Jeong-won, "Characteristics and Changes of the Legislation Process in the Korean National Assembly: Focusing on Legislation Environment and Legislation Actors," *Korean Political Science Review* 35, no. 3 (2001): 142–45

[58] Ibid., 410–11, 987–88.

(a) **New Democratic Party (NDP)**

The NDP was established in 1967 by consolidating the People's Party and New Korea Party. The NDP was the only opposition party with the potential to win the presidential election. When the NDP rallied around its presidential candidate, Kim Dae-jung, and successfully mobilized the urban electorate and Chŏlla regional voters into a formidable electoral coalition in 1971, the future defeat of the DRP in the next presidential election looked like a very real possibility.[59] This was even more obvious considering President Park was supposed to step down from the presidency in 1975 according to Korea's constitutional limitation at that time of three terms for the president. In the 1979's general election, the NDP under Kim Young-sam's leadership was encouraged because the NDP had garnered 4,861,204 votes, which was 165,209 more votes than the DRP.[60]

However, the NDP's coherence was lower than that of the DRP, one ruling entity. After the party convention in 1972, a group of NDP members split from the party and established a new party, the DUP. In the general election, where two members were selected in a constituency, the NDP recommended two candidates in 14 constituencies in the 1972 election and four constituencies in the 1979 election. This is in direct contrast with the ruling party's election strategy that the DRP recommended two candidates only in seven constituencies in the 1972 election and no constituencies in the 1979 election.[61]

The NDP's position changed from moderate Kim Young-sam's two-track strategies of dialogue and resistance (1974–1976) to Lee Chŏl-sŭng's compromising strategy of reform-through participation (1976–1979), and finally to unyielding Kim Young-sam's confrontational strategy of regime change and democratic revival (1979). The NDP finally decided to challenge President Park in 1979 because the NDP's gaining power was constitutionally impossible. The constitution abolished direct presidential elections. In addition, the chance to revise the constitution was effectively blocked because the National Assembly was transformed into a rubber stamp through President Park's appointment of a third of the assembly members with the hand-picked *Yujŏnghoe*. Furthermore, the successive promulgation of emergency decrees also hindered the opposition parties' political activities. The NDP ended up

---

59 Kim and Vogel, *The Park Chung Hee Era: The Transformation of South Korea,* 221–22. Kim Dae-jung earned 5,395,900 votes while President Park earned 6,342,828.

60 Ibid., 230.

61 Sim, "Park Chung-hee Chŏngbuhaŭi Chŏngdangkutopunsŏk(2): 1972–1979[the Analysis of Party Formation During Park Chung-hee Administration (2): 1972–1979]," 197–99.

defining its political target as nothing less than the overthrow of the *Yushin* regime. With the paralysis of the National Assembly, the NDP went to the streets to appeal directly to the public and fight the state in an all-out confrontation.[62] This standstill of the legislature took place in 1979, one year after the law for Korea's state-owned oil developer was discussed and passed in the National Assembly.

(b) **Democratic Unification Party (DUP)**

The DUP was established in 1973 when it split from the NDP in order to wage a political battle against Yu Jin-san, the NDP leader. The party focused more on its street campaign than activities in the National Assembly because the party could gain only three seats in the 1973 general election, which did not meet the 20-seat requirement to make a negotiating body in the National Assembly.[63]

1.3 Weak Presence of Interest Groups

Federation of Korean Industries (FKI)

Unlike Japan, Korea's peak organization of big business was not recognized as a veto player by the government in energy policymaking. On the contrary, the state utilized the peak organization for the policy implementation for its own purpose, creating a rather asymmetric relationship.

The FKI was established under the name of the Korea Businessmen's Association (KBA) in January 1961, just after the first President Rhee Syng-man was expelled from the presidency.[64] The KBA modeled itself after Japan's Federation of Economic Organizations (*Keidanren*), and thus represented Korea's big business (unlike Korea's Chamber of Commerce & Industry [KCCI] that represented small and medium-sized companies). Unlike the

---

[62] Park Myung-lim, "The Chaeya," in Kim and Vogel, *The Park Chung Hee Era: The Transformation of South Korea*, 397–98.

[63] Sim, "Park Chung-hee Chŏngbuhaŭi Chŏngdangkutopunsŏk(2): 1972–1979[the Analysis of Party Formation During Park Chung-hee Administration (2): 1972–1979]," 193–94.

[64] The reason behind the establishment of the KBA/FKI has been explained in a different way. Some scholars argue that the KBA was founded to protect heads of big business against the political scandal of unfair wealth accumulation. On the other hand, FKI material explains that the KBA's purpose is to overcome national economic difficulties. See Suh Je-jin, "Formation and Charateristics of Class Organization of Korean Capitalist Class: Mainly on the FKI," *Society and History* 16 (1989): 174–79; Federation of Korean Industries, "Chŏn'gyŏngnyŏn 50 Nyŏnsa: Kichŏgŭi 50nyŏnŭl Nŏmŏ Hŭimang 100nyŏnŭro [50-Year History of the Federation of Korean Industries: Beyong 50-Year of Miracle to 100-Year of Hope]" (Seoul: Federation of Korean Industries, 2011).

KCCI, the KBA was established as a purely private organization without special laws and subsequent government support in order to avoid government intervention.[65]

The organization developed as the FKI, a summit organization of large industrial conglomerates in 1968, and it served not only as a channel through which big business could voice and defend its collective interests, but it also enabled the state to control, shape, and influence those interests. From the beginning, the FKI submitted policy proposals to the government, including proposals for energy policymaking, such as its first proposal for streamlining the energy supply on October 5, 1970.[66] The Emergency Decree for Economic Stability and Growth on August 3, 1972, was planned based on the recommendation of FKI chairman Kim Yong-wan.[67] The legislation of general trading companies (GTCs or *Chonghapsansa*) in 1975 was also based on the FKI's recommendation on March 15, 1968.[68] The FKI also established a special committee in response to resources issues in 1973 and a special committee in response to energy development in 1978.[69]

However, the influence of the FKI in energy policymaking was limited during the 1970s. Unlike *Keidanren*, the FKI could not send its high officials to take on the governmental energy policymaking.[70] Advisory committees, such as the Resources Response Committee established in 1973 under the prime minister and the Energy and Resources Response Committee in 1978 under the energy minister, excluded industry insiders including an FKI staff.

Korea Petroleum Association (KPA)

Compared to Japan, it was not until 1980 that Korea established an oil industry-specific association. Though associations like Korea Petroleum

---

[65] Federation of Korean Industries, "Chŏn'gyŏngnyŏn 50 Nyŏnsa: Kichŏgŭi 50nyŏnŭl Nŏmŏ Hŭimang 100nyŏnŭro [50-Year History of the Federation of Korean Industries: Beyong 50-Year of Miracle to 100-Year of Hope]."

[66] Ibid., 933–1000.

[67] Eun-mee Kim and Gil-sung Park, "The Chaebol," in Kim and Vogel, *The Park Chung Hee Era: The Transformation of South Korea*, 285.

[68] Federation of Korean Industries, "Chŏn'gyŏngnyŏn 50 Nyŏnsa: Kichŏgŭi 50nyŏnŭl Nŏmŏ Hŭimang 100nyŏnŭro [50-Year History of the Federation of Korean Industries: Beyong 50-Year of Miracle to 100-Year of Hope]"; Duk-ryul Hong, "Study on Korea's Big Business Organization: Mainly on FKI," *Society and History* 36 (1992): 88–89.

[69] Federation of Korean Industries, "Chŏn'gyŏngnyŏn 50 Nyŏnsa: Kichŏgŭi 50nyŏnŭl Nŏmŏ Hŭimang 100nyŏnŭro [50-Year History of the Federation of Korean Industries: Beyong 50-Year of Miracle to 100-Year of Hope]."

[70] "Resources Response Committee Established," *Maeil Economic Daily*, July 12, 1973; "Appointment of 24 Members for Energy & Resources Response Committee.".

Association had existed since 1956, this organization was an association of petroleum product retailers. In 1980, the MER made an organization designed to express the industry's proposals and discuss oil policy issues. PEDCO, Korea's state-owned oil developer, and five Korean refiners joined the association.[71] Korea's oil development-specific association, equivalent to the JPDA in Japan, was established in 1991. Therefore, when the Korean government discussed the establishment of a state-owned oil developer in the 1970s, no interest group existed to express any opinions.

2 The Debate on PEDCO Law

In the aftermath of the first oil shock, the president ordered the MTI minister to establish a state-owned oil company mainly for oil stockpiling. The MTI developed the idea into a proposal to establish ENI-style oil company; however the EPB refused to support the proposal due to the difficult economic situation. This stalemate was resolved by domestic oil fever and the diplomatic problem with Japan. Without the influence of the private sector, the MER, Korea's energy ministry, could achieve its goal.

2.1 Background: The Oil Shock and the Late-Comer in the Oil Industry

The First Oil Shock in 1973

Following the first oil shock that stemmed from the Yom Kippur War in October 1973, Korea suffered from rapidly rising oil prices. In the aftermath of the first oil shock, President Park Chung-hee dispatched O Won-chol, his second economic secretary, with the heads of Korean partner refiners, such as Koo In-hwoi of Honam Refinery (currently GS-Caltex), and Kim Chong-hee of Kyung-in Energy, to Western oil majors in order to guarantee their crude oil supply to Korea.[72]

The Korean economy suffered from rapidly rising oil prices and consequent chaotic situations, called *p'atong* in Korean: shortage of oil (*yuryu p'atong*), hyper-inflation (*mulga p'atong*), and excess demand of briquettes as

[71] "Korea Petroleum Association Launched in This Month," *Maeil Economic Daily*, July 5, 1980.

[72] Won-chol O, "Wonnyukonggŭpkyosŏp Ch'ulchangboko [Debrief on the Result of Official Trip for Crude Oil Negotiation]," ed. Office of the President of Korea Secretariat (Seoul, 1973).

an alternative to oil (*Yŏnt'an p'atong*).[73] Facing this dire situation, President Park ordered his chief secretary to prepare an economic countermeasure that would be announced at the New Year's press meeting on January 14, 1974. The work had to be completed in 20 days, and therefore, the project was assigned to the task force team in *Chŏngwadae* and the announcement was in the form of an emergency decree in order to evade a time-consuming process in the National Assembly.[74] The emergency presidential decree included an austerity policy that would reduce the government budget by 50 billion won.

This unprecedented oil shock made the government recognize the importance of oil security, and thus, the policy agenda included the establishment of PEDCO. Korea's oil policy had up until this point focused on the expansion of refinery capacity to meet increasing domestic oil demand. As a result, when the MTI announced a long-term comprehensive energy plan in June 1974 as a measure to deal with the first oil shock of 1973, the proposal included plans for a state-owned oil company to be used for oil stockpiling and improving petroleum products distribution structures.[75] Securing a sustainable oil supply to Korea had become a significant issue and was included in the Long-term Energy Plan announced in June 1974.[76]

Late-Comer in Oil Industry

Since 1962, when President Park committed to economic development, the Korean economy developed at an annual rate of 8.8% of GDP until 1979.[77] Even the first oil shock did not hamper industrial development.[78] However, as seen in Table 4, Korea's absolute size GDP and GDP per capita was still not very high in 1979, when the Korean government established PEDCO. By comparison, when Japan established the JPDC in 1967, the governmental

---

[73] See Won-chol O, *Enŏchi Chŏngch'aekkwa Chungdongjinch'ul [Korean Style Economic Development: Energy Policy and Going Middle East]*, vol. 6, Hankukhyŏng Kyŏngjekŏnsŏl [Korean Style Economic Development] (Seoul: Kia Economic Institute, 1997), 383–95.

[74] The counter-measure included amendments to the budget and taxation laws; thus, the measure should have been under the consideration of the National Assembly in a normal situation.

[75] "Comprehensive Energy Plan Finalized, Nuclear Power Generation Corproation Establshed by 1980," *Maeil Economic Daily*, June 1 1974.

[76] Korea National Oil Corporation, *KNOC 25: Han'guksŏgyukongsa Isibonyŏnŭi Yŏksa [KNOC 25: 25-Year History of Korea National Oil Corporation]* (Kyunggido: Korea National Oil Corporation, 2005), 124.

[77] Based on the data in SaKong, *Korea in the World Economy*, 12.

[78] The fact that there was not minus growth after the first oil shock in terms of GDP made some economists underestimate the impact of the first oil shock on Korea's economy. For the then-Korean government officials, it was a crisis. These different views likely result from different vantage points. O., *The Korea Story: President Park Jung-Hee's Leadership and the Korean Industrial Revolution*, 260–61.

Table 4 Level of economic development and debate on SOEs

		1962	1967	1972	1977	1979
GDP (2000 constant US$)	Korea	$29.8 bil	$46.1 bil	$72.0 bil	$111.4 bil	$130.0 bil
	Japan	$817.3 bil	$1,287.6 bil	$2,054.3 bil	$2,452.9 bil	$2,723.9 bil
GDP per capita (2000 constant US$)	Korea	$1,128	$1,541	$2,163	$3,058	$3,462
	Japan	$8,528	$12,783	$19,165	$21,542	$23,507

Source World Bank WDI (World Development Index)
Note In 1967, the JPDC was established in Japan. In 1979, Korea established PEDCO, its state-owned oil company

special financier for overseas oil development, Japan's GDP per capita was almost four times that of Korea in 1979. In absolute terms, Japan's GDP in 1967 was almost ten times that of Korea's in 1979. With this level of economic development, Korea could not accumulate capital and technology, which are essential for private sector oil development (Table 4).

(a) **Lack of Capital**

During the 1970s, the Korean oil industry lacked capital and could not consider financing oil exploration. As a result, the *chaebol*, Korea's industrial conglomerates, were possible candidates for oil development in the private sector. While there was a minor personal venture in oil development, it was negligible. Oil development is one form of entrepreneurship as defined by Schumpeter: "The conquest of a new source of supply of raw materials."[79] There was Chung Woo-jin, a Korean wildcatter, but again the problem of insufficient capital blocked his efforts.[80] In this situation, GTCs, subsidiaries of the *chaebol*, were apt candidates for oil exploration in terms of capital

[79] Joseph Alois Schumpeter, *The Theory of Economic Development: An Inquiry into Profits, Capital, Credit, Interest, and the Business Cycle*, Social Science Classics Series (New Brunswick, NJ: Transaction Books, 1983), 66.

[80] Chung Woo-jin is often mentioned in Korea's literature regarding oil development. Though Chung Woo-jin, who later changed his name to Chung Sŏng-yŏp, was not an educated geoscientist, he was a wildcatter with religious faith, buoyant enthusiasm, and resolute persistence. He had begun his personal exploration in 1959 in Ŭichang, Kyungsangnanmdo, and had joined the exploration at the Pohang onshore site in the mid-1960s. He asked for help from local rich men and the government for the exploration, but his exploration failed. His constant arguing for oil development in Pohang is associated with the government-initiated exploration in 1975, but which also failed. See Dong-sung Cho, *Kukchaechawollon: Sŏgyumunchaerŭl Chungsimŭro [International Natural Resources: Mainly on Petroleum]*, 2nd ed. (Seoul: Pagyŏngsa, 1983), 238–45.

accumulation, but their efforts did not surface until after the second oil shock of 1979. Among the subsidiaries of the *chaebol*, oil refiners and GTCs could have been assigned to the task, but their situation did not permit it until the 1970s, which will be discussed later.

While Korea's old *chaebol* existed during Japanese colonial times,[81] most of the larger conglomerates appeared simultaneously with the Korean government's promotion of the heavy and chemical industry in the early 1970s. At the end of the colonial times, the Japanese owned 81.8% of the paid capital of all industrial enterprises in Korea, which included 97% of the chemical industry, 93% of metal and machinery, and over 80% of textiles and flour mills.[82] Rice milling was one of the few industries permitted for development by Korean capitalists for the purpose of maintaining an efficient supply of rice to Japan.[83] By 1945, most capitals in Korea were in the hands of the Japanese, especially the *zaibatsu*. However, even in the 1970s, Korea's *chaebol* could not easily expand their business into risky ventures like oil development, because the *chaebol* could not own banks and financial mobilization and allocation were in the hands of the state.[84] If anything, subsidiary refiners and GTCs of the *chaebol* were possible candidates for overseas oil development.

Korean oil refiners had just begun their business in the 1960s, but their financial situation became difficult due to soaring oil prices during the oil shocks of the mid-1970s. Starting in 1975, Korean refiners suffered a deficit in refining, but the loss was covered by profits in the petrochemical and lubricant business.[85] Beyond the consequences of the oil shocks, the conflict that existed between Western partners and Korean *chaebol* also contributed to this problem. Korean refiners' decision-making was controlled by Western oil companies, and most oil for refining was supposed to be supplied by Western partners with the following breakdown: 100% crude oil for Yukong, 80% for Honam, and 80% of Kyung-in Energy.[86] This is quite similar to Japan's situation in the early 1960s when the oil developed by Japanese companies could not be easily sold in Japan's domestic market. In the early 1960s, approximately 85% of oil for Japanese oil refiners was supplied by Western

[81] The term *chaebol*, a Korean translation of *zaibatsu* (the Japanese conglomerates) had been circulated as early as 1932 to refer to Korean rich men like Kim Sŏng-su and Park Hŭng-sik. Meredith Woo-Cumings, *Race to the Swift: State and Finance in Korean Industrialization*, Studies of the East Asian Institute (New York: Columbia University Press, 1991), 35.

[82] Ibid., 40.

[83] Ibid., 24.

[84] Ibid., 7.

[85] PyŏN Sang-keun, "Gulf's 18 years," in Cho, *Kukchaechawollon: Sŏgyumunchaerŭl Chungsimŭro [International Natural Resources: Mainly on Petroleum]*, 340.

[86] Ibid., 228.

oil companies, stipulated by conditions in partnership and loan contracts with Western majors.[87] The constrained supply of crude oil from Western oil companies discouraged oil development in Japan in the 1960s, and this may have also been true in Korea in the 1970s.[88]

As in Japan, Korea's GTCs had been actively investing in Korea's overseas oil development, but it was not until the early 1980s that they joined the business. The function of importation and worldwide information networks, in which the GTCs have a relative advantage, could have provided a good base upon which to create synergy for oil development.[89] However, in the 1970s, the institutional setting was not right for overseas oil development.

Korean GTCs had an inherent weakness when it came to overseas oil development. While the Korean government had learned of and introduced GTCs from Japan, the aim of Korean GTCs was quite different from Japan's. Japan's GTCs were introduced to prevent the island nation from depending too much on foreign trading companies for imports, while the prior target of Korean GTCs was to promote exports. Korea's law for GTCs, which was passed in 1975 as part of the political target to achieve $10 billion in exports, reflecting the emphasis only on export, and incentives were created only for export promotion. This outcome was different from GTC-related proposals, including the report by Sejima Ryujo, the then-Vice President of Itochu in 1972, and the report by the Korea Trade Institute of Seoul National University.[90] For GTCs, overseas oil development was an expanded function of imports, and so overseas oil development could not be a main interest until they received governmental support.[91]

Furthermore, Korea's GTC sector was in a weaker financial situation than that of Japan. In terms of capital, total assets, and sales, the average size of Japan's GTC sector was five times that of Korea in 1980, as illustrated below. The financial functions of the GTC were also different. Having a one-to-one relationship with big banks, Japan's GTC's enjoyed an almost unlimited financing supply, utilizing this to enhance their profits and control over affiliates.[92] On the other hand, Korea's GTC sector had used export credit for

---

87 "US Oil Capitals Watch Petroleum Industry Law Bill with Strong Interest."

88 In Japan, the constrained oil is called *Himotsuki* oil, which literally means "strings attached."

89 Dong-sung Cho, *Han'gukŭi Chonghapmuyŏksangsa: Chaetowa Kwalli [Korea's General Trading Company: Institution and Management]* 2nd ed., 2 vols., vol. 1 (Seoul: Bŏmmunsa, 1984), 68.

90 Ibid., 14–17. In the governmental notice on assignment of GTC, "export" appears 18 times, while "import" was used only once and "finance," "information," and "domestic sales," were not mentioned at all.

91 Cho Dong-sung describes 1981 as the year of critical situation for Korean GTC. See ibid., 24.

92 In Japan, each GTC had a close relationship with one main bank in 1978: Mitsubishi Busan with Mitsubishi Bank, Mitsui Busan with Mitsui Bank, Itochu with Taiichi Bank, Marubeni with Tokyo Bank, Sumitomo with Sumitomo Bank and Nishoiwai with Sanwa Bank.

Table 5 Comparison of Korea and Japan's GTCs

	Korea's GTC	Japan's GTC
GTC's portion in export	43.1%	52.7%
GTC's portion in import	7.0%	64.2%
Capital (average)	$23 million	$119 million
Total asset (average)	$463 million	$8,737 million
Sales (average)	$1,220 million	$23,806 million

Source Dong-sung Cho, *Han'gukŭi Chonghapmuyŏksangsa: Ponjilgwa Chŏllyak [Korea's General Trading Company: Nature and Strategy]* 2nd ed., 2 vols., vol. 2 (Seoul: Bŏmmunsa, 1984), 165
Note Data for Korea's GTC are those of 1980 and data for Japan's GTC are those of 1978

its sprawling *chaebol*,[93] and therefore, they could not enjoy favorable conditions from banks because Korean GTCs were just like any other customer for Korean banks (Table 5).[94]

Despite all of these limitations, Korea's first attempt to participate in oil development took place in 1978. Beginning in October of that year, Hyundai Yanghaeng, a Korean heavy industry company, had negotiated oil development in Indonesia with its participation in PERTAMINA's Dumai refinery construction project.[95] However, Hyundai Yanghaeng's plan did not materialize, and Korea had to wait a few more years until Choi Gye-wol, CEO of KODECO, succeeded in finding oil off the coast of Indonesia in 1982. In 1979, Korea had no overseas oil development activities, which is in contrast to Japan, where 9% of imported oil (22 million kl among 277 million kl) was Japan's equity oil.[96]

(b) **Lack of Technology**

Korea has a relatively shorter history of oil development than that of Japan and faced difficulties fostering human capital to accompany the development of oil-related technology as a result. A lack of discovery of oil in Korean territory exacerbated the situation, and unlike Japan during World War II, Korea

93 Woo-Cumings, *Race to the Swift: State and Finance in Korean Industrialization*, 164–65.

94 Dong-sung Cho, *Han'gukŭi Chonghapmuyŏksangsa: Ponjilgwa Chŏllyak [Korea's General Trading Company: Nature and Strategy]* 2nd ed., 2 vols., vol. 2 (Seoul: Bŏmmunsa, 1984), 165–67.

95 This fact is rarely mentioned in the history of Korea's oil industry but is found in the National Archive of Korea. Chan-jin Kim, "Teindonesiachawonhyŏmnyŏksaŏpch'uchinhyŏnhwang: Tongnyŏkchawonpupoko [Korea-Indonesia Resource Cooperation Projects: Ministry of Energy and Resources Reported]," ed. The Office of First Senior Economy Secretary of *Chŏngwadae* (Seoul: *Chŏngwadae*, 1979).

96 Japan National Oil Corporation, *Sekiyukōdan Nijūnensi [20-Year History of Japan National Oil Corporation]* 492–93.

had not occupied oil-rich countries such as Indonesia. At most, engineers were part of the governmental research institutes for geosciences or mining companies such as Daehan Coal Mining Corporation (DCMC), a state-owned coal company. Korea's first encounter with modern technology, such as seismic survey data, came from the work of UN Economic Commission for Asia and the Far East (ECAFE) in the late 1960s, which included the survey of offshore Pohang, a port city in the Southeastern Korean peninsula, in 1966 and the Emery Report in 1968.[97] The Education Fund, which was used for educating petroleum engineers, was established not with the national budget but with contributions from Western oil majors who had been participating in the exploration of Korea's continental shelf since 1969.[98]

Although minor attempts at oil exploration were made in 1939, Korea's first scientific onshore exploration was conducted in Haenam, Chŏllanamdo by the Korea National Geoscience Administration (a former entity of KIGAM, the Korea Institute of Geoscience and Mineral Resources) from 1959 until 1961, almost 70 years after Japan's first domestic oil production in Amase Field in 1891.[99] The exploration activity in Haenam was stopped without results. The institute was also involved in onshore exploration in the Pohang onshore site in the mid-1960s. Chung Woo-jin, who was enthusiastic about oil development, made efforts to find oil with the financial support of local areas or the help of the National Geoscience Administration. What is surprising is that even governmental institutions had not owned oil rigs for exploration in the 1960s and even in the first half of the 1970s. In 1967, 20 oil engineers were invited by the Chinese Petroleum Corporation (CPC, Taiwan's national oil company) to explore oil in Pohang and an oil rig was borrowed from the company.[100] When the KCIA launched a secret oil exploration in the Pohang onshore site in 1975, they used one of the DCMC's coal mining rigs as exploration driller rather than an oil rig.[101]

---

[97] Interview with Chang Kyung-chŏn, the then-government official of the Ministry of Trade and Industry, conducted on October 14, 2011; O., *Enŏchi Chŏngch'aekkwa Chungdongjinch'ul [Korean Style Economic Development: Energy Policy and Going Middle East]*, 250–51.

[98] Ibid., 265–66. The Gulf, which may have already known the oil potential of Korea's continental shelf, contacted the Korean government in 1968, and the negotiation for mining rights ended in 1969.

[99] Samuels, *The Business of the Japanese State: Energy Markets in Comparative and Historical Perspective*, 169.

[100] Korea Petroleum Development Corporation, *Han'guksŏgyukaepalgongsa Sibonyŏnsa, 1979–1994 [15-Year History of Korea Petroleum Development Corporation, 1979–1994]* (Seoul: Korea Petroleum Development Corporation, 1994), 143–45.

[101] Interview with industry person conducted on April 14, 2011. The interviewee stated that if there had been real oil in Pohang, there must have been a gas explosion, because the mining rig could not deal with gas that normally exists with crude oil. However, there was no gas explosion during the exploration in 1975.

Another problem was the scarcity of petroleum engineers. According to O Won-chol, few engineers had the knowledge and experience of oil exploration in the 1970s, but he mentions Dr. Kim Young-il as one reliable engineer. Dr. Kim was Vice President of Kyung-in Energy and had majored in geology and participated in Alaskan oil development.[102] Myung In-sung and Kim Youn-soo, who worked abroad for oil development, were also invited to the Pohang oil development by the KCIA.

Part of the reason for this lack of experienced engineers was the distribution of Korea's underground resources. The only fossil fuel Korea had was anthracite coal, and, therefore, petro-engineers were not in high demand. Education for petro-engineering was provided in a piecemeal fashion by the chemical engineering departments or industrial chemistry departments at Korea's universities.[103]

Other efforts to foster education in petroleum geology consisted mainly of a US$200 thousand fund collected starting in 1969 from Western oil majors that had joined the oil exploration in Korea's continental shelf. This fund, jointly managed by the mining bureau of the MTI and Western majors, was used for scholarship, to fund the Petroleum Industry Development Center, and to conduct overseas study tours for National Assembly members.[104] Along with the educational opportunities created by the fund, on-the-job training on the oil rigs also provided Korean engineers with valuable experiences.[105] Given this, it is understandable why it was "almost impossible" to develop offshore oil fields in the 1970s, and why Korean companies depended on foreign engineers or companies for their first successful oil ventures in Indonesia and Yemen in the 1980s.[106]

102 O., *Enŏchi Chŏngch'aekkwa Chungdongjinch'ul [Korean Style Economic Development: Energy Policy and Going Middle East]*, 322.

103 Sang-gil Shin, "Enŏjipumun Chŏnmunillyŏgŭi Hyoyulchŏgin Hwakpopangan Yŏnku [Study on Efficient Securing of Experts in Energy Sector]," (Kyunggido: Korea Energy Economics Institute, 1998), 17.

104 O., *Enŏchi Chŏngch'aekkwa Chungdongjinch'ul [Korean Style Economic Development: Energy Policy and Going Middle East]*, 319.

105 Interview with Chang Kyung-chŏn, the then-government official of the Ministry of Trade and Industry, conducted on October 14, 2011.

106 Korea National Oil Corporation, *KNOC 25: Han'guksŏgyukongsa Isibonyŏnŭi Yŏksa [KNOC 25: 25-Year History of Korea National Oil Corporation]*, 161.

2.2 The Establishment of the Korea Petroleum Development Corporation

The Inception and Development of the Proposal

The first oil shock in 1973 was the catalyst for establishing PEDCO and, more significantly, for general changes to Korea's energy policy. *Chŏngwadae* proposed PEDCO, and the MTI developed it after the first oil shock took place. O Won-chol, the then-Second Senior Presidential Secretary for the Economy at *Chŏngwadae* proposed to President Park the idea of a state-owned oil company for the purpose of oil stockpiling. President Park accepted his proposal and asked Mr. Jang Ye-jun, the MTI Minister, to draw up plans for a state-owned oil company.[107]

For President Park, securing a steady oil supply was a critical issue. Before the oil shocks, political needs had already encouraged the government to pursue an autonomous oil supply. At that time, Korean policymakers believed that the US government had influenced the Korean government's decision-making by threatening an oil supply cut.[108] The Korea Oil Storage Corporation (KOSCO), which was the only supplier of petroleum products in Korea in the 1950s, had been controlled by the US government through the finance of the US Agency for International Development (USAID) to KOSCO.[109] Thus, securing an autonomous oil supply by producing oil products in Korean refineries became a priority for the Korean government, especially after General Park Chung-hee, who gained power in 1961, began a government-initiated economic development. The policy of securing an autonomous oil supply could be applied to PEDCO, too.

While President Park Chung-hee was an anti-communist who supported private ownership of capital and opposed *dirigiste* ideas, state-owned companies were not prevented outright. He had advocated *myŏlgong t'ongil* (destroy communism, unify the nation) and utilized it for his political victory in elections by contrasting it with Kim Dae-jung's three-stage peaceful unification with North Korea.[110] Also, his support for private ownership is stated in *Our Nation's Path*, which includes President Park's ideas following the May 16th military coup:

---

[107] Interview with O. Won-chol, conducted on July 5, 2011.

[108] Ibid.

[109] The oil had been provided by Exxon, Shell, Caltex, and so on.

[110] Hyug-baeg Im, "The Origins of the Yushin Regime," in Kim and Vogel, *The Park Chung Hee Era: The Transformation of South Korea*, 245. In the 1971 presidential election, President Park even accused Kim Dae-jung, an influential opposition candidate, of being a communist sympathizer to play on public anxiety about the North Korean threat.

> Private ownership of production should be unconditionally encouraged except in instances where it is necessary to control it to stimulate national development and protect the interests of the people in general. In other words, private ownership is entitled to protection, which must be expanded for the ideological requirements of a free society, unless it operates against the construction of a sound social order.[111]

President Park also followed the official position of the previous constitution, which changed substantially in terms of economic ownership in 1954.[112] Originally, Korea's first constitution, established in 1948, had leaned toward socialist ideology.[113] It also stipulated the public ownership of a wide range of economic activity[114] and contained open-ended provisions allowing nationalization of any industries "related to the public welfare" or "required for vital and urgent needs for national defense or livelihood of the people."[115] In addition, political reality forced President Park to choose not SOEs, but the *chaebol* as the main driver of economic development.[116] The members of the military that gained power after the May 16th coup had rural backgrounds and personal experiences with poverty.[117] The fact that corruption under President Rhee Syng-man had involved government officials running businesses for personal interests placed political constraints on members of the Park regime. Therefore, they preferred to avoid being perceived as businessmen.[118]

---

[111] Park Chung-hee, *Our Nation's Path; Ideology of Social Reconstruction*, 2nd ed. (Seoul: Dong-A Pub. Co., 1964), 229.

[112] Jones and SaKong, *Government, Business, and Entrepreneurship in Economic Development: The Korean Case*, 147.

[113] Ibid., 144.

[114] (Article 87) Significant transportation, communication, finance, insurance, electricity, water service, gas and companies that have the characteristic of providing public goods are to be state-owned or public-owned. "Taehanmin'guk Hŏnbŏp [the Constitution of the Republic of Korea]," ed. The Committee for Drafting Constitution and Government Organization Law (1948).

[115] Ibid., Article 88. This flavor of socialist ideology reflected the reality of the time. When the first constitution was drafted in 1948, most companies were under state control. During colonial times, infrastructural properties, such as railroads and communications, were owned by the Japanese colonial regime. Most companies were Japanese, but all these assets were transferred to the Korean government in1945 when Korea was liberated from Japan. See Yong-oak Shin, "Sovereignty Principle and Economy Orders of the Constitution of Korea Made in 1948," *The Journal for the Studies of Korean History* 17 (2004): 230–31.

[116] Kim, *Big Business, Strong State: Collusion and Conflict in South Korean Development, 1960–1990*, 117–20.

[117] Huntington and Harvard University. Center for International Affairs, *Political Order in Changing Societies*, 241.

[118] Kim, *Big Business, Strong State: Collusion and Conflict in South Korean Development, 1960–1990*, 119.

However, the establishment of SOEs during President Park's term can also be seen pragmatically. President Park's attitude toward state-owned companies was that they were necessary evils. This attitude was similar to that commonly held in high-growth "rightist" regimes that depended on public enterprises, such as Spain, Brazil, and Taiwan.[119]

The high level of state-owned company activity can be seen as President Park's pragmatic approach to economic development.[120] One reason President Park permitted the expansion of state-owned companies was to overcome market failure. For him, the intervention of state-owned companies was necessary to counteract private markets from controlling the market. He believed that national interests and public welfare could be protected by state-owned companies, but also that publicization can be justified only when fair and sound coordination of order and structure of economic forces fails.[121]

In terms of value added, the portion of Korean public enterprise was as high as in countries such as India and Pakistan where socialist goals were well pronounced,[122] but this did not mean the state excluded businesses entirely. It is fair to say that the Korean government had not sought to compete with the private sector, but rather wanted to undertake new industrial investments only when necessary either to avoid creating a powerful private monopoly or to introduce a large-scale activity where private capital had not been available or able to do the job.[123]

The largest share of value added from public enterprises was from manufacturing, where projects involved significant amounts of capital, long gestation periods, and high investment risks, challenges not easily overcome by Korea's nascent private sector. In addition, Korea's state-owned companies were established in strategic sectors of the economy. This included the Daehan Oil Corporation (*Yukong*), Korea's first oil refiner, and Korea Electric Power Corporation (KEPCO), Korea's only utility provider. The far higher input–output interdependence of state-owned companies compared with the Korean economy as a whole or the manufacturing sector alone shows how President Park strategically utilized the state enterprise sector.[124] In addition,

---

119 Leroy P. Jones, *Public Enterprise and Economic Development: The Korean Case* (Seoul: Korea Development Institute, 1975), 130.

120 Ibid., 139. SaKong, *Korea in the World Economy*, 30.

121 Park, *Our Nation's Path; Ideology of Social Reconstruction*, 229–30.

122 SaKong, *Korea in the World Economy*, 27.

123 Gilbert T. Brown, *Korean Pricing Policies and Economic Development in the 1960s* (Baltimore: John Hopkins University Press, 1973), 78.

124 SaKong, *Korea in the World Economy*, 30. Input interdependence is calculated by the ratio of intermediate purchases to the total value of output. Output interdependence is estimated by the ratio of intermediate sales to the total value of output. A high input and output interdependence indicates a sector's high degree of linkage with the rest of the economy.

the government intended to promote technology development by setting up a state-owned company to finance the venture and provide human capital, the latter of which was trained at *Yukong* and spread to other industries, such as refining, fertilizer, and petro-chemicals.[125] This was also true of PEDCO. Leroy P. Jones's comment provides the essence of this explanation: "Where received doctrine is tempered by a pragmatic concern for 'ends,' ideological predilection for a particular 'means' may well play a subsidiary role."[126]

MTI and EPB

The establishment of PEDCO required a 100 billion won capital injection, equivalent to US$237 million in 1974.[127] However, the EPB hesitated to provide the budget for 1975 due to its austerity policy and budget constraints.[128] In the dire situation just after the first oil shock of 1973, the austerity policy in the emergency presidential decree, which should be prior to the president's previous instructions, was maintained until the end of 1974. This austerity policy prevented the EPB minister from considering the budget for the establishment of PEDCO.[129] In addition, one possible reason for neglecting PEDCO is other prior budget demands. The HCI promotion program planned in *Chŏngwadae* was not backed by a workable financial program of the relevant ministry.[130] Therefore, the HCI program contracted overall budget management, and the heavy defense budget, which was approximately 35% of the general budget or 5–6% of GDP, added further budget constraints.[131]

In spite of the lack of budget support by the EPB, the MTI advanced the PEDCO project, which was supposed to undertake oil exploration in the domestic area, direct oil purchases bypassing Western oil majors, refining, and the control of petroleum product sales.[132] While the MTI minister reported

[125] Cho, *Kukchaechawollon: Sŏgyumunchaerŭl Chungsimŭro [International Natural Resources: Mainly on Petroleum]*, 215.

[126] Jones, *Public Enterprise and Economic Development: The Korean Case*, 130.

[127] Based on the annual average of exchange rate in WDI (World Development Index), 1$ = 404.4725 won.

[128] "Ministry of Trade and Industry Considers Establishing PEDCO in 1974," *Dong-A Ilbo*, July 24, 1974.

[129] O., *The Korea Story: President Park Jung-Hee's Leadership and the Korean Industrial Revolution*, 307–13.

[130] Duck-woo Nam, "Korea's Economic Takeoff in Retrospect," in Sung-yeung Kwak, *The Korean Economy at a Crossroad: Development Prospects, Liberalization, and South-North Economic Integration* (Westport, Connecticut: Praeger, 1994), 10.

[131] Ibid., 6, 10.

[132] "Ministry of Trade and Industry Considers Establishing PEDCO in 1974."

Table 6 Korea's three refiners at the end of 1974

	Yukong	Honam Refinery Company	Kyung-In Energy Company[134]
Ownership	Korean government: 50% Gulf: 50%	Lucky Group: 50% Caltex: 50%	Hanhwa Group: 50% Union Oil: 50%
Refining Capacity	215 thousand b/d	160 thousand b/d	60 thousand b/d
Annual Profit (post-tax)	1,881 million won	2,940 million won	△ 1,891 million won
Composition of Board of Executive	Korea: 4, Gulf: 4 (Gulf had casting vote)	Lucky: 5, Caltex: 5 (unanimity)	Hanhwa: 3, Union: 3 (Union had casting vote)
Year of Establishment	October 13, 1962	May 19, 1967	September 20, 1968
Beginning of Operation	April 1, 1964	May 19, 1969	June 11, 1973

Source Cho, *Kukchaechawollon: Sŏgyumunchaerŭl Chungsimŭro [International Natural Resources: Mainly on Petroleum]*, 214–230

the plan to establish PEDCO at the presentation of MTI's annual plan on January 29, 1975, President Park commented only on the importance of energy conservation, not on PEDCO. It seems that his silence was interpreted as a "go-ahead" from the perspective of the MTI. After the presentation, the MTI suggested capital investment from three refiners in Korea in order to finance the establishment of PEDCO.[133] Considering PEDCO was supposed to intervene downstream through direct oil purchases, refining, and distribution of petroleum products, there was a conflict of interest between the refiners and PEDCO. Therefore, this financing method was improbable, and the MTI was forced to depend on the EPB6.

With a strong push by the MTI, the PEDCO draft bill passed in the economic ministers' meeting in May 1975. However, this did not mean that the EPB minister decided to support the plan. In addition, whether existing Korean refiners would give up their portion of market shares and distribution networks was also a problem.[135] Since July 1975, the MTI had been

[133] "PEDCO Will Be Established in 1975," *Maeil Economic Daily*, Jan. 31, 1975.

[134] Unlike the other two Korean refiners, Kyung-In could not receive enough crude oil from Union Oil California, an independent US oil company. In 1980, five thousand b/d was provided. Thus, Kyung-In had no choice but to diversify the sources of oil supply, which included Khafji oil, Japan's first successful overseas oil venture. See Cho, *Kukchaechawollon: Sŏgyumunchaerŭl Chungsimŭro [International Natural Resources: Mainly on Petroleum]*, 218–19.

[135] "Improving Demand and Supply of Oil," *Maeil Economic Daily*, May 24, 1975.

discussing, albeit in vain, with the EPB and the MOF whether a supplementary budget for the year of 1975 or a general budget for the year of 1976 was possible for PEDCO.[136] The National Assembly focused more on the development of domestic energy resources, such as coal and nuclear power, and oil did not command much attention.[137] Also, PEDCO's oil stockpiling and oil development function could not make headway due to budget constraints, an issue connected to the sharp increase in the defense budget.[138] This shows that it was still difficult to get budgetary support from the EPB without the President's clear support, because the EPB had a mechanism to prioritize all presidential orders, on which the ministries depend, within the limit of financial resources.[139]

Fake Oil Discovery in 1975 and PEDCO

President Park's decision to fabricate an oil discovery in 1975 in Pohang garnered attention for the PEDCO proposal. Though the Korean government's first attempt to discover oil at the Pohang onshore site from 1964 until 1968 failed,[140] the KCIA secretly resumed the exploration in Pohang in May 1975. A project company called Dongshin Industrial Corporation was established with members of the KCIA, the Institute of Geoscience and Minerals, and the Daehan Mineral and Resources Corporation.[141] At the site,

[136] Sung-ho Lee, "Bright Beginning, Dull Finish in 1975: The Establishment of PEDCO," *Kyunghyang Shinmun*, Dec. 16, 1975.

[137] The National Assembly Secretariat, *Budget Review Report for the Fiscal Year of 1976* (Seoul: The National Assembly Secretariat, 1975), 64.

[138] While the defense budget for 1976 increased by 52.2% compared to that of the previous year, the total budget increased only 28.8%. The portion of the defense budget for 1976 increased from 29.2% for 1975 to 34.5% for 1976. See Korea National Assembly, Special Committee on Budget, *7th Meeting: Examination Reference Material*, Interpellation, Nov. 2, 1975.

[139] Interview with a Korean government official of Ministry of Strategy and Finance conducted on July 20, 2011.

[140] Despite the fact that one oil rig was borrowed and 22 petroleum engineers were invited from Chinese Petroleum Corporation (CPC, Taiwan's national oil company), the exploration of three wells was not successful. See Korea Petroleum Development Corporation, *Han'guksŏgyukaepalgongsa Sibonyŏnsa, 1979–1994 [15-Year History of Korea Petroleum Development Corporation, 1979–1994]*, 144.

[141] Interview with a person who was involved in the project, conducted on April 14, 2011, and Kap-che Cho, "1976nyŏn P'ohangsŏgyupalgyŏnŭn Kajjayŏtta [Pohang Oil Discovery in 1976 Was Bogus]," *Monthly Chosun* March 2000, 192–213.

liquid resembling crude oil was found, but it was a false alarm.[142] Nevertheless, President Park announced that oil was discovered in Pohang at the New Year's press conference in January 1976 as follows.[143]

> It seems that since the beginning of this year [1976] a rumor of an oil discovery in Korea has become a big topic. In addition, more attention was drawn to oil due to a recent Japanese newspaper article saying that there seems to be a big oil reserve in Zone VII, which was also known to Korea. [Clearing throat] I mean, I remember that took place in early December in the end of last year[1975], and it is true that Korea's first oil was found near Korea's Gulf of Young-il. As a result of a long period of surveys and exploration by Korean engineers, gas and oil was found in one of three or four wells. Surely, [3-second pause] the amount of oil is small, but we found the first oil in Korea. The oil was found near a 1,500m-depth. Approximately a few drums of oil was extracted, and the result of a chemical analysis done by the Korea Institute of Science and Technology (KIST) showed that this is good quality oil... As we always tell that God helps those who help themselves, I would like to say that God might give us a nice gift if we together work hard with diligence, self-help, and cooperation, and if we do our best in order to establish a rich and powerful homeland and to make a better country. I would like to say that everybody should please be patient and wait until the exploration is finished.[144]

The fever that spawned from this news conference added momentum to the MTI's oil policy, including the founding of PEDCO. Oil fever spread throughout the country through exaggerated news articles,[145] and even to Japan through news articles by Japanese correspondents.[146] The Vice Minister of MTI stated cautiously that the MTI would consider the establishment of PEDCO on the condition that the Pohang site would

---

[142] According to O Won-chol, the liquid was tested by Caltex and was identified as gasoline, which might be used for lubrication. O., *Enŏchi Chŏngch'aekkwa Chungdongjinch'ul[Korean Style Economic Development: Energy Policy and Going Middle East]*, 300–2.

[143] Selig S. Harrison describes the situation as if there had been a real oil discovery. Considering real facts were not released until O. Won-chol, the then—Second Senior Economic Secretary to President Park, revealed the truth in 1997, we cannot fault Harrison for the inaccurate description. See ibid., 293–308; Selig S. Harrison and Carnegie Endowment for International Peace, *China, Oil, and Asia, Conflict Ahead?* (New York: Columbia University Press, 1977), 137.

[144] Daehan News, "Pakchŏnghŭi Taetongnyŏng Sinnyŏn Tŭkpo [President Park Chung-hee's New-Year Special News]" (Korea, 1976).

[145] President Park's announcement was treated specially by Korean newspapers. The examples are: "Happy News of Dragon Dream That Excited the Beginning of the Year. Oil Discovered!" *Kyunghyang Shinmun*, Jan. 15, 1976, "Placid Full-Scale Exploration in the Dream of Oil Rich Nation," *Dong-A Ilbo*, Jan. 16, 1976, "Buoyant Expectation, Oil Development," *Maeil Economic Daily*, Jan. 16, 1976.

[146] "Japanese News Papers Reported with Interest in Oil Discovery. Stimulate Continental Shelf Development," *Dong-A Ilbo*, Jan. 16, 1976.

be commercially viable.[147] Considering that a Korean government official visited the JPDC to consult on Korea's state-owned oil company in 1976, it seems that the MTI continued its efforts to launch PEDCO.[148] One month after the press conference, the Association of Gas Stations proposed an oil fund, with assets totaling 100 billion won. The function of this oil fund would include supporting domestic oil exploration and PEDCO.[149]

It is likely that the conference was held and the oil exploration in Pohang was not suspended in an effort to boost the environment for Korea's continental shelf development, including the Joint Development Zone between Korea and Japan, which was subsequently related to PEDCO. On January 16, 1976, the MTI summoned representatives of the Western oil companies involved in oil exploration in the Korean continental shelf and strongly urged them to resume exploration activities, which had been halted since July 1975. In addition, President Park may have wished to indirectly push Japan to ratify the Korea–Japan Continental Shelf Joint Development Agreement, which was signed in January 1974. In a *Yomiuri Shinbun* article dated January 26, 1976, the news of Pohang was recognized as adding momentum to a Joint Development Zone between Korea and Japan.[150] Though the KCIA's Pohang oil exploration was stopped without positive results in April 1977—two months before the ratification of the agreement by Japan's Diet—the official suspension of oil exploration was not made by the government until March 19, 1982.[151] This delay was in part because the Korean government did not want to lose the opportunity to ratify the agreement with Japan. The event was meaningful from the perspective oil development industry because a real oil rig, with the ability to dig 5,000 m deep, was imported from the United States after President Park's New Year's announcement on this alleged discovery of oil in 1976.[152]

[147] Here are some examples: "Petroleum Institution Will Be Established," *Maeil Economic Daily*, Jan. 17, 1976. and "4 Western Oil Companies Inactive in Continental Shelf Development," *Dong-A Ilbo*, Jan. 17, 1976.

[148] According to a then-JPDC staff member, a Korean government official visited the JPDC to consult on Korea's state-owned oil company with Mr. Akinobu Tsumura, the then-Director General of Research and planning of the JPDC in 1976. Interview with Tatsu Kambara, a then-staff member of Research and Planning Bureau of the JPDC, conducted on June 16, 2011.

[149] "Oil Fund Will Be Established," *Maeil Economic Daily*, Feb. 11, 1976.

[150] "Korea Might Develop Joint Development Zone Based on the Result of UN Law of the Sea Conference," *Dong-A Ilbo*, Jan. 27 1976.

[151] "Energy Ministry Concluded No Oil in Pohang," *Maeil Economic Daily*, Mar. 20, 1982.

[152] Cho, "1976nyŏn P'ohangsŏgyupalgyŏnŭn Kajjayŏtta [Pohang Oil Discovery in 1976 Was Bogus]."

Joint Development Agreement Between Korea and Japan and PEDCO

On May 30, 1970, the Korean government announced Korea's seven offshore mining zones, based on the natural prolongation principle, where Korea's continent is connected up to the Okinawa trench. Korea's Zone VII and a part of Zone V overlapped with the continental shelf that Japan considered as its maritime boundary based on a median-line approach. Japan responded with a threat to suspend its economic aid to Korea, which was at the time quite essential for Korea's economic development; however, President Park decided not to retract the announcement and Japan requested a meeting on the topic. When a working-level discussion to resolve the dispute failed to make any progress from November 1970 until February 1972, Japan's then-Foreign Minister Ohira proposed to Kim Jong-pil, Korea's Prime Minister, that Korea and Japan develop the overlapping area together.[153] The Korean government accepted Minister Ohira's proposal, and the draft of the agreement was finalized in July 1973. The agreement was signed between Mr. Kim Dong-jo, the then-Foreign Minister of Korea, and Mr. Ushiroku, the then-Japanese Ambassador to Seoul, on January 30, 1974.[154]

Although the Korean National Assembly ratified the agreement on December 7, 1974, Japan's Diet did not ratify it at that time. In Korea, the ruling DRP, aggressively ratified the agreement at the end of 1974's regular session. The resolution was passed in the Committee on Foreign Affairs with only one member's question on December 16, 1974, and in the Plenary Session the next day.[155] In Korea, the unicameral system made this quick procedure possible, but this was not the case for Japan. Though Japan's LDP

---

153 According to Chang Kyung-chŏn, the then-government official of Ministry of Trade and Industry, Korea presented a case based on a previous work of Oda Shigeru, a prominent Japanese international lawyer. Faced with this potent argument by Korea, Japanese authorities changed its hawkish stance and pursued joint development. Oda Shigeru was involved in the North Sea Continental shelf delimitation case of International Court of Justice (ICJ) as a lawyer representing the Germany, and Germany team's argument of the principle of natural prolongation was accepted in the ICJ. Later, he became Judge of the ICJ. Interview with Chang Kyung-chŏn, the then-government official of the Ministry of Trade and Industry, conducted on October 14, 2011.

Mainichi Shinbun reported that former Prime Minister Kishi and Yasugi Katsuo proposed the idea to Kim, Jong-pil when they visited Seoul for a Korea-Japan Cooperation Executive Meeting in 1973. It seems that this proposal was followed by Foreign Minister Ohira's formal proposal. Hyŏng-su Jung, "Kishi Said Delayment of Ratification Ridiculous. Agreement Proposed by Japan," *Dong-A Ilbo*, May 2, 1977.

154 Later, the then-Korean Foreign Minister Kim, Dong-jo was appointed as the first chairman of PEDCO in 1979.

155 "Reason to Pass Late in the Night," *Kyung-hyang Shinmun*, Dec. 17, 1974. "90th National Assembly Regular Session Ended," *Maeil Economic Daily*, Dec. 18, 1974.

had a majority in the Lower House then,[156] the agreement could not be ratified due to tenacious arguments from opposition parties. Despite consistent calls by the Korean government for Japan to endorse the agreement, the situation in Japan did not change.

This stalemate provided even further impetus for the establishment of PEDCO. On April 25, 1977, *Yujŏnghoe* demanded that the MTI minister establish PEDCO for Korea's solo exploration and development in Zone VII in case Japan's Diet did not ratify the agreement that had already been signed in January 1974.[157] The demand coincided with the Korean government's decision to develop Zone VII independently if Japan's Diet did not ratify the agreement during the regular session that was scheduled to end on May 28, 1977. In order to show Korea's seriousness on the issue, the government expressed its intention to declare a 200-mile exclusive fishery zone and to borrow an offshore oil rig for the independent exploration of Zone VII from foreign countries. *Yujŏnghoe*'s request for PEDCO was raised in the middle of this diplomatic warfare.[158]

Ironically, the incident related to Korea and Japan's joint development of the continental shelf provided the MTI with momentum to overcome the financing issue. The Japanese failure to ratify the agreement led the MTI to push for PEDCO. Even after Japan's Diet ratified the agreement on June 8, 1977, the Korean government continued to prepare for the establishment of PEDCO as Korea's counterpart to Japanese companies that explored Zone-VII and other domestic areas.[159]

National pride may have also factored into the founding of PEDCO.[160] In the agreement, while Japanese oil companies—Nihon Sekiyu, Nishi Nihon Sekiyu and Teikoku Sekiyu—were to be involved in the joint development zone from the Japan side, no Korean companies were to be involved in the

[156] From December 10, 1972 until December 4, 1976, the LDP occupied 271 seats of 491 total. This also shows that Japan's Diet is an effective veto-player in Japan's political institutions. See Gerald L. Curtis, *The Japanese Way of Politics*, Studies of the East Asian Institute (New York: Columbia University Press, 1988), 255.

[157] "Yujŏnghoe Asked Gov't to Establish PEDCO," *Kyughyang Shinmun*, Jan. 17, 1976.

[158] "Gov't Noticed Japan Its Decision to Develop Zone-VII Independently," *Kyughyang Shinmun*, Jan. 17, 1976. Japan faced two different pressures from its neighbor countries. On the one hand, Korea urged Japan to ratify the agreement. On the other hand, China protested against the Japanese government's plan to get the agreement ratified in the Diet, because China thought the agreement infringed upon China's sovereignty. Finally, the agreement was ratified on June 9, 1977, and the instrument of ratification was exchanged in 1978, when Japan made Special Measure Law for the Development of Petroleum and Natural Gas, which accommodates Japan's domestic offshore oil development legally.

[159] "Gov't and Ruling Party Deceided to Establish PEDCO," *Maeil Economic Daily*, Aug. 12, 1977.

[160] Interview with Chang Kyung-chŏn, the then-government official of the Ministry of Trade and Industry, conducted on October 14, 2011.

development from the Korea side. Instead, foreign oil companies—Texaco, Shell, Gulf and KOAM—were slated to be involved in the development for Korea.[161] Due to a lack of technology and engineers, no Korean company could serve as a counterpart to Japanese oil companies in the joint development project. Lee Chŏl-sŭng, an opposition party member, noted that the situation was a national disgrace during the National Assembly interpellation on the ratification of Korea–Japan Continental Shelf Joint Development in December 1974.[162]

Quick Process in the National Assembly

Once the draft of the PEDCO bill was confirmed, the procedures, including the deliberation in the National Assembly, were processed quickly. The task to establish a public corporation was transferred to the MER, which had been newly established in January 1978.[163] The MER initiated the effort to make a draft law and secure a budget for PEDCO.[164] The draft bill was passed in the Economic Ministers' meeting on August 31, 1978, and again in the Cabinet meeting on September 26, 1978.

The PEDCO Law was passed swiftly in the regular session in 1978, with its budget for 1979. At the beginning of the regular session of 1978 on October 4th, President Park expressed his intentions for policy in 1979, including the establishment of PEDCO.[165]

> Especially, in the new year [1979], [the government] will found the Korea Petroleum Development Corporation, which will take responsibility for the stockpiling of oil and petroleum products, development of oil resources, importation of crude oil and so on. Regarding Korea-Japan joint-development of the continental shelf, its full-scale operations, such as drilling and exploration in the next year [1979], will be started.

For President Park, the policies of 1979, associated with the biggest budget in the country's history, were significant as Korea prepared to enter a new decade. In this context, the establishment of PEDCO would seem to fit under

---

[161] PEDCO, *Han'guksŏkyukaepalgongsa Sibonyŏnsa 1979–1994 [15 Year-History of PEDCO, 1979–1994]* (Seoul: PEDCO, 1994), 149.

[162] *11th Committee Meeting.*

[163] Korea Energy Economic Institute, *Enŏchi Chŏngch'aek Pyŏnchŏnsa [the History of Energy Policy]* (Kyunggi-do: Korea Energy Economics Institute, 2006), 1033.

[164] The Ministry of Energy and Resources was severed from the Ministry of Trade and Industry in early 1978.

[165] "Full Text of President Park's Administrative Policy Speech," *Maeil Economic Daily*, Oct. 4, 1978.

"*Charip kyŏngjeŭi Ch'okchin* [Promoting self-supporting economy]," which was one of the four policy targets for 1979.[166]

In the Standing Committee on Commerce and Industry on Oct. 23, 1978, Kim Tong-young, a member of the NDP, objected to PEDCO. He argued it would be better to give the initial projected budget of 50 billion won to the private sector for oil stockpiling. However, the Energy Minister answered that the government has national security concerns regarding oil stockpiling due to the strategic importance of petroleum.[167] A subcommittee was formed to deliberate on seven bills including the PEDCO bill, composed of two members from each negotiating body: two from the DRP, two from *Yujŏnghoe*, and two from the NDP.[168] Therefore, it was to be expected that the subcommittee agreed on the government's bill on PEDCO as submitted and reported their decision to the Committee on Commerce and Industry on October 26, 1978.[169] After a slight correction of the text of the bill in the Legislation and Judiciary Committee, the PEDCO bill was passed along with 54 other bills in the Plenary Session on November 13, 1978.[170]

Although the bill was successfully passed, MER officials were disappointed by the size of the budget allocated for implementing PEDCO. The MER initially requested 80 billion won, but the Cabinet only granted 50 billion won, an amount that was further cut by the EPB and finalized at one billion won.[171] It was decided that the necessary funds would come from the Petroleum Industry Fund later.[172] However, the measure could still be described as a success, because once the state-owned company was established, the budget would have to be supplied by the government in some fashion. At the very least, the Petroleum Industry Fund, which was stipulated in the law in 1977, would supply the needed money for oil stockpiling and overseas oil development.[173]

---

166 "Efficient Solution in Entering the 1980s: Backgroud and Meaning of New Year's Administrative Policy Blueprint," *Kyung-hyang Shinmun*, Oct. 4, 1978.

167 Korea National Assembly, Standing Committee on Commerce and Industry, *14th Committee Meeting*, Interpellation, Oct. 23, 1978.

168 Ibid.

169 Korea National Assembly, Standing Committee on Commerce and Industry, *15th Committee Meeting*, Interpellation, Oct. 26, 1978.

170 "Budget for New Year is 4,533.8 Billion Won," *Maeil Economic Daily*, Nov. 14, 1978. The Assembly members seem to have prepared for the 10th general election for December 12, 1978.

171 "Energy Ministry Disappointed by Budget Cutting," *Kyung-hyang Shinmun*, Sept. 20, 1978.

172 Comment by Energy Minister Jang Ye-joon in *14th Committee Meeting*.

173 Interview with Seo Joo-seok, the then-official of Ministry of Energy and Resources, conducted on September 19, 2011 via e-mail. The fund for oil stockpiling has accumulated since October 1979, and the fund for oil development has been collected since June 1983. See PEDCO, *Han'guksŏkyukaepalgongsa Sibonyŏnsa 1979–1994 [15 Year-History of PEDCO, 1979–1994]*, 530–39.

The function of PEDCO, as stated in the final version of the law, included most functions of a typical vertically integrated oil company: Oil development in the domestic and overseas area; stockpiling, transportation and sales of crude oil, natural gas, and petroleum product; investment, loan, and equipment rental to the relevant companies; and any other business the government assigned.[174] When the preparation team designed the structure of PEDCO, it studied other countries' cases, such as the JPDC and US oil companies. Interestingly, the JPDC's original function was to be a special financier for the private sector's oil E&P, but from 1967 to 1970, it was also acting as a state-owned oil company with E&P functions. While the PEDCO preparation team studied the case after the JPDC had split into two distinct entities, they continued to regard the JPDC as amalgamated with JAPEX, an oil development company.[175]

3 Summary and Conclusion

The small number of veto players in Korea's political system contributed to the establishment of PEDCO, which was a decisive outcome from the non-existence of a state-owned oil company in Korea. Korean policy regarding the establishment of a state-owned oil company in the 1970s included seven veto players: the President, MTI/MER, and the EPB, the DRP, *Yujŏnghoe*, the National Assembly and the Standing Committee on Commerce and Industry. Compared to Japan, Korea did not have any interest group veto players in the 1970s because the FKI was not included in *shingikai*-style committee.

Furthermore, the number of effective veto player becomes even smaller when the absorption rule is applied. Unique characteristics of the *Yushin* Constitution and Korea's decision-making norm affect the distance among veto players. The preferences of the DRP and *Yujŏnghoe*, two ruling entities, are included in those of the president. In a majority-based decision-making norm, the Standing Committee and the National Assembly were controlled by the president through majority members from the two ruling entities, and thus the preferences of the legislature and its committee are also included in those of the president. Therefore, the number of effective veto players is reduced to three.

---

[174] Article 19 in "Sŏgyukaepalkongsabŏp [Korea Petroleum Development Corporation Law]," ed. Ministry of Energy and Resources (1978).

[175] Interview with the then-study team member conducted on April 7, 2011. This may have been due to outdated information or selective interpretation of history.

The decisive outcome in the policymaking process followed MTI/MER's original intention, which was to establish a state-owned oil company. As an agenda setter, the MTI finally achieved its goal. While the EPB opposed it due to budget constraints caused by internal and external challenges, the establishment of an off-budget Petroleum Industry Fund relieved the burden for the EPB, who then distributed the minimal funds to launch PEDCO. Korea's National Assembly had been in line with the President's policy, and, therefore, the PEDCO bill could be easily passed in the legislature. This outcome also verified the relevance of veto player theory that states the outcome passed through a small number of veto players tends to be more public-regarding than private-regarding.

Historical events and the level of economic development should be considered as important factors affecting the preferences of veto players. External events affecting the realization of PEDCO includes the first oil shock in 1973 and Japan's delayed ratification of Korea–Japan Continental Shelf Agreement. During the discussion, PEDCO was supported as a national symbol in oil development. While Korea developed quickly, it was still a developing country with $3,000 of GDP per capita (constant value in 2000) in the 1970s and the level of economic development was illustrated by the lack of capital and technology, both essential in overseas oil development. In John Kingdon's terms, a policy window opened when a problem was recognized with external shocks. The idea of PEDCO as a policy solution to the shock was available and the political circumstances were positive for a change.[176]

It is worth noting that Japan's response to the oil shock was different than Korea's. While the oil shock in 1973 provided momentum for the Korean government to make a state-owned oil company, the same external shock only made Japan's private sector's overseas oil development proactive. The JPDC's financial support for Japan's overseas oil development increased sharply from 26 billion yen in 1973 to 71 billion yen in 1974, and finally to 81 billion yen in 1974.[177]

This chapter explains why Korea was able to establish a state-owned oil company by showing the positive correlation between a small number of veto players and the decisiveness of policy change. In the previous two chapters, veto player theory helped guide the contrast between this development outcome in Japan and Korea. The next two chapters will use veto player theory to explain Japan and Korea's cases of establishment/strengthening of a state-owned oil company in the early twenty-first century.

---

[176] John W. Kingdon, *Agendas, Alternatives, and Public Policies* (Boston: Little, Brown, 1984).

[177] Japan National Oil Corporation, *Sekiyukōdan Nijūnensi [20-Year History of Japan National Oil Corporation]* 522.

4

The Attempt to Establish a Japanese State-Owned Oil Company in the Early Twenty-First Century

Examining Japan in the early twenty-first century provides further support for the main argument of this book that a state-owned oil company cannot be established when there are a number of political actors who can block the initiative. Through qualitative analysis, this chapter will show how the interaction among veto players affected Japan's attempt to realize a state-owned oil company in the twenty-first century. In addition, the case study shows that even a strong political leader could not fulfill its preference due to a high number of veto players, implying the strong correlation between the number of veto players and the level of policy resoluteness.

In Japan, the idea to establish a state-owned oil company was conceived during the debt-ridden JNOC debacle during the Koizumi Administration. As a countermeasure to the abolishment of the JNOC, METI (the successor ministry of MITI) proposed a state-owned oil company to deal with important oil and gas assets inherited from the defunct JNOC. However, METI's attempt to establish a state-owned oil company failed.

The final outcome was to privatize one of the better-performing subsidiaries of the JNOC and have the privatized company inherit JNOC's assets. This obstruction was mainly due to a disagreement among ministries and councils of the ruling party in Japan's policymaking. This final outcome was invented to evade a complex political and legislative power struggle among many veto players. Also, the establishment of JOGMEC (Japan

S.-i. Oh, *Overseas Energy Investment of Korea and Japan*,
https://doi.org/10.1007/978-981-99-0285-9_4

Oil, Gas, and Metals National Corporation), was a similar public institution to previously existing JNOC, which shows another example of policy resoluteness.

1 Japan's Political Institutions in the Early Twenty-First Century

Compared to the 1960s, the prime minister's power has been strengthened thanks to formal and informal change. However, there was not a substantial change in terms of the number of veto player in Japan's policymaking system, and thus the policy change is resolute rather than decisive.

1.1 Institutional Veto Players: The Prime Minister, the Ministry, and the Diet

Strong Prime Minister

Prime Minister Koizumi Junichiro, who was inaugurated as Japan's prime minister in 2001, dramatically expanded the power of the executive branch in Japan, making it stronger than ever before. While Japan is a parliamentary democracy with a political system resembling that of the United Kingdom, the role of the prime minister and the members of Cabinet had been limited in significance in policymaking for most of the country's postwar history. In fact, the prime minister could direct his ministers only when backed by a unanimous Cabinet decision, forcing him to build a consensus among different groups within the ruling parties and the government so as to ensure Cabinet approval. Due to an unclear delineation of powers, policies were mainly initiated not by the prime minister, but by working-level bureaucrats. This bottom-up policy process unfolded at the same time as another bottom-up process, the prior review process in the LDP, the ruling party.[1]

Japan's "Un-Westminster" system is contrasted with the central feature of the British parliamentary system—a strong executive government.[2] In the Westminster system, substantive policy debate is conducted by the

[1] Tomohito Shinoda, *Koizumi Diplomacy: Japan's Kantei Approach to Foreign and Defense Affairs* (Seattle: University of Washington Press, 2007).

[2] This term is mentioned by Aurelia George Mulgan. According to Mulgan, the Japanese parliamentary system is different from the Westminster model because the power of the executive is undermined by the ruling party and the bureaucracy. Mulgan, "Japan's 'Un-Westminster' System: Impediments to Reform in a Crisis Economy," 76.

cabinet under the prime minister, while the prime minister takes charge of crafting policy. Policymaking authority is essentially top-down from the prime minister to the Cabinet, the majority party and parliament. The majority party members act only as a sounding board, at most as a source of minor dissent from the Cabinet. Prior party appraisal is not always required and the budget is submitted to the party as a *fait accompli*.[3] This process is sometimes called "presidentializaion," defined as: "The movement over time away from collective to personalized government, movement away from a pattern of governmental and electoral politics dominated by the political party towards one where the party leader becomes a more autonomous political force."[4]

This model of "president-style prime minister" had been attempted by Prime Minister Nakasone Yasuhiro. Believing that postwar Japanese prime ministers had more potential power than British prime ministers, he aimed to change the policy process by promoting more top-down political strategies. He established advisor councils composed of civilians in order to counter the political parties and the bureaucracy; appointed influential political figures to important posts such as heads of the Ministry of Foreign Affairs (MOFA), MOF, MITI, and Japan Defense Agency (JDA) to nurture future prime ministers; and appointed more competent staff to strengthen the *Kantei*.[5] However, Japanese leaders with presidential aspirations like Prime Minister Nakasone were still beholden to the factional coalitions that put them in office.[6] In addition, the effectiveness of these strategies depended on a consistent level of support and interest from each successive prime minister. The absence of such strategic continuity resulted in Japan's pronounced lack of effective leadership as seen in the country's delayed response to the first Gulf War. This became a central issue in national politics in the 1990s, and Prime Minister Hashimoto Ryutaro, who served as Minister of Finance during the Gulf War, strove to enhance policy leadership within the executive branch.[7]

The Administration's reform-related bills of July 1999 strengthened the power and function of the Cabinet Secretariat, the prime minister's supporting body. The Cabinet Office (*Naikakufu*) was established, into which the Prime Minister's Office (*Sorifu*), the Okinawa Development Office, and

[3] Ibid., 76.

[4] Anthony Mughan, *Media and the Presidentialization of Parliamentary Elections* (Houndmills, Basingstoke, Hampshire, New York: Palgrave, 2000), 7; Presidentialization in Germany since 1949 and in Israel since 1996 has a legal basis, while that of Britain is based on customary practice.

[5] Shinoda, *Koizumi Diplomacy: Japan's Kantei Approach to Foreign and Defense Affairs*, 25–26.

[6] Mulgan and Australian National University. Asia Pacific School of Economics and Management, *Japan's Failed Revolution: Koizumi and the Politics of Economic Reform*, 51.

[7] Shinoda, *Koizumi Diplomacy: Japan's Kantei Approach to Foreign and Defense Affairs*, 48–62.

the Economic Planning Agency were merged. The Defense Agency and the National Public Safety Commission were also within the sphere of the Cabinet Office.[8] The role of Chief Cabinet Secretary (CCS) was expanded from that of a policy mediator and spokesperson for the prime minister to a post responsible for planning and drafting, Furthermore, the CCS would also be the acting prime minister in the event of a crisis.[9] Additional staff was provided to the prime minister's office with unlimited private secretaries, up to five special advisors—two more than before 2001—and the opening-up of *Kantei* posts to non-bureaucrats.[10] Ad-hoc offices for special issues were allowed to be created in the *Kantei* for further coordination of inter-ministerial policymaking, and the number of officials at the secretariat swelled to 655, three times more than before 2001. The reforms also enhanced the institutional capacity of the prime minister by clarifying his authority to propose important, basic policies. The revised law also provided the *Kantei* with the power to plan and draft concrete proposals under the direction of the Cabinet and the prime minister.[11] Also, the Council on Economic and Fiscal Policy (CEFP or *Keizai zaisei shimon kaigi*) was established and came to play a leading role in invigorating prime ministerial direction in the area of economic management. The CEFP reviewed and discussed the basic principles of economic and fiscal policies, including budget formulation, and deliberated on important items of economic and fiscal policy. It was considered a new venue to implement reforms, separate from the web of special interests that had previously hindered such efforts.[12]

In the Koizumi Administration, this institutional strength was utilized through special State Ministers who were attached to the Cabinet office. They supported prime minister-led policymaking in the Cabinet. In Koizumi's Cabinet, Takenaka Heizo, Ishihara Nobuteru and Kōnoike Yoshitada took charge of economic and fiscal policy, administrative reform, and special structural reform, respectively. They were not bound to specific ministries but were assigned specifically to Koizumi's program of structural reform. Special

[8] Stockwin, *Governing Japan: Divided Politics in a Resurgent Economy*, 148.

[9] Shinoda, *Koizumi Diplomacy: Japan's Kantei Approach to Foreign and Defense Affairs*, 64–66; Mulgan and Australian National University. Asia Pacific School of Economics and Management, *Japan's Failed Revolution: Koizumi and the Politics of Economic Reform*, 179.

[10] Shinoda, *Koizumi Diplomacy: Japan's Kantei Approach to Foreign and Defense Affairs*, 70.

[11] Ibid., 76–77.

[12] Yu. Uchiyama, *Koizumi and Japanese Politics Reform Strategies and Leadership Style* (Hoboken: Taylor & Francis, 2010), 26–27.

organizations like the Secretariat for Promotion of Special Public Institution Reform in the Cabinet office supported their roles in cross-ministerial work.[13]

In addition to the strengthened *Kantei*, Prime Minister Koizumi also exercised more effective policymaking leadership through the media. Koizumi's used his twice-a-day meeting with the media, during the day for newspapers and the evening for TV, to disseminate his statements. He also made good use of the sensational and dramatized reporting of weekly tabloid magazines and sports newspapers to enhance his populist style. His "one-phrase (*wanfurēzu*) politics," which uses easily understood terms and simple assertions to convey messages, was well-suited to capturing the attention of listeners and narrowing the psychological sense of distance. This media strategy added his personal power to institutional change aimed at strengthening executive power.[14]

However, Koizumi was not able to impose his policy stance on traditional veto players due to the predominance of the dual policy making system in the government and the LDP. Thus, the country faced a chronic policymaking stalemate. The pattern of confrontation over policies and legislation shifted from the ruling camp versus the opposition parties to the prime minister versus the ruling camp.[15]

As an attempt to strengthen executive power, Koizumi wanted the executive branch to be the center of policymaking, as it was in Britain. In keeping with this objective, Koizumi himself circumvented the LDP's prior approval requirement for some of the bills presented to the Diet by the Administration. Koizumi deployed the CEFP as a key mechanism in pursuit of his reform agenda. Its membership included all economics ministers and other key figures like the President of the Bank of Japan.[16] While the CEFP was strongly supported by Prime Minister Koizumi and Takenaka Heizo, the Minister of State for Economic and Fiscal Affairs, it could not ignore the traditional review system completely. In the early phase of his administration, Koizumi announced his policies, which included reviewing road-specific funds, privatizing the Japan Highway Corporation, and reducing grants to local prefectures, with no explanation to the LDP. However, Takenaka Heizo had to explain CEFP's policies to the LDP and relevant ministries.[17] Only ten

[13] Ibid., 132–33. The secretariat was established on Jan. 6, 2001.

[14] Ibid., 6–12.

[15] Mulgan and Australian National University. Asia Pacific School of Economics and Management, *Japan's Failed Revolution: Koizumi and the Politics of Economic Reform*, 193.

[16] Stockwin, *Governing Japan: Divided Politics in a Resurgent Economy*, 143–44.

[17] "Pressure to Kantei, LDP PARC's Hearing of CEFP," *Asahi Shinbun*, June 6, 2001.

among 150 staff members were recruited from outside the government, and this prevented the CEFP from operating as a tool of reform.[18] Furthermore, according to Masaru Takagi, a professor of economics at Meiji University, the council turned into a rubber stamp for policies drafted by the MOF and other ministries. As seen in the case of Kazuhiko Takeshima, a former Finance Ministry bureaucrat who played a key role in the CEFP as an assistant to the Deputy Chief Cabinet Secretary, the influence of the Finance Ministry in the CEFP's policymaking was widely recognized,[19] thereby lowering the credibility of the CEFP in the eyes of other ministries.[20]

To formalize the executive's authority in decision-making, in November 2001 Koizumi instructed the LDP's National Strategy Headquarters (*Kokka Senryaku Honbu*) to consider ways in which the existing decision-making process could be restructured to give the Cabinet sole authority in implementing policy.[21] This meant that the veto power of the LDP PARC and the Executive Council would have been removed from Japan's policymaking process.

In March 2002, the LDP's National Strategy Headquarter drafted the principles of a new decision-making system: leadership by the Cabinet, centering on the prime minister; removal of the power of bureaucrats in the process; and curbing the ability of party lobbies to influence policy. The proposal called for the eradication of the current practice of gaining ruling party approval of bills before the Cabinet submits them to the Diet, although parties would still have some input into Cabinet decisions in the form of policy coordination by State Ministers. These positions would have been occupied concurrently by the chairmen of the policy affairs research councils of the ruling coalition. Committee chairmen in those councils would concurrently sit as either senior vice ministers or parliamentary secretaries.[22]

In July 2002, the effort to make a new decision-making system came up with a draft recommendation to abolish the LDP's prior screening system. This proposal also included eliminating the practice of forcing members to abide by party decisions in order to diminish the power of the prior screening system and to reduce *zoku*'s influence; abolishing the Executive Council's

---

18 Mulgan and Australian National University. Asia Pacific School of Economics and Management, *Japan's Failed Revolution: Koizumi and the Politics of Economic Reform*, 197–98.

19 Yosuke Naito, "Bureaucrats Victors as Diet Draws to Close," *Japan Times*, July 27, 2002.

20 A METI official also described the CEFP as a puppet of the MOF. Interview with a METI official who is versed in the JNOC reform issue, conducted on June 22, 2011.

21 Mulgan and Australian National University. Asia Pacific School of Economics and Management, *Japan's Failed Revolution: Koizumi and the Politics of Economic Reform*, 201.

22 Ibid., 202. "Diminishing Role Lowers Political Power, from MITI to METI," *Asahi shinbun*, May 31, 2002.

practice of requiring a unanimous vote on policy issues in order to prevent certain Diet politicians from obstructing specific policy issues; allowing the same person to serve as both LDP PARC Divisional Chairman and deputy minister of the related ministry in order to unify the policymaking system of the LDP and the Cabinet.[23] However, the proposal was dropped in the face of strong opposition from the LDP, resulting in Koizumi's failure to overhaul the decision-making process, the most important element in his structural reform program.[24]

Under LDP rule, Japan's prime minister was selected by a vote in the Diet as well as a grass roots vote, which is close to the American primary election system. In addition to the LDP Diet members, three rank-and-file members from each of the party's prefectural chapters were also allowed to vote in a so-called popular election.[25] For Koizumi, it provided a chance for being the prime minister, but it also created a problem. In spite of lukewarm support from party members, he was elected as party president with a substantial margin after receiving 123 votes of the 141 available from the prefectural branches. This result then affected the voting of the LDP Diet members, who were reluctant to support Koizumi's radical approach, but who received pressure to do so from their constituencies. This path to victory thus compounded Koizumi's difficulties in the policymaking process, where the party's support is critical for passage of legislation.

Ministerial Veto Player: The METI and the Reform Minister

(a) Ministry of Economy, Trade, and Industry (METI)

MITI, the predecessor to METI, had been described as the "Headquarters of Japan, Inc." and the "Vision Office", but this role gradually changed in the 1990s due to the globalization of the Japanese economy and the increasing capacity of the private sector in terms of information, expertise, and resources. In the tide of deregulation, MITI's regulatory power largely decreased. In response, MITI attempted to extend its jurisdiction from its traditional policy areas to new issues like the environment, consumer protection, and the movie industry through a broad interpretation of its Establishment Law. MITI had

23 Mulgan and Australian National University. Asia Pacific School of Economics and Management, *Japan's Failed Revolution: Koizumi and the Politics of Economic Reform*, 204–5.

24 Ibid., 240.

25 Ibid., 45.

also taken on the role of promoting international cooperation.[26] In 2001, there was a major restructuring of the government bureaucracy. Twenty-two ministries and agencies of the central government were reduced to thirteen. In this process, the MITI changed to METI by giving some of its functions to the new Ministry of Environment.[27]

As a result, its power diminished in a number of ways. Deregulation meant a decrease in the ministry's regulatory power over individual industries, which meant that special interests no longer required the influence of *zoku* Diet members to deal with METI. This, in turn, resulted in a decrease of benefits for those Diet members. In addition, the *zoku* politicians who were supportive of METI decreased in the Koizumi Administration. Of five former MITI ministers following the LDP's return to power in 1994, one passed away, and three failed to be elected in the 2000 general election. The only survivor was Mr. Horiuchi, who had been clashing with MITI bureaucrats.[28] In the Koizumi Administration, when the *Kantei* became the center of the policymaking, METI could not dispatch any bureaucrat to the CEFP, a powerful organization that could initiate policies in the *Kantei*.[29]

The steady erosion of MITI's power through deregulation and the globalization of Japanese manufacturing had a galvanizing effect on its enthusiasm for energy development projects over the past decade. Energy security has been one of the few enduring national imperatives in which the government has a self-evident role that also lies unambiguously at the heart of MITI's organizational mission.[30]

The ANRE under METI, the successor to MITI, currently deals with Japan's energy policies, including overseas oil and gas development.[31] In response to the new international oil market situation of the 1970s, the ANRE was established by Nakasone Yasuhiro, the then-MITI Minister, and Morozumi Yoshihiko, the then-MITI Administrative Vice Minister, in July 1973 as a part of MITI's most significant organizational restructuring since its inception in May 1949.[32] The establishment of the agency can be seen as

---

26 Kohno Masaru, "A Changing Ministry of International Trade and Industry," in Jennifer Ann Amyx and Peter Drysdale, *Japanese Governance: Beyond Japan Inc.*, Politics in Asia Series (London, New York: RoutledgeCurzon, 2003), 96–111.

27 Stockwin, *Governing Japan: Divided Politics in a Resurgent Economy*, 146–49.

28 "Diminishing Role Lowers Political Power, from MITI to METI."

29 Ibid. In the Koizumi Administration, one Assistant Vice Minister and one General Director were sent from the Ministry of Finance.

30 Calder, "Japan's Energy Angst and the Caspian Great Game," 17.

31 See Article 4, 16 and 17 in *METI Establishment Law*, 108 (July 18, 1999).

32 The establishment of the ANRE took place just before the first oil shock in Oct. 1973. Ministry of International Trade and Industry, *Tsūshōsangyō Seisakushi [the History of International Trade and Industry Policy]*, 17 vols., vol. 15 (Tokyo: Tsūshōsangyōchōsakai, 1991), 510–21.

a revival of old energy agencies, such as the Coal Agency (*Sekitanchō*), which existed under the MCI from December 1945 until the establishment of MITI in 1949, and the Agency of Natural Resources (*Shigenchō*) that existed from the establishment of MITI in 1949 until August 1952.[33]

In the ANRE, overseas oil development is addressed by the Department of Natural Resources and Fuel, which is known to be more powerful than the agency's other two departments, the Department of Energy Conservation and Renewable Energy and the Department of Electricity and Gas Industry.[34] The agency utilizes the oil and gas self-sufficiency ratio, which is the ratio between the oil and gas developed by Japanese companies and total oil and gas imports, for its quantitative policy target. While the numeric target was abolished around 2000, it was revived in Japan's comprehensive energy strategy of 2006, at which time the government decided to aim for 40% by 2030.

The ANRE has advisory committees (*shingikai*), such as the Petroleum Subcommittee under the Comprehensive Resource and Energy Investigation Committee, which deals with the issues pertaining to overseas oil development. As of April 2002, there were 25 members who attended the *shingikai*.[35]

(b) **State Minister for Administrative Reform and Regulatory Reform**

Ishihara Nobuteru, a reform-minded young LDP Diet member, became State Minister for Administrative and Regulatory Reform, the Chief of the Headquarters for Administrative Reform in the *Kantei*. He was one of the "ministers for special missions" to assist the prime minister.[36] The secretariat operated under the assistant CCS for internal affairs to address three major issues: (1) special public corporations, (2) the public system and (3) administration-commissioned public service.

While the Reform Minister Hashimoto, the first State Minister for the public sector reform under Mori, focused on changing the civil servant system, Ishihara Nobuteru, under Koizumi fixated on special public corporation (*tokushu hojin*). The secretariat swiftly drafted a bill proposing the Special

33 Ministry of International Trade and Industry, *Tsūshōsangyōsesakushi [the History of International Trade and Industry Policy]*, 17 vols., vol. 17 (Tokyo: Tsūshōsangyōchōsakai, 1991), 204–46.

34 Interview with a METI official, conducted on Aug. 19, 2010.

35 Ministry of Economy Trade and Industry, "Member List of Petroleum-Subcommittee under Comprehensive Resources and Energy Investigation Committee" (Tokyo 2002).

36 The state minister is appointed at the prime minister's discretion, and their role is to implement key policies in a prompt way beyond established ministerial jurisdiction and vested interests, which is quite similar to Korea's presidential committee system. See Mulgan and Australian National University. Asia Pacific School of Economics and Management, *Japan's Failed Revolution: Koizumi and the Politics of Economic Reform*, 73–74.

Public Institutions Reform Promotion Headquarters, which would review the fate of all public corporations. After the bill passed in June 2001, 77 special corporations, including JNOC, became the target of drastic reforms.[37]

Legislative Institutional Veto Players: The Diet and the Committee

The distinct feature of the Japanese bicameral system is that unlike that of Britain, the Upper House has autonomous power to decide what bills will pass, while the prime minister has no measure with which to keep the Upper House in check. As seen in the September 2005 general elections, the prime minister may dissolve the Lower House and eliminate any enemies in the ruling party. However, the constitution does not give the prime minister the same power vis-à-vis the Upper House. Since the LDP had never regained an overall majority in the upper house between 1989 and 2007, the prime minister needed to conciliate the Upper House members who disagreed with him. This has been the case not only for opposition members but also for LDP members of the Upper House. For example, Aoki Mikio, the leader of the LDP's Upper house members, resisted Koizumi on the issue of road construction reform. At the end of 2006, three months after Abe succeeded Koizumi's prime ministership, Aoki Mikio forced Abe to allow opponents of postal reform to rejoin the party. While the Upper House had been generally compliant with the Lower House in the past, its increasing tendency toward independence continues to complicate the legislative process.[38]

Also, the Diet committee system, inspired by the US Congress, produces still one more level of deliberation that legislation must pass. Time pressure, caused by a three-day-a-week meeting tradition in the committee, is normally accompanied by the difficulty and urgency of the issues, and this makes the committee visible as a veto player.[39] Therefore, the Diet has four veto players—the Upper House, the Lower House, and the respective Standing Committee for Economy, Trade, and Industry.

In 1994, Japan's Lower House election system changed entirely from a single, non-transferrable vote in multi-member constituencies (SNTV in MMCs) to a combination of single-member districts and proportional representation (PR). In other words, three hundred constituencies elect one

[37] Shinoda, *Koizumi Diplomacy: Japan's Kantei Approach to Foreign and Defense Affairs*, 80–81.

[38] Hiroshi Kaihara, "The Advent of a New Japanese Politics: Effects of the 1994 Revision of the Electoral Law," *Asian Survey* 47, no. 5 (2007): 740–42.

[39] Kent E. Calder, *Japan's Stealth Reform: The Key Role of Political Process*, Asia–Pacific Policy Paper Series (Washington, DC: The Edwin O. Reischauer Center for East Asian Studies, 2005), 6–7.

member each, and the remaining 200 seats are allocated in party-list proportional representation from eleven regional constituencies. Each vote receives two votes, one for the local constituency and one for the regional (PR) constituency.[40]

The electoral reforms of the Lower House in 1994 also shifted the relative importance of policy interference for individual Diet members. New electoral rules weakened the incentive for policy specialization that corresponded with the interests of a narrow political support base focusing on well-established ties with specific industry groups. On the other hand, it strengthened the incentive to maximize benefits for the constituency as a whole, through activities such as pork barreling in order to win a plurality.[41]

1.2 Partisan Veto Players: The LDP Councils and Coalition Partners

The new political reality in Japan contributed to the increasing number of veto players. Following the 1996 general election that implemented the new electoral system, the LDP continued to be the ruling party and the largest party in the Diet. However, the LDP needed to form a coalition government with other political parties in order to form a working majority in the Diet (Table 1).

Making a coalition government increases the number of veto players more than before by the number of coalition partners.[42] At the same time, compared to the 1960s, the diminished practice of consensus-based decision-making removed the veto player status from opposition parties, thereby decreasing the number of veto players. Political parties outside the LDP were either brought into coalition arrangements with the LDP or saw their confrontational impact blunted by arrangements made in their favor. The Democratic Party of Japan (DPJ) was the largest party outside the LDP-centered coalition, but it could not reach an internal consensus on whether to enter this coalition or grow into a party capable of substituting the LDP.[43]

[40] Stockwin, *Governing Japan: Divided Politics in a Resurgent Economy*, 167–68.

[41] Mulgan and Australian National University. Asia Pacific School of Economics and Management, *Japan's Failed Revolution: Koizumi and the Politics of Economic Reform*, 151–55.

[42] Tsebelis, "Decision Making in Political Systems: Veto Players in Presidentialism, Parliamentarism, Multicameralism and Multipartyism," 302.

[43] J. A. A. Stockwin, "Reshaping of Japanese Politics and the Question of Democracy," *Asia–Pacific Review* 9, no. 1 (2002): 56.

Table 1 General election results and the LDP's need to form a coalition

Parties	General election date	
	June 25, 2000	Nov 9, 2003
LDP	**233 (48.5%)**	**237 (49.4%)**
CGP	**31 (6.4%)**	**34 (7.08%)**
CP	**7 (1.4%)**	**4 (0.83%)**
DPJ	127 (26.4%)	177 (36.9%)
JSP/SDP	19 (3.9%)	6 (1.25%)
JCP	20 (4.2%)	9 (1.9%)
LP	22 (4.6%)	-
Others	21 (4.4%)	13 (2.7%)
Total seats	480	480

Source Stockwin, *Governing Japan: Divided Politics in a Resurgent Economy*, 278–79
Note The number of coalition parties is in bold face. The number in parenthesis is the percentage of the party among total number of Lower House seats. The debacle regarding the JNOC took place mainly between April 2002 and 2003. The LP was merged with the DPJ between September and October 2003. The parties of the coalition government are expressed bold

One Ruling Party: The LDP

The new political reality did not change the number of veto players in the LDP, because the LDP maintained the status of a ruling party and the role of councils were retained. The LDP's one-party dominance ended in August 1993. Following nine months as a minority party, the LDP asserted its dominance by forming another ruling coalition with the assistance of small parties in 1994. Unlike other parliamentary systems, as the majority party, the LDP must approve all legislation including budgets, a well-established convention that produces a separation of powers between the ruling party and the executive branch. The LDP's considerable de facto power over government policy meant the party was not subordinate to the executive branch as is customary in other parliamentary systems.[44] Koizumi wanted to change this old order in the LDP, and attempted to do so by bypassing the LDP councils in some regards, such as reforming the postal service. However, the roles of the LDP PARC and the Executive Council could not be easily removed, and thus there existed two veto players in the LDP.

While the electoral reform in 1994 weakened the faction, the reform did not change the significance of PARC in policy making. Since the change of the electoral law from a multi-member district system to a single member

[44] Mulgan and Australian National University. Asia Pacific School of Economics and Management, *Japan's Failed Revolution: Koizumi and the Politics of Economic Reform*, 40, 130.

system, the influence of factions in the LDP decreased, although most LDP Diet members still belonged to a faction. Both the revision of the electoral law in 1994 and the strengthening of the prime ministerial authority by the Hashimoto Cabinet in 1998 were handed down to Koizumi. When Koizumi organized his first Cabinet, he eschewed LDP custom by deliberately ignoring inputs from the various factions, the Hashimoto faction in particular, which was then the largest faction in the LDP.[45] He also instructed Mr. Horiuchi, the Chair of the LDP Executive Council, to disregard the tradition that every faction sends a vice executive to the Executive Council.[46] However, this did not change the number of veto players in the LDP, but their composition.

In addition, electoral reform did not affect the important function of PARC. In the new electoral system, the LDP members had to respond to the need for a wider and more diverse constituency. Furthermore, not being able to bring benefits to his or her district made the LDP vulnerable. Therefore, LDP members were required to become familiar with diverse policy areas in order to meet the needs of a wider constituency.[47]

Coalition Partners: CGP and CP

The coalition of the LDP with two small parties, the Clean Government Party (CGP) and the Conservative Party (CP), had a legislative majority in both houses of the Diet. This, in theory, provided Koizumi with the guaranteed passage of Cabinet-sponsored bills.[48] The impact of the coalition rule on policy making was considerable, especially when Koizumi's approval rate was not high enough to neglect coalition partners. During the Koizumi Administration, many inter-party coalition negotiations were conducted by the secretaries-general and policy chairs of the three parties. Their meetings were followed by gatherings of the three coalition party leaders, who formally adopted the policy.[49] Therefore, coalition parties can also be categorized as veto players.

---

[45] Stockwin, *Governing Japan: Divided Politics in a Resurgent Economy*, 192. Some Japanese observers interpret his confrontation with the Hashimoto faction as stemming from their different factional roots, the Fukuda faction and Tanaka faction, which had long-standing animosity in the LDP. Uchiyama, *Koizumi and Japanese Politics Reform Strategies and Leadership Style*, 13–14.

[46] "Listen to the New Three High Officials of the LDP," *Asahi Shinbun*, Apr. 29, 2001.

[47] Ellis S. Krauss and Robert Pekkanen, *The Rise and Fall of Japan's LDP: Political Party Organizations as Historical Institutions* (Ithaca, NY: Cornell University Press, 2011), 200–01.

[48] Mulgan and Australian National University. Asia Pacific School of Economics and Management, *Japan's Failed Revolution: Koizumi and the Politics of Economic Reform*, 61.

[49] Ibid., 159–60.

(a) **Clean Government Party (CGP)**

Formed in 1964, the CGP governed in coalition with its dominant partner, the LDP. The CGP is the only significant party to be founded by a religious group, the Nichren Buddhist sect, in the post-World War II era. It was formed in 1964 and attained remarkable success in general elections. In the 1960s, the CGP drew its support base from the less educated, less well-off, and less well-organized sections of big city populations, and its policy orientation was moderate conservative.[50]

As a junior partner in a coalition government, the CGP could exert only limited influence on government policy. However, it maintained some influence because the LDP needed the CGP to keep its majority status in the Upper House which has the constitutional right to block bills.[51] The party had been communicating with the LDP through Hashimoto faction members, and the resulting gap between the Hashimoto faction and the Koizumi Administration made intra-coalition communication rather rough.[52] For example, when the CGP and the Hashimoto faction collectively asked for the dismissal of the agriculture minister, Koizumi did not accept it.[53]

(b) **CP (Conservative Party)**

The CP is a breakaway party from the Liberal Party (LP). When Mr. Ozawa wanted to split from the coalition in April 2000, nearly half the Liberal Party members opted to remain in the government and formed the Conservative Party.

Opposition Parties

(a) **DPJ (Democratic Party of Japan)**

The DPJ was founded in September 1996, and the right wing of the JSP was its central element at its inception. Two years later, the DPJ was re-launched with members of the collapsed New Frontier Party (NFP), many

---

[50] Stockwin, *Governing Japan: Divided Politics in a Resurgent Economy*, 198.

[51] Ibid., 197.

[52] Mulgan and Australian National University. Asia Pacific School of Economics and Management, *Japan's Failed Revolution: Koizumi and the Politics of Economic Reform*, 160.

[53] "Withering Koizumi Cabinet Having Historical Mission Takes Risk of Life by Dissolving Diet in July," *Asahi shinbun*, Apr. 29, 2002.

of whom were conservative. The inclusion of NFP members diluted the old socialist element within the party, creating difficulties in making the party ideologically coherent.

In policy, the DPJ had long been an advocate of cuts in wasteful public works spending, as well as policies such as deregulation and decentralization, all of which overlapped with Koizumi's structural reform agenda. Partly due to this similarity, the DPJ often operated like a de facto ruling coalition member. For example, Mr. Hatoyama, the then-President of the DPJ, showed his support for Koizumi's reform efforts on June 6, 2001, just prior to the July 2001 Upper House election. This unusual support was likely based on the expectation that the reforms would split the LDP and restructure the political landscape in favor of the DPJ.[54] Facing internal opposition from the LDP and the CGP on his reform effort, Koizumi utilized the DPJ's cooperative stance to check opposing forces in the LDP and the CGP. However, after Koizumi accepted the LDP's request to dismiss Foreign Minister Tanaka Makiko, the DPJ shifted its efforts toward toppling the Koizumi Cabinet because of Koizumi's compromising attitude toward LDP resistance forces.[55]

(b) **JCP (Japan Communist Party)**

The JCP is the oldest party in Japan, formed in 1922 and legalized in 1945 by the American Occupation. It gained significant support during the Occupation period, from the late 1960s until the late 1970s, and then again in the late 1990s. The JCP never joined a coalition arrangement with other parties, and it holds to a policy of running candidates, even in unwinnable seats, throughout the country, often splitting the opposition vote.[56] While the JCP supported the abolition of the JNOC and private sector-driven oil development, it opposed the privatization of the Housing Finance Corporation and the Japan Scholarship Foundation due to the high costs of real estate and education. Also, it argued the need for an immediate prohibition

---

[54] "Quick Survey 2: Names of DPJ Members Not Resisting Koizumi's Reform," *Weekly Asahi*, June 22, 2001.

[55] Mulgan and Australian National University. Asia Pacific School of Economics and Management, *Japan's Failed Revolution: Koizumi and the Politics of Economic Reform*, 118–19. Interestingly, Tanaka Makiko joined the DPJ with her husband in 2009 and at the time of this writing is working, as Chair of the Standing Committee on Foreign Affairs. See "Ex-Foreign Minister Tanaka to Join DPJ," *Kyodo News International*, Aug. 15, 2009. The Democratic Party of Japan, "Members Profile," http://www.dpj.or.jp/english/member/?detail_4032=1.

[56] Stockwin, *Governing Japan: Divided Politics in a Resurgent Economy*, 198–99.

of *amakudari*.[57] Overall the party's stance was rather confrontational toward the Koizumi Cabinet.[58]

(c) JSP (Japan Socialist Party) or SDP (Social Democratic Party)

The JSP had been the largest opposition party to the LDP. The socialists were central to the movement in its defense of the 'peace clause' of the constitution, and in opposition to rearmament and enhanced security relations with the United States. Between 1993 and 1994, the JSP was a key player in the coalition governments and supported Prime Minister Murayama Tomiichi, the first socialist prime minister in 46 years. Between 1994 and 1996, the JSP composed a coalition with the LDP and the New Party *Sakigake*.[59] In 1996, the leftist legislators of the defunct JSP formed the SDP. In Koizumi's reform of special public corporations, the party had argued in favor of abolishing the JNOC,[60] but opposed the abolition of the Housing Finance Corporation.[61]

(d) LP (Liberal Party)

The LP was established by Ozawa Ichiro in January 1998 after the collapse of the New Frontier Party. The party pursued a free market economy, Japan as a "normal state," and internationalism. Between January 1999 and April 2000, the LP was part of the LDP collation. When Ozawa pulled the party from the coalition, almost half of the members chose to remain in the coalition government, and the rest merged with the DPJ in 2003.[62]

1.3 Interest Groups: *Keidanren*, the PAJ, and the JPDA

Keidanren: Weakened Role in Energy Policy

During the economic downturn of the 1990s, the Federation of Economic Organizations (*Keidanren*) gradually relinquished its role as a pooling body for donations from its member companies to the party. The splintering of

[57] "Japan Communist Party's Proposal on Special Corporation Reform," *Asahi shinbun*, July 19, 2001.

[58] Mulgan and Australian National University. Asia Pacific School of Economics and Management, *Japan's Failed Revolution: Koizumi and the Politics of Economic Reform*, 119.

[59] Stockwin, *Governing Japan: Divided Politics in a Resurgent Economy*, 198.

[60] Comment by Ooshima Reiki, Diet member of Democratic Socialist Party in Japan House of Representative, Standing Committee on Economy, Trade and Industry, *22nd Committee Meeting*, Interpellation, June 12, 2002.

[61] "Cautious on Abolition of Housing Finance Corporation," *Asahi shinbun*, Sept. 15, 2001.

[62] Stockwin, *Governing Japan: Divided Politics in a Resurgent Economy*, 201.

the conservative side of politics followed by the continuous dissolution and reform of parties and coalition arrangements had complicated *Keidanren*'s political funding. In addition, due to economic recessions, the financial base of businesses, in general, had deteriorated. Faced with these difficulties, two major business federations, *Keidanren* and the Japan Federation of Employers Associations (*Nikkeren*), merged into *Nippon Keidanren* (Japan Business Federation) in an effort to reduce expenses for their member firms.[63]

While *Keidanren* continued to represent big business interests overall, industrial sector groupings and powerful individual corporations became more important to the exercise of political influence.[64] This trend also applied to Japan's overseas oil development. Overall, *Keidanren* dealt with big issues, such as climate change around the year 2000, and industry-specific issues like overseas oil development, the latter of which became a specific issue dealt with industrial organizations such as PAJ and JPDA.[65] However, *Keidanren* still sent its staff to the related *shingikai*. As of April 2002, Wada Ryūkō, the Secretary of General of *Keidanren*, was sent to the Petroleum subcommittee under Comprehensive Resources and Energy Investigation Committee in METI.[66]

Petroleum Association of Japan (PAJ)

The PAJ, which had represented refiners, now became supportive of JNOC's role because refiners became involved in overseas oil development to secure oil supply and expand business opportunities. Therefore, compared to the 1960s, the preference of the PAJ converged with that of the JPDA, which represented oil developers. In the case of Cosmo Oil, the company had secured 17% of the company's total crude requirements from the projects it owned, the majority of which were in the Middle East, by 1998. Keiichiro Okabe, President of Cosmo Oil Co. and Chairman of the PAJ in the late 1990s, frequently emphasized the importance of oil production as the "head of the oil business."[67] Mitsubishi Oil successfully found oil at the Rang Dong field

[63] Mulgan and Australian National University. Asia Pacific School of Economics and Management, *Japan's Failed Revolution: Koizumi and the Politics of Economic Reform*, 230–31.

[64] Curtis, *The Japanese Way of Politics*, 111–12.

[65] See Japan Business Federation, "Policy Proposals Environment and Energy," http://www.keidanren.or.jp/en/policy/index07.html.

[66] Ministry of Economy Trade and Industry, "Member List of Petroleum-Subcommittee under Comprehensive Resources and Energy Investigation Committee."

[67] Mr. Okabe describes oil production as the head, refining as the body, and marketing as the legs of the business. See "Japan/JNOC/Oil Development-2: Equity Crude Target Intact," *Dow Jones Newswire*, Aug. 21, 1998.

off the coast of Vietnam in 1994, and at the time of this writing is about to start commercial production under a 25-year production-sharing contract (PSC) with Petro Vietnam, a Vietnamese state-owned oil company.[68] Mr. Okabe was also a member of the Petroleum subcommittee under Comprehensive Resources and Energy Investigation Committee as of April 2002.[69] As a lobby, the PAJ was claimed to provide a total 798 million yen to the LDP from 1990 until 2002.[70]

Japan Petroleum Development Association (JPDA)

The JPDA has made efforts to enhance the government's support for the private sector. For the JPDA, oil development is a national mission to secure a sustainable energy supply. While the JPDA is an association of private companies, its element of public service can be attributed to the fact that most JPDA chairs were from MITI.[71]

Mr. Wagasugi Kazuo, the JPDA Chairman, was a member of the Petroleum subcommittee under the Comprehensive Resources and Energy Investigation Committee as of April 2002.[72] As a lobby, the JPDA was claimed to provide a total of 104.5 million yen to the LDP from 1990 until 2002. The JPDA was criticized by Shiokawa Tetsuya, a Diet member from the JCP, when it was discovered that the beneficiaries of these contributions included the debt-ridden Japan Oil Development Corporation (JPDC) was criticized by.[73]

The JPDA's primary objective was to secure a sustainable energy supply, and it maintained that the private sector needed the government's support to take high risks in overseas oil development. Therefore, it did not support the abolition of the JNOC system.[74] However, its stance on the establishment of a state-owned oil company inheriting JNOC's oil assets was different from that of METI. While it could not deny the will of METI directly, the

68 Mika Watanabe, "Japan Oil Chief: Indus Still Needs Japan Natl Oil Corp. Aid," *Dow Jones Newswires*, July 15, 1998; "Japan/JNOC/Oil Development-2: Equity Crude Target Intact."

69 Ministry of Economy Trade and Industry, "Member List of Petroleum-Subcommittee under Comprehensive Resources and Energy Investigation Committee."

70 Comment by Shikawa Tetsuya, Diet member from JCP at Japan House of Representative, Standing Committee on Economy, Trade and Industry, *23rd Committee Meeting*, Interpellation, June 26, 2002.

71 Examples are Mr. Hidaka, Mr. Takamiza, Mr. Wada, Mr. Katzuya, Mr. Konaga and Mr. Wagasugi.

72 Ministry of Economy Trade and Industry, "Member List of Petroleum-Subcommittee under Comprehensive Resources and Energy Investigation Committee."

73 Comment by Shiokawa Tetsuya, Diet member from JCP at *23rd Committee Meeting*.

74 Japan Petroleum Development Association, *Wagakuni Sekiyu Tennengasukaihatsu No Genjō to Kadai [the Situation and Issues of Our Country's Oil and Natural Gas Development]* (Tokyo: Japan Petroleum Development Association, 2001), 47.

JPDA asserted the importance of private sector-centered oil development. In a meeting to hear from private companies in November 2002, the JPDA stated:

> If the state intervenes strongly, this would undermine the efficiency in management by making the responsibility of management ambiguous. [A core company] should be made through autonomous consolidation in the private sector.[75]

2 Background: Economic Downturn and Japan's Overseas Oil Development

The idea of the founding of a Japanese state-owned oil company for overseas oil development gained momentum against the backdrop of Japan's economic downturn and the unsatisfactory performance of the JNOC. The METI thought the idea might counter the dissolution of the JNOC.

2.1 Japan's Financial Crisis and Economic Downturn

Following the asset bubble of 1985–1990, Japan experienced an economic crash. The Nikkei Stock Index plunged, and land prices began to fall beginning in early 1992, both of which affected the ability of corporate and individual borrowers to repay loans. This left Japanese banks with massive amounts of bad debts, totaling U$ 1 trillion by 1998, at which point the government initiated measures to deal with the crisis.[76]

The financial crisis had severely affected people's lives. Beginning in 1998, the number of suicides surged to over 300,000 a year, the major portion of which was due to financial problems, such as bankruptcy or an inability to repay loans.[77] Homelessness, which had been rare in Japan, became more easily observed due to the prolonged economic stagnation and high unemployment rates. Taxpayers suffering from harsh economic conditions could not be generous when they felt public corporations were using their tax

---

[75] "Consolidation through the Private-Oriented for Making a Core Company is General Tendency," *Asahi shinbun*, Nov. 29, 2002.

[76] Amyx, *Japan's Financial Crisis: Institutional Rigidity and Reluctant Change*, 1–5.

[77] Ibid., 5. Japan's life insurers had been generous in the 1970s and 1980s, but after the financial crisis, they became strict. Life insurers, such as Alfac Japan and DIY Life Insurance, a unit of Nissan Fire & Marine Insurance, told customers that in suicide cases they would not pay a yen for three years. See Brian Bremner, "Sorry, We're Not Paying," *Business Week*, June 3, 2002.

money inefficiently and accumulating enormous debts.[78] As mentioned by members like Tanaka Keishu in a Diet session, it was unfair for debt-ridden public organizations to evade bankruptcy, while small and medium companies collapsed and faced losing their properties.[79] JNOC in particular was singled out by the government as an example of everything that was wrong with Japan's depressed economy, and many policymakers demanded reform of the ailing giant in favor of private sector-driven overseas oil development.[80]

2.2 Japan's Overseas Oil Ventures and JNOC's Performance

Since the establishment of the JPDC (JNOC's predecessor) in 1967, Japanese companies had become actively involved in overseas oil development. Especially after the oil shocks of the 1970s, Japanese investment focused more on quantity than efficiency. This quest for oil led the JPDC (or JNOC) to rack up enormous debts. From the organization's founding in 1967 until the end of March 1997, it had spent a total of 372 billion yen on loans and investments in unsuccessful exploration projects. In spite of this enormous investment, Japan had yet to achieve its policy target of 30% of oil imports coming from *Hinomaru* oil, the oil developed by Japanese companies. While some leaders in politics and business, called *Shigen-ha* (resource faction), initiated oil ventures in the 1970s, the mega-projects that would have provided Japan with enough oil to achieve the 30% target did not work well.[81] As a result, Japan's self-sufficiency ratio stayed at around 12–15% in the early twenty-first century.

In addition, the one-project and one-company system and irresponsible management among many shareholders had lowered efficiency and obstructed the accumulation of technology and expertise. Plummeting oil prices after the oil shocks of the 1970s and the appreciation of the yen also worsened the financial health of oil developers. In this situation, many of

[78] Interview with a Japanese industry person, conducted on Aug. 17, 2010.

[79] Comment by Tanaka Keishu, Diet member from Democratic Party of Japan, in Japan House of Representatives, Standing Committee on Economy, Trade and Industry, *24th Committee Meeting*, Interpellation, June 28, 2002.

[80] "Japan Reforms—Only to Remain Much the Same," *Petroleum Intelligence Weekly*, Dec. 3, 2001.

[81] Journalist Tatasawa Kōji mentioned in the Diet interpellation that mega-projects in Iran, the UAE, and the Soviet's Tumen had been expected to provide a total of 100 million kl of oil out of 250 million kl of oil imports by 1978, but this plan did not proceed as anticipated. For more information on the resource faction, see Japan House of Councillors, Standing Committee on Economy, Trade and Industry, *24th Committee Meeting*, Interpellation, July 16, 2002. Junichiro, *Tanaka Kakuei Fujirareta Shigen Senryaku: Sekiyu, Uran, Sosite Americatono Toui [Tanaka Kakuei's Sealed Resource Strategy: Oil, Uranium and Battle with the U.S.]*, 138–44.

Table 2 MITI's oil self-sufficiency ratio target and results

Planned year	Target	Result
1967	30% of oil by 1985	10.7% by 1985
1978	1.5 MMBD by 1990	0.45 MMBD by 1990
1983	1.2 MMBD by 1995	0.69 MMBD by 1995
1993	1.2 MMBD by the beginning of twenty-first century	0.58 MMBD

Source Masanari Koike, Gento Mogi, and Waleed H. Albedaiwi, "Overseas Oil-Development Policy of Resource-Poor Countries: A Case Study from Japan," *Energy Policy* 36, no. 5 (2008): 1765
Note MMBD means Million Barrels per Day

JNOC's loans could not be paid back, and the *Yomiuri Shinbun* estimated that JNOC lost a total of more than 340 billion yen (U$2.7 billion) by investing in and offering loans to oil development companies (Table 2).[82]

JNOC's accumulated deficiency became a target of accusation in the Diet interpellation of the Lower House auditing committee on December 3, 1997, right in the middle of the economic downturn. The deficit of JNOC had not been overtly publicized, and even most MITI officials (except those in the oil division) did not know the facts. Most work related to the JNOC was decided at the hand of the oil division chief, and the MITI minister had not been informed about the organization's financial straits. Unorthodox bookkeeping also disguised the reality of JNOC's finances.[83]

In this situation, the then-MITI Minister Horiuchi calculated JNOC's deficiency without the assistance of MITI officials, who had been reluctant to help him.[84] While JNOC had announced earned profits of about 707 billion yen in December 1997, which would have covered the 372 billion yen in losses,[85] Mr. Horiuchi's analysis revealed that the JNOC had accumulated a 1.3 trillion yen deficit. While MITI Minister Horiuchi attempted to significantly change the JNOC after his analysis, he had no choice but to compromise with the MITI bureaucrats. In spite of opposition from

[82] The Diet members quoted *Yomiuri Shinbun*'s article on JNOC's enormous deficit in November 1997. "Newsbrief: Japan," *Platt's Oilgram News,* Nov. 18, 1997.

[83] Interview with a METI official who is versed in the JNOC reform issue, conducted on June 22, 2011.

[84] The relationship between the then-MITI Minister and the Ministry's bureaucrats was not always comfortable. Mr. Horiuchi is known to have had negative experiences with government bureaucrats when he was President of the railway and resort company Fuji Kyuko Corp, a history which may have affected his relationship with MITI career officials.

[85] "JNOC Says Failed Exploration Cost 372 Billion Yen," *Reuters,* Dec. 3, 1997.

Mr. Osamu, MITI's Administrative Vice Minister, Horiuchi pressured Kunio Komatsu, the then-JNOC Chairman, to resign and instructed the JNOC to stop paying his pension.[86] The JNOC Reassessment Committee was established in July 1998 in MITI, and Minister Horiuchi submitted the reassessment report to the Audit and Administrative Surveillance Committee of the Lower House on July 27, 1998, a few days before he stepped down from the ministership. The report called for stricter auditing standards to uncover any undisclosed bad loans by the JNOC to its affiliates and prompted the state-run company to set aside adequate loan-loss reserves. By submitting the report, Mr. Horiuchi intended to pressure MITI to complete a reconstruction plan for the corporation by September, regardless of the forthcoming change in government.[87]

Despite the debacle of 1998, few industry sources expected the JNOC to be abolished or the Japanese government to dramatically lower its involvement in overseas energy development. On the contrary, representatives of Japanese companies on the Petroleum subcommittee called for a bigger government role in overseas oil development projects, arguing that it was necessary to ensure an equal footing with foreign competitors.[88] Keiichiro Okabe, President of the PAJ, also supported JNOC's financial role because he determined that Japanese companies could not enter overseas oil development projects without the JNOC's support.[89]

This overall supportive stance is understandable considering JNOC's role as financier and facilitator in Japan's overseas oil development. For Japan's large firms that had been severely affected by the economic downturn, ventures in the Caspian Sea, Siberia, or elsewhere provided major opportunities to generate profits through oil and gas development, equipment supply and financing, and construction material sales that could not easily be made in the depressed domestic market. The private companies actively involved in overseas oil and gas ventures included oil developers, general trading companies, steel makers, and utilities.[90] For example, the JNOC had been playing

---

[86] Horiuchi Mitsuo, "Tsushōsangyōshō No Chibu Sekiyukōdan Wo Kokuhatsusuru [Denounce the JNOC, MITI's Disgrace]," *Bungeishunzū*, November 1998, 110–20.

[87] Japan National Oil Corporation, *Sekiyukōdanshi (1997 <Heisei9> Nendo-2004 <Heisei16> Nendo) [the History of JNOC (1997-2004)]* (Tokyo: Japan National Oil Corporation, 2005), 12; "1,405-B.-Yen Losses Feared for JNOC—MITI Horiuchi," *Jiji Press*, July 28, 1998, Mika Watanabe, "Japan's MITI Submits National Oil Restructuring Plan to Diet," *Dow Jones Newswire*, July 28, 1998.

[88] Mika Watanabe, "JNOC/Debt/Japan Oil Investments-3: Gov't Role Needed," *Dow Jones Newswires*, June 10, 1998. In the late 1990s, the Caspian Sea presented a new opportunity for Japanese oil companies and they needed major capital investments.

[89] Mika Watanabe, "Japan Oil Chief: Indus Still Needs Japan Nat'l Oil Corp. Aid."

[90] Calder, "Japan's Energy Angst and the Caspian Great Game," 12–17.

a role as a facilitator of the South Caspian project for a Japanese consortium composed of Itochu Oil Exploration Co., Ltd., Indonesia Petroleum Exploration, Ltd. (INPEX), JAPEX and Teikoku Oil.[91] The JNOC also participated in planning for a multi-national project within Russia, China, South Korea, Mongolia, and Japan, all of which aimed at tapping the Kovyktinskoye field, a huge gas reserve in Siberia.[92]

From the perspective of the Japanese government, Japan's efforts for overseas oil development were vital, though the performance of Japan's overseas oil development since the establishment of JNOC was not satisfactory. Japan, whose GNP was over 15% of the global total in the 1990s, had no significant domestic oil or gas fields. At that time, Japan imported 99% of its oil, more than 5.7 million barrels per day.[93] The Asian energy market of the coming decade, beyond the financial crisis in the late 1990s, looked ever more precarious for Japan, despite low oil prices. Japan was expected to face more active competitors for available energy sources, such as China and Korea. This national anxiety was compounded by Japan's high dependence on the Gulf region and the possibility of future competition with China for its traditional sea lanes.[94]

This situation made the role of the JNOC visible. New MITI Minister Yosano, the successor to Mr. Horiuchi, said in his inaugural speech on July 31, 1998, that JNOC's mission to provide financial support to private companies for overseas oil exploration remained crucial. The reassessment plan, which called for the liquidation of 27 project companies and an equity sale of 17 companies to cover its losses, was endorsed in the LDP and submitted to the audit and administrative surveillance committee in the Lower House on September 28, 1998.[95] In the loosened environment following Mr. Horiuchi's departure as a minister, on September 3, 1999, the JNOC paid a 46-million yen allowance to former JNOC Chairman Komatsu in defiance

[91] Peter Lardner, "Azeri Leader's Japan Visit May Unlock Caspian Oil," *Reuters*, Jan. 23, 1998.

[92] Keun-wook Paik, "Natural Gas Expansion in Korea," in Ian Wybrew-Bond, Jonathan P. Stern, and Oxford Institute for Energy Studies, *Natural Gas in Asia: The Challenges of Growth in China, India, Japan and Korea* (Oxford, New York: Oxford University Press, 2002), 218–26; Kiyoshi Takenaka, "Talks on Long-Delayed Siberia Pipeline Begin," *Reuters*, Jan. 14, 1998. The participants were mainly from public entities: Russia's Department of Fuel and Energy, Siberia-Far East Oil Company (Sidanco), the Petroleum Authority of Mongolia, China National Petroleum Corp (CNPC), Korea Gas Corp (KOGAS), Japanese trading house Sumitomo Corp., as well as JNOC. For details on natural gas development in East Siberia and its geopolitical implications, see Keun Wook Paik, *Sino-Russian Oil and Gas Cooperation: The Reality and Implications* (Oxford: Oxford University Press, 2012).

[93] Kent E. Calder, "Fueling the Rising Sun," *Harvard International Review* 19, no. 3 (1997): 24–28.

[94] Calder, "Japan's Energy Angst and the Caspian Great Game," 10–12.

[95] "MITI/Nat'l Oil-2: Submits Restructuring Plan to Diet," *Dow Jones Newswires*, Sept. 28, 1998. "MITI Gets LDP Nod for Reconstruction Plan for JNOC," *Dow Jones Newswires*, Sept. 28, 1998.

of Mr. Horiuchi.[96] While many felt that the JNOC issue was resolved, it remained latent until the launching of the Koizumi Administration in April 2001.

3 JNOC and the State-Owned Oil Company Debacle

3.1 Inception of a Proposal for a State-Owned Oil Company

The proposal for Japan's state-owned oil company was born during the revived debate on the JNOC after Koizumi became Prime Minister in April 2001. Following a reshuffle of the heads of veto players on April 25, 2001, Mr. Horiuchi, newly named chair of the LDP Executive Council, asked to include the JNOC in the list of the special public corporations that would be reassessed with the possibility of abolition to be endorsed by Koizumi's Administrative Reform Headquarters on June 22, 2001. This took place despite strong advocacy for the continued existence of JNOC by METI and the Petroleum subcommittee, an advisory group (*shingikai*) under METI.[97] The JNOC, which had not previously been considered a target of Koizumi's reforms, thus became one main issue of his policy agenda.[98]

Horiuchi's visit to Koizumi on July 6, 2001, dramatically altered the fate of the JNOC. During the meeting, Horiuchi suggested that Koizumi abolish the debt-ridden JNOC for the cause of administrative reform.[99] Horiuchi's suggestion was modeled after the privatization of Nippon Telegraph and Telephone Corporation (NTT) in the 1980s when the government

---

[96] "Sacked Oil Agency Chief Returns Quarter of Allowance," *Japan Energy Scan*, Sept. 6, 1999.

[97] "METI and Advisory Council Untied to Support That Jnco is Needed for the Nation," *Asahi shinbun*, June 21, 2001. "'Consider Abolition of JNOC' Former METI Minister Drive Old Enemy into Corner," *Asahi shinbun*, June 22, 2001.

[98] In the Weekly Asahi's survey of May and June 2001 on Diet members' attitudes toward Koizumi's reforms, specific items mentioned as part of the agenda were: the use of road-specific funds for general purposes; the privatization of three businesses of Japan Post (mailing, postal saving, and easy insurance); the reduction of grants to local prefectures; and the privatization of the Highway Corporation. See "Quick Survey: Names of LDP Members Resisting Koizumi's Reform," *Weekly Asahi*, June 15, 2001. And "Quick Survey 2: Names of DPJ Members Not Resisting Koizumi's Reform."

[99] Mr. Horiuchi had already written another article, "Dissolve the JNOC," in the magazine *Bungeishunzū* in May 2000. During Koizumi's reform of the Highway Corporation, the magazine was utilized as an information conveyer to the public. In *Bungeishunzū* of July 2003, Katagiri Sachio, a reformist within the Japan Highway Public Corporation, disclosed the existence of financial statements showing Japan Highway Public Corporation had more than 617.5 billion yen of debt. See Uchiyama, *Koizumi and Japanese Politics Reform Strategies and Leadership Style*, 45.

established a new joint-stock company to sell shares in order to cover accumulated losses.[100] The dissolution of the JNOC, which was already known to have enormous deficits due to unsuccessful overseas oil development, was an attractive target for Koizumi's slogan of reform. In addition, unlike other public corporations, the JNOC was hardly involved in domestic economic activities, such as highway construction, and thus the subsequent absence of specific political constituencies in local areas did not pose a burden for Koizumi's strategy for the upcoming election for the House of Councillors on July 29, 2001. Though the abolition of the JNOC was under METI's jurisdiction, the agreement between Prime Minister Koizumi and Mr. Horiuchi was made during the absence of METI Minister Hiranuma, who was visiting the Middle East to negotiate the extension of concession of the AOC's Khafji oil field and the development of the Azadegan field in Iran.[101] The decision was communicated to Mr. Hiranuma by an international phone call from Mr. Horiuchi. In an interview with the press on July 6, 2001, Prime Minister Koizumi confirmed that the first shot in his fight to overhaul debt-ridden public corporations would be to abolish the JNOC.

Prime Minister Koizumi's stance on the JNOC issue is in line with his credibility as a reformer who prefers small government and emphasizes the need to eliminate wasteful government expenditure. He had previously published four books on the need to reform various parts of the bureaucracy and privatize the postal services. He had already campaigned for the presidency of the LDP on a similar platform of privatizing the postal services in 1995.[102] In line with the firm reformist posture shown in his first statement as prime minister on April 26, 2001, it was to be expected that he would not support a state-owned oil company.[103] Meanwhile, the policy preference of Mr. Horiuchi, Chair of the LDP Executive Council, was inconsistent and thus likely based on emotion rather than political philosophy. This is apparent in his opposition to the postal service reform later, so his willingness to abolish the JNOC seemed to be rooted in his personal experience trying to accomplish this task as MITI Minister.

For METI officials, the decision did not make sense. They had understood that Mr. Horiuchi had been persuaded to accept MITI's plan to renovate the JNOC. The oil companies with bad assets had been dissolved one by

100 "LDP Executive Council Chair Horiuch Shows Intention to Abolish JNOC Possibly in This Year," *Asahi shinbun*, July 12, 2001.

101 This visit was largely seen as a counter-measure to Horiuchi's initiative on the JNOC reform.

102 Mulgan and Australian National University. Asia Pacific School of Economics and Management, *Japan's Failed Revolution: Koizumi and the Politics of Economic Reform*, 49.

103 Cabinet Secretariat, "Statement by Prime Minister Junichiro Koizumi at the First Meeting of the New Cabinet," http://www.kantei.go.jp/foreign/koizumispeech/2001/0426setuji_e.html.

one as planned in 1999. In addition, the loss of the AOC's oil concession in Saudi Arabia's section of the Neutral Zone in February 2000 was a challenge to MITI's energy security policy. The AOC's failure to renew its 150,000 b/d worth of drilling rights made it nearly impossible for Japan to attain its 30% oil self-sufficiency target.[104] Thus, METI was seeking an alternative to the Neutral Zone in Iran's Azadegan oil field, which required massive capital investment, possibly from the JNOC.[105] Faced with this situation, the Diet strengthened the role of JNOC on June 13, 2001, by: (1) endorsing a revision of JNOC Law to enable it to provide capital to Japanese companies for acquiring existing oil fields, which was added to JNOC's funding for exploration, and (2) passing legislation that stipulated JNOC's key role in securing a stable oil supply in emergencies.[106] This situation was diametrically opposed to the new political confusion surrounding JNOC in 2001 and 2002.

METI opposed the idea of dissolving the JNOC, but it could not persuade the Prime Minister and Mr. Horiuchi. After the LDP's success in the Upper House election of July 2001, Koizumi was politically more powerful than before, and the Japanese media were very supportive of his initiative.[107] METI officials did not want to be seen as "forces of resistance" (*teikō seiryoku*) to Koizumi's reform, so they had no choice but to accept the abolition of the JNOC.[108] With a push from Mr. Horiuchi, the JNOC was included among

[104] In August 2000, the Japanese government abolished its long-standing numerical target of oil development to raise the proportion of crude oil imports from Japanese-owned overseas projects to 30% of total domestic crude needs. This decision was made after the AOC's failure to renew the Khafji oil concession in February 2000.

[105] "Iran: Japan Looks to Iran to Replace Assets Lost in Neutral Zone," *Middle East Economic Digest*, Nov. 10, 2000.

[106] "Role of Japan National Oil Necessary: METI's Hirose," *Kyodo News*, June 14, 2001.

[107] According to AFP, Japanese newspapers supported Koizumi's initiative. An Asahi editorial dated July 8, 2001, states "The reform of the public sector centering on public corporations is an eminent task for the Japanese economy. We have argued that we have to start the reform with elimination, or scaling back in our view...Anyway, the era of *hinomaru* oil has already ended. Since the government has decided to withdraw from its full commitment to developing foreign oil resources, much of the rationale for having the JNOC has been eliminated." On the issue of the CEO of Idemitsu Oil criticizing METI for its approach to the JNOC issue, the *Asahi Shinbun* repeatedly published the same article with only a few words changed on Sept. 22 and Oct. 4. See "Koizumi Seeks to Dissolve State Oil Firm," *Agence France-Presse*, July 10, 2001 and "Editorial: Abolition of JNOC is Model of Special Corporation Reform," *Asahi shinbun*, July 8, 2001. See also "No Need of Tax Input in Oil Development, President of Idemitsu Showed Real Intention Publicly (Osaka)," *Asahi shinbun*, Sept. 22, 2001; "'For Oil Development, Support Private Business Rather Than Tax Input,' President of Idemitsu Criticized METI," *Asahi shinbun*, Oct. 4, 2001.

[108] "METI Hiranuma: OK to Abolish Japan Nat'l Oil Corp-Kyodo," *Reuters*, July 12, 2001.

seven public corporations to be dissolved prior to others,[109] and the list was endorsed by the CEFP on September 21, 2001.

During this time, the JNOC did not have a voice in the debate,[110] and the voices of the JPDA and private oil developers that supported JNOC were not well-received.[111] Only *Keidanren*'s support for the reform was pronounced. Takashi Imai, Chairman of *Keidanren*, urged the state-run JNOC to be dissolved or scaled down. For him, the time when Japan should explore its own oil fields and secure its concessions has come to an end.[112]

While METI accepted the inevitability of the abolition of JNOC as an organization, they did not give up the function and the assets of JNOC. They began to persuade veto players by separating these functions and assets from the corporation itself. It was in this pursuit that the idea of state-owned oil company to manage JNOC's assets appeared. METI Minister Hiranuma reached out to veto players such as Reform Minister Ishihara Nobuteru, LDP Executive Council Chairman Horiuchi Mitsuo, the LDP PARC Chairman Aso Taro, and the other policy chiefs of the tripartite coalition.[113] Mr. Hiranuma tried to convey the ministry's stance that the government should continue its commitment to financing risky oil development projects and maintain JNOC's technological expertise and oil stockpiling. This move was supported by private sector actors like Fumiaki Watari, President of the leading oil distributor Nippon Mitsubishi Oil Corp., and Keiichi Okabe, President of the PAJ.[114] In the fall of 2001, METI also made efforts to create a favorable environment by consulting with experts in international politics and economics, such as journalist Tahara Soichiro and Professor Iokibe Jin, both of whom were not directly involved in oil policy.[115] These efforts succeeded in a limited fashion. Mr. Horiuchi accepted the need for

---

[109] Japan National Oil Corporation, *Sekiyukōdanshi (1997 <Heisei9> Nendo-2004 <Heisei16> Nendo) [the History of JNOC(1997-2004)]*, 13. The seven corporations were four highway-related corporations, City Corporation, Housing Finance Corporation, and the JNOC.

[110] The JNOC asked a former JNOC staff member to write an article to defend the JNOC, but the request was withdrawn. Interview with Tatsu Kambara, a former JNOC staff, conducted on Aug. 19, 2010.

[111] This was confirmed by many Japanese industry insiders during an interview in June 2011. For the opinion of the JPDA, see Japan Petroleum Development Association, *Wagakuni Sekiyu·Tennengasukaihatsu No Genjō to Kadai [the Situation and Issues of Our Country's Oil and Natural Gas Development]*, 47.

[112] "Bureaucracy Overhaul Key to Reforming Public Firms," *Kyodo News*, July 9, 2001.

[113] "Lead: Ministers Agree on State Involvement in Oil Projects," *Kyodo News*, July 17, 2001.

[114] "Nippon Mitsubishi Oil Head Urges Gov't to Keep JNOC Functions," *Kyodo News*, July 18, 2001. Mika Watanabe, "Japan Oil Chief: Indus Still Needs Japan Nat'l Oil Corp. Aid."

[115] "Mr. Tahara Criticized METI Minister Japan Does Not Have Energy Strategy and JNOC Should Be under Review," *Asahi shinbun*, Aug. 1, 2001. "Consultations to Be Held before Dissolution of Oil Corporation," *BBC Monitoring Asia Pacific-Economic*, July 25, 2001.

government support for the private sector's oil development with a limit of 50% of the cost and the condition that public funds be used only on projects that could not be implemented solely by the private sector, such as Azadegan.[116] On the other hand, Mr. Horiuchi continued to oppose the use of Petroleum Special Account for oil development and oil stockpiling and asked the Administrative Reform Secretariat to include his opinion, which became another flash point with METI.[117]

On October 29, 2001, METI proposed its plan for JNOC reform, and the idea of establishing a "special company" (*tokushu kaisha*) was included.[118] METI's reform plan broadly consisted of the following points: (1) R&D and oil stockpiling would be managed by a special public corporation that integrated the JNOC and the Metal Mining Agency of Japan (MMAJ); (2) the private sector would be supported with subsidies and loan guarantees; and (3) public oil stockpiling companies would be abolished, and management would be entrusted to the private sector. Also, METI intended to establish a company that would inherit JNOC's oil and gas assets for approximately 100 companies invested by the JNOC.[119] However, METI's original idea was to utilize the situation to make a Japanese oil major (*wasei meja*) in the form of a state-owned oil company, similar to French Total or KNOC.[120]

In August 2000, the interim report of the development committee in the Petroleum Advisory Group of METI highlighted the need for a core company group (*Kakutekina Kaisha gurupu*), meaning integrated among upstream and downstream oil companies, and combined with energy-related companies.[121] According to METI Minister Hiranuma, a core company group would have effectively been the same as a Japanese oil major, maintaining and expanding its oil and gas development for itself and eventually

---

116 "LDP Executive Council Chairman Horiuchi Accepted Support for Oil Development Conditionally," *Asahi shinbun* Aug. 3, 2001.

117 "Mr. Horiuchi Asked Reform Headquarter to Stop the Expenditure of Special Account to Oil Stockpiling and Development," *Asahi shinbun,* Aug. 4, 2001.

118 According to Johnson's category on Japan's public company, a narrowly defined national policy company was used to refer to the special company. They were joint stock companies found by special laws. The examples are the Imperial Oil Company established in 1941, and the Japan Petroleum Exploration Company established in 1955. See Johnson, *Japan's Public Policy Companies*, 53–54.

119 "METI's Proposal: JNOC Will Be Integrated with MMAJ, Financing Will Be Subsidy," *Asahi shinbun,* Oct. 30, 2001.

120 Interview with a METI official who is versed in the JNOC reform issue, conducted on June 22, 2011. This fact is not well known in academia. For example, Llewelyn Hughes mentions the company is just a special company without stating the intended function of the company. See Hughes, "The Limits of Liberalism? Long-Run Petroleum Prices and Government Intervention in Petroleum Markets in Japan, France, and the United States."

121 Development division under Petroelum Advisory Committee, "Interim Report of Basic Policy Subcommittee" (Tokyo, 2000), 9. This report also mentioned the need to abolish the self-sufficiency target and a flexible application of the duty to import equity oil to Japan only in a time of emergency.

growing to equal Western oil majors.[122] Interestingly, this idea was similar to an opinion of an advisory group to Reform Minister Ishihara. The advisory group, called the Administrative Reform Execution Conference Meeting (*gyōkakudankōbyōgikai*), suggested that the government establish a wholly owned joint-stock company that would become a leading shareholder of project companies. The wholly owned company would also be a holding company that would manage four currently healthy companies and one company created through a merger among small-size developers. As of 2001, oil development assets owned by the JNOC amounted to 1.4 trillion yen, which was similar to the Italian ENI, a quasi-oil major.[123] Therefore, the reasoning went that the holding company would be a quasi-oil major and open to the market.[124] Additionally, METI's special company proposal was similar to that of the Ministry of Land, Infrastructure, Transport and Tourism (MLIT). As a response to Koizumi's instruction to reform four highway corporations in August 2001, the MLIT in the next month proposed to combine Japan Highway Public Corporation, the Metropolitan Expressway Public Corporation, and Hanshin Expressway Public Corporation into a new public corporation that would be privatized after ten years as a government-owned special company (*tokushu kaisha*).[125] METI started to strengthen its commitment to establishing a state-owned oil company, and the controversy among veto players also resumed.

3.2 First Quarrel on a Proposal Before Finalizing the Reform Plan

As soon as METI Minister Hiranuma reported METI's plan on October 30, 2001, Mr. Horiuchi, Chairman of the LDP's Executive Council, criticized the plan as being equivalent to handing over the duties of the JNOC to two separate entities while leaving the huge debts of the oil explorers intact. In response to METI's proposal, he submitted his alternative plan to Ishihara Nobuteru on November 9, 2001. Mr. Horiuchi accepted 50% of loans for oil exploration, but argued for a reassessment of the Petroleum Special Account from a zero-base. He also proposed to abolish the JNOC and to establish

---

[122] Japan House of Representative, Standing Committee on Economy, Trade and Industry, *26th Committee Meeting*, Interpellation, July 3, 2002.

[123] "JNOC's Consolidated Financial Statement Shows That JNOC's Asset is More Than 4 Trillion Yen, Quasi-Major Level," *Asahi shinbun*, Aug. 31, 2001.

[124] "Administrative Reform Execution Conference Meeting Suggeted Abolition of JNOC and Idea to Integrate Oil Developers," *Asahi shinbun*, Aug. 23, 2001.

[125] Uchiyama, *Koizumi and Japanese Politics Reform Strategies and Leadership Style*, 40.

a three-year-term liquidation organization that would sell JNOC's assets to pay off accumulated debt, which would obstruct METI's plan to establish a special company to inherit the JNOC's assets. He opposed METI's idea of a holding company due to a concern that the holding company would encroach upon private businesses as a rival.[126] During Reform Minister Ishihara's review of the two proposals, on November 20, Koizumi requested that Ishihara Nobuteru clearly mention "full abolishment" of the JNOC in the final plan he would submit to the Prime Minister later that month.[127]

The stance of Mr. Aso Taro, the then-LDP PARC Chairman, is an interesting point in the political battle among veto players. Mr. Aso was more supportive of METI's idea than other heads of veto-player institutions. In the LDP presidential election of 2001, he had already clashed with Mr. Koizumi by arguing for an expansionist budget over Koizumi's structural reforms.[128] While he accepted Koizumi's offer of the chair of the LDP PARC,[129] he was reluctant to support Koizumi's structural reforms as seen *Asahi Shinbun* survey incident of May 30, 2001. For Asahi's survey request to the LDP Diet members for questioning whether they support Koizumi's reform, the LDP PARC, whose Chair is Mr. Aso, did not help the newspaper's survey. Instead, the PARC instructed LDP members not to answer the first time, and later PARC sent the members sample answers, which were based on Mr. Aso's decision.[130] At the same time, the LDP PARC was neglected by the prime minister in making important policies.[131] Later, in Horiuchi's description of the negotiation among main veto players, Mr. Aso used METI's proposal as his own.

3.3 The Compromise Among Veto Players

On November 22, 2001, veto players in the government and the LDP, such as State Minister for Administrative Reform and Regulatory Reform Ishihara, METI Minister Hiranuma, the LDP's PARC Chairman Aso, and the LDP's Executive Council Chairman Horiuchi, reached a compromise on the JNOC abolition scheme. On the same day, the JNOC reform plan with

[126] "LDP Executive Council Chair Proposed to Establish 3-Year Term Liquidation Organization for JNOC Reform," *Asahi shinbun*, Nov. 10, 2001.

[127] "Japanese Premier Calls for 'Full Abolishment' of JNOC," *Platts Commodity News*, Nov. 21, 2001.

[128] "Stitched Confrontation among 3 Coalition Partners, Discuss the Structure," *Asahi Shinbun*, Apr. 16, 2001.

[129] Originally, METI Minister Hiranuma received the offer on the LDP PARC Chairman from Koizumi, but he could not accept it due to the stance of his faction.

[130] "Quick Survey: Names of LDP Members Resisting Koizumi's Reform."

[131] "Widening the Gap between Kantei and LDP, Mr. Aoki's Nail," *Asahi Shinbun*, May 30, 2001.

those of six other corporations was endorsed by the coalition leaders—Prime Minister Koizumi, New Kōmetō Chief Kanzaki Takenori and Conservative Party leader Noda Takeshi—and the plan was endorsed by the LDP and the Cabinet on November 27, 2001. According to the plan, the JNOC would be integrated with the MMAJ, another public corporation, and become an independent administrative agency modeled after Britain's agency system.[132] Under this scheme, the function of loans for risky oil exploration, which had been blamed as the main cause of JNOC's enormous deficit, would be abolished.[133] Also, the equity investment in oil development companies would be limited to 50%, and METI would take direct charge of the national oil stockpiling scheme.[134] The plan was broadly in line with METI's claims that the government should maintain its involvement.

It is notable that METI's suggestion on a special company was accepted. In the agreed-upon scheme, JNOC's assets, such as stakes in subsidiaries, would be transferred to a special company the government would set up in three years, and the JNOC would sell bad assets in two or three years. The special company would oversee the 104 oil development firms now under the supervision of JNOC, and an official from the private sector would be president of the special company.[135] This scheme meant METI's bid for "total control," as well as an earlier scheme to consolidate viable upstream projects around the few profitable operators, such as JAPEX, INPEX, and Japan Oil Development Corporation (JODCO), faded out.[136] At the same time, the private sector worried that firms unaffiliated with the new company would face stiff competition from the special company which would also inherit information on the private sector. For that reason, observers anxiously waited to see how much of its authority the new special company would use.[137]

While the reform plan written by Special Corporation Reform Headquarters was endorsed in the Cabinet on December 19, 2001, the Cabinet decision had extremely complex characteristics, called a "settlement of jewel

---

[132] In Britain, the executives at British agencies are recruited from the general public, and, moreover, they hire employees with efficient management skills rather than good contacts. Not surprisingly, many more new employees in the agencies came from the private sector.

[133] JNOC's financial function of extending loans to assist oil-field developers would be transferred to the Japan Bank for International Cooperation (JBIC), a public corporation.

[134] "Gov't, Coalition Finalize Plans to Disband State-Run Oil Firm," *Kyodo News*, Nov. 22, 2001.

[135] "Japan to Abolish National Oil Firm," *Jiji Press English News Service*, Nov. 22, 2001; "Gov't, Coalition Finalize Plans to Disband State-Run Oil Firm."

[136] "Japan Reforms—Only to Remain Much the Same."

[137] "Koizumi Reforms; Old Guard Champions Vested Interests," *Daily Yomiuri*, Dec. 21, 2001.

beetle color" (*tamamushi-iro no ketchaku*).[138] In reality, the plan for rationalization of a special public corporation could not include all the details for 146 corporations in its 27 pages.

In the section concerning the establishment of a special company for JNOC's reform, the plan mentions shortly:

> The development-related properties currently owned by the JNOC, will be dealt with adequately by clearing off what is to be cleared off and disposing of what is to be disposed of after carrying out a strict evaluation of assets. In addition, an organization for the purpose of liquidation like the sales of assets will be established with (approximately three-year) time limit and deal with disposal. Waiting for the termination of the liquidation company, a special company will be established and be privatized.[139]

In spite of the Cabinet's authorization of the agreement, Mr. Horiuchi had already begun to show his different interpretation of the agreement from other veto players, and mentioned that METI's effort to establish a state-owned oil major (*kansei meja*) was not acceptable.[140] The controversy on the JNOC reform remained latent without a clear resolution.

4 A New Debacle on the Reform Before the Legislation

4.1 METI's Draft Bill and the LDP's Obstruction

In 2002, METI drew up three bills dissolving the JNOC and establishing a new agency and a new special company, but met objections from Mr. Horiuchi, chairman of the LDP Executive Council and a key advocate of total abolition of the JNOC. On April 15, 2002, Horiuchi visited Koizumi and

---

[138] "JNOC, Difficult Mediation between Dissolution and Privatization, Burden of Jewel Beetle Color Decision," *Asahi shinbun*, Apr. 17, 2002. The term *tamamushi-iro* is used in the context of Japanese officialdom, when politicians or bureaucrats use language that is unclear. The term comes from a bug that has iridescent wings that glow lengthwise with different colors depending upon the light angle, one cannot be sure exactly which color it is. Therefore, it gave rise to the expression *tamamushi-iro* (tamamushi color), in reference to a convoluted statement that can be interpreted in more ways than one. See Kaori O'Connor and Mitsubishi Sh*oji Kabushiki Kaisha, *Japanese Business Language: An Essential Dictionary* (London KPI: Distributed by Routledge & Kegan Paul, 1987), 193.

[139] Special Corporation Reform Headquarter, "Tokushu Hojin Nado Seiri Gōrika Keikaku [Plan for Disposal and Rationalization of Special Corporations]" (Tokyo: Special Corporation Reform Headquarter, 2001), 2–3.

[140] "Confrontation on the JNOC Reform Reappear, LDP Horiuchi Argues to Abolish Loan Gurantee," *Asahi shinbun*, Dec. 13, 2001.

asked him to let METI revise the draft bills, claiming METI's draft bills were against the Cabinet decision and would water down Koizumi's reform initiative. Koizumi agreed. Horiuchi also told Koizumi that allowing the JNOC to sort out its own assets would hinder the drive toward reform as it would leave as many as 50–60 of the roughly 100 oil development companies, including loss-making ones, intact under a special company to be set up under one of the bills.[141]

The next day, Koizumi gave METI Minister Hiranuma an order to stay in line with the Cabinet decision, instructing him to discuss the matter with relevant parties, the LDP Executive Council Chairman Horiuchi, Reform Minister Ishihara, and the LDP PARC Chairman Aso, who were the main veto players in the JNOC issue.[142] Koizumi did not adjudicate the final outcome, meaning he did not force the change, but rather allowed the veto players to compromise. For Koizumi, this renewed political infighting posed a serious impediment to his reform initiative. Once he had failed in the JNOC reform, which was seen as the earliest indication of his reform initiative, many anticipated a "domino effect" for subsequent reforms, like that of the Japan Highway Corporation.[143] Moreover, the failure in the scheduled signing-off of the JNOC reform bills at the Cabinet meeting on April 26th, the first anniversary of Koizumi's premiership, tainted his reform initiative.[144]

As seen in the following chart, there was a large gap in the interpretations of the Cabinet of December 2001 between METI and the LDP Executive Council. Reform Minister Ishihara sided with Mr. Horiuchi and urged METI to revise the bills.[145] Horiuchi estimated that sales of healthy assets would clear 1.3 trillion yen of JNOC's bad assets and the remaining 400 billion yen of surplus could be returned to the National Treasury. METI's plan to transfer 60 companies among 80 JNOC-related companies was not acceptable. In addition, he did not oppose the idea of a Japanese oil major, but he did not support a government-led Japanese oil major (Table 3).[146]

In spite of Koizumi's order, METI had no plan to revise the bills, because METI thought the bills were in line with the Cabinet's decision. Instead,

---

141 "Koizumi Seeks Quick Dissolution of State Oil Firm," *Kyodo News*, Apr. 16, 2002.

142 Ibid. Koizumi's reactive response to Mr. Horiuchi's appeal shows that his own weakness in detail was an obstacle to his implementation of a reform initiative.

143 "JNOC, Difficult Mediation between Dissolution and Privatization, Burden of Jewel Beetle Color Decision."

144 Shinichi Kishima, "Japan PM's Reform Battle Heats up as Bills Stall," *Reuters News*, Apr. 25, 2002. At the Cabinet meeting, it was also scheduled to approve post service reform bills, which did not receive formal consent of the ruling LDP in a break from a longstanding protocol.

145 "Ishihara Takes Issue with METI's JNOC Dissolution Plans," *Kyodo News*, Apr. 23, 2002.

146 "Listen to Horiuchi, LDP Executive Council Chairman, Arguing the Dissolution of JNOC," *Asahi shinbun*, Apr. 20, 2002.

Table 3 Different interpretations of the cabinet decision

Cabinet decision	Interpretation	
	METI	LDP Executive Council
Risk money is limited to equity investment	Remove loan function, but maintain loan guarantee	Remove loan guarantee
Clarify the time to dissolution of JNOC in 2003	JNOC exists until March 2004	Immediate dissolution of JNOC
File and sell JNOC's assets with serious evaluation	Yuushikisha Iinkai (literally 'knowledgeable peoples' committee) decides criteria for filing and sales[147]	In principle, sell all assets of JNOC
Facilitate liquidation of JNOC for 3 years and then establish a special company	JNOC implements its own liquidation after METI minister decides basic criteria based on report from special committee	JNOC becomes corporation to be liquidated, and reform headquarters-related ministers meeting will authorize the sales plan

Source "JNOC, Difficult Mediation between Dissolution and Privatization, Burden of Jewel Beetle Color Decision." *Asahi Shinbun*, Apr. 17, 2002

METI Vice Minister Katsusada Hirose visited Koizumi and attempted to persuade him to accept METI's draft bills.[148] If METI disbanded the JNOC in one year and let a residuary firm sort out and dispose of its assets in the JNOC-affiliated oil development companies, it would lead to the sale of healthy assets, leaving any successor company with only loss-making affiliates. Considering METI planned to develop the company, which was initially wholly owned by the government but would be privatized later, into a Japanese oil major (*wasei meja*), Horiuchi's idea was not acceptable to METI.[149] In addition, METI worried that Western oil majors or Chinese state-owned companies would buy the healthy assets Japan had owned.[150] In this situation, METI's will to maintain the current draft bills was so strong

[147] This committee's composition implies that the participation of the industry would be minimal.

[148] Mulgan and Australian National University. Asia Pacific School of Economics and Management, *Japan's Failed Revolution: Koizumi and the Politics of Economic Reform*, 145.

[149] "METI Has No Plan to Revise Bills for JNOC Abolition," *Kyodo News*, Apr. 17, 2002.

[150] Interview with a METI official who is versed in the JNOC reform issue, conducted on June 22, 2011.

that METI even considered retracting the JNOC reform bills. The retracting would have hurt Koizumi's reform initiative.[151]

Mr. Aso, the LDP PARC Chairman, tried to bridge the differences between the two sides, which encouraged METI.[152] Other players also joined the debate. On April 25, 2002, Cabinet Secretary Fukuda Yasuo proposed a compromise to establish a special company at a suitable moment after JNOC is dissolved, but this effort was in vain.[153]

Finally, the veto players reached an agreement on May 1, 2002, via international phone call,[154] though a few points remained unclear. The JNOC would be dissolved in March 2005, and the JNOC itself would sell its assets. Also, the JNOC's functions of financial support and R&D would be transferred to a newly-established independent administrative agency. However, the contents related to the function of loan guarantee and a special company, such as criteria for dealing with assets and the characteristics of the special company, remained topics for future discussion.[155]

This imperfect agreement can be attributed to the political situation at the time. Facing disappointment about his reform initiative, Prime Minister Koizumi vulnerably operated without a strong personal base of support inside the LDP.[156] For him, it seemed to be more realistic to realize a partial success in his reform rather than a thorough but costly victory. Since the enactment of the Anti-Terrorism Law in late 2001, the Koizumi Cabinet's approval rating, as surveyed by *Asahi Shinbun*, had declined, especially after the dismissal of the popular foreign minister Tanaka Makiko on January 29, 2002.[157] The media and commentators had also begun to raise suspicions about his reform efforts, with one Japanese journalist describing Koizumi as a "sheep in wolf's clothing" in the *Yomiuri Shinbun* in December 2001. This implied he appeared to be a radical reformer but was actually a tame underperformer. Niwa Uichiro, the President of Itochu, commented that if

---

151 "METI Did Not Agree with Kantei's Proposal. The Final Conclusion after Holidays," *Asahi shinbun*, May 1, 2002.

152 "Ishihara Takes Issue with METI's JNOC Dissolution Plans." Mr. Aso's stance was mentioned by METI minister Hiranuma in a meeting with reporters. This means Mr. Aso was compassionate toward METI, a fact confirmed by a METI official. Interview with a METI official who is versed in the JNOC reform issue, conducted on June 22, 2011.

153 "Fukuda's Proposal Did Not Make an Agreement," *Asahi shinbun*, Apr. 26, 2002.

154 The settlement between the reform minister and METI minister was made over the telephone, because the METI minister was visiting Washington, D.C., at the time. "Lead-State-Run Oil Firm to Be 'Virtually' Dissolved in Fy 2003," *Kyodo News*, May 2, 2002.

155 It seems that loan guarantees were permitted later. METI submitted the JOGMEC bill with the addition of the function of loan guarantee as Clause 3, Article 11.

156 Mulgan and Australian National University. Asia Pacific School of Economics and Management, *Japan's Failed Revolution: Koizumi and the Politics of Economic Reform*, 106.

157 Shinoda, *Koizumi Diplomacy: Japan's Kantei Approach to Foreign and Defense Affairs*, 136–39.

Koizumi were a manager of a private company, he would be fired due to zero outcomes in his first year.[158] Gerald Curtis labeled Koizumi as "Mr. NATO" (No Action Talk Only).[159]

In addition, the Koizumi Administration's efforts to exploit a sense of economic crisis to facilitate its radical reform program did not work well. The much-anticipated "March-crisis" of 2002 in Japan's financial system did not materialize, and stock prices rose modestly. This situation dissipated the government's sense of crisis and urgency. While the economic crisis did open a window of opportunity for reform in Japan by making the public more receptive to the Koizumi phenomenon, it did not produce a political crisis, where the power structure would have been sufficiently undermined to allow the LDP politicians and bureaucrats to become more committed to Koizumi's reforms.[160]

The submission of the bill for the establishment of a special company was delayed after the disposal of JNOC's assets, and the other two bills (calling for the abolishment of the JNOC and the establishment of the JOGMEC) were to be submitted in the regular session of 2002.[161] From the perspective of the Reform Minister, abolishing one public corporation by establishing two public entities did not make sense politically, and METI had to accept it.

However, what METI gave up was not its goal to establish a Japanese state-owned oil major but submitting the draft bill in this Diet session. While the prime minister would be consulted in the process, METI would be involved in the detailed work of selection and sales of assets through the special committee under the METI Minister's advisory group. Therefore, METI maintained the possibility of keeping good assets for a special company. As shown in an additional clause in the Law for Dissolution of JNOC and MMAJ, METI was still eager to establish a special company:

> (Establishment of Joint-Stock Company) Article 3. The government will deliberate on the necessary measures to …establish a joint-stock company

---

[158] Mulgan and Australian National University. Asia Pacific School of Economics and Management, *Japan's Failed Revolution: Koizumi and the Politics of Economic Reform*, 20.

[159] "Koizumi Branded 'Mr. NATO'—No Action, Talk Only," *Kyodo News International*, Dec. 5, 2001.

[160] Mulgan and Australian National University. Asia Pacific School of Economics and Management, *Japan's Failed Revolution: Koizumi and the Politics of Economic Reform*, 96–97.

[161] "Dissolution of JNOC Delayed One More Year, Special Company Not Included for Legislation," *Asahi shinbun*, May 2, 2002.

succeeding rights and duties of the JNOC through the government's full investment and will privatize as early as possible.[162]

The joint-stock company was not a special corporation (*Tokushū hōjin*), but rather a state-owned company (*Kokuei Kaisha*).[163]

Based on this agreement, the LDP Executive Council endorsed the two draft bills on May 15th. The Cabinet authorized the bills on May 17th and the draft bills were submitted to the Diet on the same day.[164] Now, the Lower and the Upper Houses discussed the matters on the JNOC reform, including a special company issue.

4.2 Discussion of the Proposal in the Diet

In the 2002 Diet regular session from June 9th to July 5th, the Diet members of the Standing Committee on Economy, Trade, and Industry of the Lower House asked broad questions on energy security and the JNOC issue. Diet members and the METI Minister showed a different perspective on the level of the government's intervention and support. For the question of why METI's stance in 2002 was different from that of 2001 when METI made efforts to strengthen the role of the JNOC, the METI Minister answered ambiguously by saying that the abolition was in line with Koizumi's reform. Suzuki Yasutomo, a Diet member of the DPJ, raised doubts about the need to support the private sector financially. During Diet interpellation, he illustrated this point by showing an *Asahi shinbun* article that quoted the President of Idemitsu Oil as saying the private sector did not need financial support. The ANRE Administrator Kawano Hirobumi responded that there was a strong demand from the private sector for the government's financial support for overseas oil development.[165] In the discussion of JNOC's reform, socialist party members voiced concerns on *amakudari* and the JNOC's political contribution to the LDP.

Many questions centered on the problem of disposing of JNOC's properties and subsequent issues of a special company that METI had been pushing.

---

[162] Ministry of Economy Trade and Industry, "Sekiyukōdanhō Oyobi Kinzokukōkyōzigyōdanhō No Haishi Tōnikansuru Hōritsu [Law Related to Abolition of Law for Japan National Oil Corporation, Law for Metal Mining Agency of Japan, Etc.]," ed. Japanese Diet (2002).

[163] *24th Committee Meeting*. Comment by Kumashiru Akihiko, Vice Minister of Cabinet Ministry.

[164] Japan National Oil Corporation, *Sekiyukōdanshi (1997 <Heisei9> Nendo-2004 <Heisei16> Nendo) [the History of JNOC (1997-2004)]*, 14.

[165] *23rd Committee Meeting*. As of April 2012, Kawano Hirobumi is President of Japan Oil, Gas and Metals National Corporation (JOGMEC), the independent administrative agency integrating the JNOC and the MMAJ, which was the outcome of the JNOC reform.

Still, both the asset disposal plan and the timing of the establishment of a special company and its privatization were not decided at that time, instead being entrusted to the future. Some issues were publicized in the committee session. The METI Minister stated that there would be four categories of asset disposal: (1) bad assets will be disposed of in principle; (2) if a decent asset is not considered a core asset for a special company, it will be sold; (3) if the assets are under exploration or development, and if the government needs to show its involvement to oil-producing countries, the asset will be transferred to a special company; and (4) if the asset is bad but is expected to have higher value due to a close connection with other projects or businesses, it will be passed on to a special company.[166] METI's submission of a bill for the establishment of a special company was anticipated in three years when the JNOC would officially be abolished.[167]

The attitude on a special company was diverse. On the one hand, more reform-oriented Diet members from the DPJ, such as Ubukata Yukio, Tanaka Keishū, and Nagatsuma Akira, opposed the establishment of a special company, because they thought this would make it meaningless to abolish the JNOC.[168] Nagatsuma Akira urged for immediate privatization of the special company, if the company were to be established. Also, Suzuki Yasutomo, a Diet member from the DPJ, conveyed some of the private sector's concerns about the special company.

On the other side of the reformists were Diet members who were more concerned with energy security, such as Takemoto Naokazu, a Diet member from the LDP, and Gotō Shigeyuki and Nakayama Yoshikatch, Diet members from the DPJ. These members supported the idea of a Japanese oil major like the French Total Fina Elf or Italian ENI, because the integration of assets would generate more value than would separate assets. Mr. Gotō thought this was the "last chance" to make a Japanese oil major (*wasei meja*).[169] Mr. Takemoto supported the government's effort to promote a vertically integrated major oil company with substantial financing capacity from the perspective of resource diplomacy (*shigen gaikō*).[170]

While the request to hear from Mr. Horiuchi was not accepted, Reform Minister Ishihara attended the last day of discussions in the Lower and the

---

[166] Ibid. Comment by METI Minister Hiranuma Takeo.

[167] Ibid. Comment by METI Vice Minister Ooshima Yoshihiko.

[168] Ibid.; *24th Committee Meeting*; *26th Committee Meeting*.

[169] *23rd Committee Meeting*; *26th Committee Meeting*. This split of opinions in the opposition party had already taken place during the discussion on the Petroleum Industry Law in 1962, when the idea of a state-owned wholesaler was raised.

[170] *24th Committee Meeting*.

Upper House committees on July 5th and July 18th, respectively.[171] Asked about his stance on the disposal of the JNOC's assets, he repeatedly emphasized the "minimization of people's burdens" rather than having to make a choice between immediate sales of all assets and keeping all healthy assets through a special company.[172]

Also, the Lower House committee invited energy experts, such as Toichi Tsutomu, the then-Managing Director of the Institute of Energy Economics Japan (IEEJ), Arai Mitsuo, *Yomiuri Shinbun* Editor, and Isono Akira, President of Teikoku Oil, on July 2, 2002. All of them supported the need for loan guarantees and a special company, both of which were controversial issues. They estimated that establishing a capable upstream company was a priority task. In that regard, they were negative on sales of the JNOC's assets, because that would have removed a chance to create a Japanese oil major.[173] The Upper House committee invited energy experts, such as Okabe Keiichiro, President of both the PAJ and Cosmo Oil, Prof. Morita Nobuo from Waseda University, and Tatesawa Kōji, a freelance journalist. While the former two experts were sympathetic to JNOC's role and supportive of METI's direction, Mr. Tatesawa was skeptical of the performance of the JNOC and METI bureaucrats. Mr. Tatesawa maintained that a government-led Japanese oil major would not work, an argument that was well received by DPJ member Honda Ryōichi, who also opposed METI's state-owned company idea.[174]

In the final committee meeting at the Lower House, the bills were passed by a majority with no revisions on July 5th though explicit opposition was registered by Tasso Takuya, Shiokawa Tetsuya, and Ooshima Reiko, the members from the LP, the JSP, and the SDP, respectively. One of the main reasons was that there was no clear picture for the creation of a special company, and it appeared to give the government "carte blanche." With similar reasoning, Ogata Yasuo, the Upper House member from the JCP, opposed the bills in the final committee meeting at the Upper House, but the bills passed with majorities in the committee and the Plenary Session with no revisions on July 18th. Finally, while successful repayable loans were abolished, government financial support for the private sector's overseas oil development survived by establishing JOGMEC instead of abolishing

[171] The request to hear from Horiuchi was made by Shiokawa Tetsuya, a Diet member from the Japan Communist Party on June 23, 2001.

[172] Japan House of Representatives, Standing Committee on Economy, Trade and Industry, *27th Committee Meeting*, Interpellation, July 5, 2002; Japan House of Councillors, Standing Committee on Economy, Trade and Industry, *25th Committee Meeting*, Interpellation, July 18, 2002.

[173] Japan House of Representative, Standing Committee on Economy, Trade and Industry, *25th Committee Meeting*, Interpellation, July 2, 2002.

[174] *24th Committee Meeting*.

the JNOC. The establishment of a special company remained unclear and postponed for future discussion.

4.3 Evading Veto Players Through INPEX

The possibility of establishing a special company became more realistic when the bills to dissolve the JNOC and MMAJ passed, though the situation facing METI was not favorable. The political presence of veto players, such as Koizumi, Horiuchi and Ishihara, diminished the possibility of successful legislation for a special company that METI had planned.[175] In addition, the three-year time limit set for the dissolution of the JNOC was estimated as too short to complete the procedures needed to create the special company.[176] In addition, the private sector was not supportive of a government-led special company. In the subcommittee hearing under Comprehensive Energy Examination Committee, METI's *shingikai*, on November 29, 2002, a majority of the 23 participants, including representatives of the JPDA, Shin Nippon Oil, Teikoku Oil, Mitsubishi Corporation, Mitsui & Co. and Mizuho Corporate Bank, opposed METI's plan. Only Daiwa Securities expressed support, saying it would be unclear whether an internationally competitive oil company could be realized by the private sector.[177] Therefore, METI decided to choose a healthy subsidiary of the JNOC as the core company (*chūkakutekina kaisha*). In order to evade a complex political procedure, the company would be a private company based on Commercial Law, not a special company based on the specific law that would have been submitted to the LDP and the Diet.[178]

INPEX Corporation, led by Matsuo Kunihiko, a former METI official, was chosen as the core company because it had shown better performance among all of JNOC's subsidiaries. Originally, a March 2003 report of the subcommittee suggested integration of INPEX Corporation, JODCO,

---

[175] Koizumi maintained his premiership from Apr. 26, 2001 until Sept. 26, 2006. Horiuchi maintained the Chair of LDP Executive Council from April 25, 2001 until Sept. 27, 2004, and Ishihara continued to serve as Reform Minister until Sept. 22, 2003, at which point he became MLIT Minister. Liberal Democratic Party of Japan, "Liberal Democratic Party—Terms in Office of Top Party Leaders"; Cabinet Secretariat, "Junichiro Koizumi Administration," http://www.kantei.go.jp/foreign/rekidaisouri/koizumi_e.html.

[176] JNOC Asset Evaluation & Liquidation Review Subcommittee, "Sekiyukōdanga Hoyūsuru Kaihatsukanrenshisan No Shorini Kansuru Hōshin [Plan Related to Disposal of Development-Related Asset the JNOC Possesses]," ed. Comprehensive Resource and Energy Examination Committee (Tokyo, 2003), 14.

[177] "Consolidation through the Private-Oriented for Making a Core Company is General Tendency."

[178] Interview with a METI official who is versed in the JNOC reform issue, conducted on June 22, 2011.

SODECO (Sakhalin Oil & Gas Development Corporation) and their subsidiaries. However, the shareholders of SODECO, such as JAPEX, Itochu, and Marubeni, opposed its integration with INPEX. In particular, JAPEX, one of the shareholder companies of SODECO, needed SODECO's assets to receive a high value in its listing on the Tokyo Stock Exchange in November 2003.[179] Therefore, JNOC's subsidiaries—JODCO, INPEX Southwest Caspian Sea, INPEX Zau, and INPEX ABK—were combined to form INPEX Corporation through the exchange of stocks between the JNOC and INPEX Corporation.[180] In May 2004, the exchange of stock was completed, and as a result, INPEX Corporation became the holding company for JNOC's subsidiaries except for SODECO.[181]

In the process of privatization, a golden share was applied instead of giving the JNOC's assets to INPEX Corp. to prevent the following national energy security-related reasons: (1) myopic profit-seeking behavior like selling important assets; (2) foreign capital from controlling the core company; and (3) the core company from veering off the policy track of securing a sustainable energy supply.[182] Applying a golden share was suggested by Matsuhara Yoshitake, a Lower House member from the LDP, during interpellation for the JNOC reform bills in 2002 and in METI's subcommittee meeting in November 2002.[183] Ironically, this partial governmental intervention in the management of the core company was the lesson learned from the JNOC's failure, because a lack of attention to the oil developers invested by the JNOC was one reason for JNOC's huge debt accumulation.[184] As mentioned by

179 Ibid.; Subcommittee, "Sekiyukōdanga Hoyūsuru Kaihatsukanrenshisan No Shorini Kansuru Hōshin [Plan Related to Disposal of Development-Related Asset the JNOC Possesses]," 10.

180 Unofficially, the integration of the AOC, JAPEX and INPEX, which used to be a project company of JAPEX, had been mentioned, but it was not realized. One possible reason circulated in the industry was the rivalry between the presidents of JAPEX and INPEX, who entered MITI in the same year. Nagazima Ikuo, *Shigengaikō Rensenrenpai: Azadeganyuden No Satetsu [Consecutive Defeats in Resource Diplomacy: Set-Back of Azadegan Oil Field]* (Tokyo: Yosensha, 2009), 224–26.

181 Japan National Oil Corporation, *Sekiyukōdanshi (1997 <Heisei9> Nendo-2004 <Heisei16> Nendo) [the History of JNOC (1997-2004)]*, 34–35.

182 Ibid., 32. Golden share is a type of share that gives its shareholders veto power over changes to the company's charter. A golden share holds special voting rights, giving its holder the ability to block another shareholder from taking more than a ratio of ordinary shares. Ordinary shares are equal to other ordinary shares in profits and voting rights. These shares also have the ability to block a takeover or acquisition by another company. Unlike Japan, Korea has not adopted the golden share system as of November 2012. For more information on golden share, see Investopedia U.L.C., "Dictionary: Golden Share," http://www.investopedia.com/terms/g/goldenshare.asp#axzz1jpNkr6bf.

183 *24th Committee Meeting*; "Consolidation through the Private-Oriented for Making a Core Company is General Tendency."

184 Subcommittee, "Sekiyukōdanga Hoyūsuru Kaihatsukanrenshisan No Shorini Kansuru Hōshin [Plan Related to Disposal of Development-Related Asset the JNOC Possesses]," 13.

METI Minister Hiranuma in the 2003 Diet interpellation, METI had the will to make a good core company with appropriate supervision.[185]

Since the rebirth of INPEX as the core company, INPEX has played a leading role in Japan's overseas oil development. As of March 2011, it has approximately 4.13 billion BOE of reserves and produces 423 thousand BOE per day, which accounts for approximately half of Japan's overseas oil and gas production. Furthermore, through asset acquisition and aggressive exploration as an operator, like in Ichthys in the Australian offshore and Abadi in the Indonesian offshore, the production output of INPEX is expected to reach one million BOE/day by 2020.[186]

While the residue of the government's major share ownership for INPEX through the JNOC still exists, INPEX is made up of basically private companies based on the Commercial Law. METI's share portion, inherited from JNOC's majority share ownership on INPEX, decreased from 53.96% after the consolidation in 2004 to 18.9% after increasing its stock in 2011.[187] In addition, METI's special-class share, or golden share, became not only an effective countermeasure to prevent control over the business or hostile takeovers for speculative purposes but also a good tool for international negotiation and credits.[188] For INPEX, while national energy security is considered one of its management aims due to INPEX's special relationship with METI, profit-maximization is the most significant aim of INPEX like other Japanese oil developers.

5 Summary and Conclusion

In terms of establishing a state-owned oil company, Japan in the early twenty-first century shows that the interaction among many veto players produces a resolute outcome compared to METI's intended agenda. In the early twenty-first century, Japan had fourteen veto players: the prime minister; the reform minister; METI; two LDP councils of the LDP; the other two coalition parties; the two Diet chambers; the standing committee in each chamber; *Keidanren*; the PAJ, and the JPDA. Compared to the MOF in the case of the

[185] Japan House of Representatives, Standing Committee on Economy, Trade and Industry, *10th Committee Meeting*, Interpellation, Apr. 2, 2003.

[186] INPEX Corporation, "Annual Report 2011: Firm Commitment to Sustainable Energy" (Tokyo, 2011), 32–33.

[187] Ibid., 121.

[188] INPEX Corporation, "Corporate Governance," http://www.inpex.co.jp/english/company/governance.html#gov04.

1960s, making a coalition government increased the number of veto players, and the Reform Minister appeared as a regulator during the time of Koizumi's reform. This case also counts the last three interest groups as veto players mainly because they were included in *shingikai*, the formal channel for the interest groups to affect policy outcome, and they played roles in the discussion. On the other hand, opposition parties were not veto players because they were paralyzed due to the majority-based decision-making system in the Diet.

Applying the absorption rule, the preferences of the coalition partners, the reform minister, and the LDP Executive Council are included in those of the prime minister in this case. In a majority-based decision-making system during Koizumi's prime ministership, the opposition parties had limited influence in legislative decision-making. Therefore, the preference of the Diet and committees are included in those of the ruling coalition. However, the PARC and Executive Council in the LDP still have different preferences. Among interest groups, the preferences of the JPDA and the PAJ are similar in that they supported a state financier and opposed the idea of a state-owned company. As a result, the number of effective veto players becomes five.

The resolute outcome in the decision-making process differed from METI's original intention, which was to establish a state-owned oil company. The initiative of METI, the agenda setter for the establishment of state-owned oil company, was obstructed, compromised, and transferred into the privatization of one subsidiary of JNOC to make a sizeable oil developer. At the same time, the Prime Minister and LDP Executive Council, both agenda setters for the JNOC reform, could only abolish the JNOC and its lucrative success repayable loan, and they had to permit METI to establish a new public institution to support the private sector's overseas oil development. These results also verify the applicability of veto player theory that an outcome passed through many veto players is more private-regarding than public-regarding. This outcome confirmed the relevance of veto player theory that the outcome passed through many veto players tends to be more private-regarding than public-regarding.

While this case supports the veto player approach, it also reveals interesting aspects of Japan's political economy. In the decision-making of collective veto players, Japan's previous consensus-based decision-making gradually diminished over time. The outcome also showed that the private sector's preference affects veto players' preferences and behaviors, as well as how the media was involved in politics. At the same time, this chapter shows that the presence of a strong executive itself does not guarantee swift and decisive outcomes if there are many veto players, such as the LDP councils. Japan's unique review system is one noticeable difference from the Westminster Parliamentary system, and it hinders Japan's prime minister from making decisive policy changes. Prime Minister Koizumi inherited a strengthened *Kantei* and attempted to change the policymaking system, but he could not achieve the expected outcome, instead settling for one that saved face. The outcomes in the early twenty-first century in Japan will be compared in the next chapter to the case of Korea in 2008, when the Korean government attempted to strengthen its existing state-owned oil company.

5

A Business-Friendly President and the Strengthening of Korea's State-Owned Oil Company

While Korea became an industrialized capitalist country in the early twenty-first century, the country depended more on SOEs for its overseas oil development. While both Japan and Korea experienced strong political leadership during this time, only Korea could realize the strong demiurge role in overseas oil development. However, given the weak presence of KNOC in the 1980s and 1990s, the state's deeper involvement in overseas oil development cannot be seen as path-dependent. This chapter explains why Korea's approach in overseas oil development was much more successful than Japan's, even in the twenty-first century. Examining the case of Korea in 2008 further strengthens the main argument of this book that a state-owned company is more likely to be established or strengthened when there are a small number of veto players. Through qualitative analysis, this chapter will show how the number of veto players affected Korea's realization to expand its state-owned oil company in the twenty-first century.

The Korean government made efforts to strengthen its state-owned oil company, KNOC, when it faced an oil price hike in the early twenty-first century. One key policy was an effort to increase the country's oil self-sufficiency ratio through overseas oil development, and the state needed to directly fund large investments to increase E&P assets and the capacity of KNOC. For the Korean government, Korean state-owned oil companies needed to be sizeable to compete with gigantic Western oil majors and state-owned oil companies of newly industrialized countries, such as China. The

S.-i. Oh, *Overseas Energy Investment of Korea and Japan*,
https://doi.org/10.1007/978-981-99-0285-9_5

government succeeded in amassing substantial funding in the budget process of the unicameral legislature.

KNOC's reinvigoration was possible due to a small number of veto players in Korea's policymaking process. The Korean president who is mandated to run only the single five-year term encouraged the government to improve the energy security situation in such a short period, regardless of the burden of massive debt.[1] Also, relevant ministries and his majority party in the National Assembly supported this direction. It is important to note that interest groups were not consulted in this decision-making. Ironically, this policy that strengthens an SOE was implemented predominantly when a "business-friendly" president, a president who supports the private sector first, was in power.

1 Korea's Political Institutions in the Early Twenty-First Century

Korea's recent 9th revision of the constitution in 1987 did not accompany the change in the number of veto players. The people of Korea most desired a direct, rather than indirect, presidential election, and the amendment did not create a substantial transformation in Korea's political structure.[2]

1.1 Institutional Veto Players: The President, the Ministry, and the National Assembly

Strong President

While Korea's 1987 Constitution had replaced the president's dominant power with a more balanced parliamentary system, the constitution still reinforced certain characteristics of presidentialism. While the 1987 Constitution abolished the president's power to dissolve the National Assembly and limited

---

[1] Interview with Keun-Wook Paik, senior research fellow at Oxford Institute for Energy studies (OIES), University of Oxford, conducted on Oct. 7, 2012.

[2] "Meet Aspiration of People General, New Era of Constitutional Politics," *Kyung-hyang Shinmun*, Aug. 31, 1987. In this process of the amendment, a small number of assembly members showed different opinion from the absolute majority in the assembly. For example, Lee Chul-sŭng argued the need to change the political structure from a presidential system to a parliamentary one. See "Resolved Political Storm by Pursuing for Desire of People General," *Kyung-hyang Shinmun*, Oct. 13, 1987.

each president to a single five-year term, the power of the president was not diminished much. Many rights of the president were maintained, and direct presidential elections afforded more legitimacy to the elected president than the previous indirect presidential elections.[3]

In the 1987 Constitution, electoral laws were made less favorable to the president's power than in previous constitutions (i.e., the *Yushin* Constitution). The president is elected through direct election by the people with terms fixed at five years with no opportunity for reelection. While this is a countermeasure to prevent long periods of power, it also has some weak points. With the inevitable lame duck period in the latter part of the presidency, the president only has a short amount of time in which to institute policy or make changes, causing presidents to rush into policymaking and implementation.

Unlike Japan, coalition governments are rare in Korea's presidential system, making it more difficult to increase the number of veto players. In spite of multiple candidates in Korea's presidential election, it maintained a one-round presidential election rule, lessening the need for coalitions with other parties. This is in contrast to other political systems where only the candidate with the absolute majority vote gains power, such as in Japan's parliamentary system and Russia's two-round presidential election. The only exception was President Kim, Dae-jung's coalition government with Mr. Kim Jong-pil's Liberal Democratic Federation (LDF), called the "DJP Coalition." In this coalition, the prime minister from the LDF had the real power, such as the appointment of economic ministers, but the coalition ended due to the conflict surrounding the dismissal of the unification minister in September 2001, the fourth year of the Kim Dae-jung Administration.[4] Considering the presidential election system and history of Korean politics, it is less likely in Korea than in Japan to increase the number of veto players in policymaking due to fewer coalition governments.

Compared to the *Yushin* Constitution, the president can still submit draft bills and veto laws passed by the National Assembly. The president can also appoint the prime minister and other Cabinet members, as well as CEOs and

---

[3] Heo, *Hankuk Hŏnbŏmnon [Korean Constitution]*, 760–62.

[4] Media purport that the real reason of the conflict in 2001 was that the then-President Kim Dae-jung's de facto withdrawal of political promise with Kim Jong-pil to revise Korea's constitution from a presidential system to a parliamentary system. See Youn-jong You, "Today's Bookmark: Kim Dae-Jung and Kim Jong-P'il Agreed on Single Presidential Candidate in 1997," *Dong-A Ilbo*, Nov. 2, 2004.

supervisors of state-owned companies.[5] Ministers take part in the meetings of the Cabinet, which is a delegation rather than a decision-making institution, and ministers are able to maintain their status as members of the National Assembly like in a parliamentary system. The president also has the right to submit regular budgets and supplementary budgets, which require the authorization and inspection of the National Assembly.

The Presidential Transition Committee has been formed since 1993 and became the de facto policymaking office for the president-elect.[6] The activities of this committee were not limited to studying administrative works and included important decision-making like the restructuring of the government. In the case of President-elect Lee Myung-bak, the Committee Chair announced the plan on January 16, 2008, and the plan was submitted by Ahn Sang-soo, a ruling party member, to the National Assembly on January 21, 2008. The government restructuring plan was passed after a slight compromise between parties, despite the opposition party having more seats than the ruling party at that time.[7] Considering the power of the Committee, it is natural to regard the President-elect's guidance, such as his emphasis on the role of a state-owned oil company, as an order of the highest power for ministries.

Chŏngwadae, Korea's Office of the President, is the control tower of policymaking. As in previous administrations, important policy agendas, such as the privatization of state-owned companies and the "Green Growth" sustainable development policy, were managed by *Chŏngwadae* or presidential committees in the Lee Myung-bak Administration. Important issues in each ministry must be approved by *Chŏngwadae* before announcement and implementation, and relevant senior secretaries were responsible for examining each policy. The senior secretaries report ministry policy to the president; therefore, ministries are dependent on senior secretaries' favorable comments. While President Lee downsized the regular staff from 533 to 427, with one chief of secretariat, seven senior secretaries, one spokesperson, as well as 47

[5] For KNOC, President Lee appointed Kang Young-won, who had been CEO of Daewoo International, as CEO in August 2008. The president appointed Choi Sŏng-ryong, who had worked at the Preparation Committee for Presidential Inauguration as an auditor of KNOC, and Ryu Je-hyun, who also worked at his Presidential Transition Committee, as Mr. Choi's successor.

[6] The predecessor of the Presidential Transition Committee is President Roh Tae-woo's Inauguration Preparation Committee in 1988, and the first Presidential Transition Committee began in 1993 for President Kim Young-sam. See Kwang-Soo Kim, "On Presidential Transition in Korea," *Public Law Journal* 10, no. 3 (2009): 317–18.

[7] Ibid., 336. Thanks to the restructuring, the jurisdiction of Ministry of Commerce, Industry, and Energy, which dealt with energy issue, was expanded to the IT industry, and the name was changed to the Ministry of Knowledge Economy.

secretaries and aides at the beginning of his administration,[8] this did not weaken the influence of *Chŏngwadae*. Approximately half of *Chŏngwadae* officials are "aces" seconded from ministries.[9] In addition to its status as the office of the "president" that is the highest authority in the Korean government, *Chŏngwadae* controls bureaucrats by intervening in personnel management of each ministry. For example, *Chŏngwadae* controlled promotions within ministries and state-owned companies directly.[10]

One recent phenomenon is the establishment of presidential committees directly under the president, such as the Presidential Committee of Future and Vision (PCFV), to delegate the president's important tasks. This has mainly taken place since the Roh Moo-hyun Administration in order to make comprehensive political visions beyond ministerial sphere of influence and overcome bureaucratic rigidity.[11] However, the presidential committees have been criticized as being beyond the control of the legislature and having an inappropriate role in policymaking. Also, the president generally lends greater support to the committee rather than the ministries.[12] The active role of independent non-cabinet presidential committees is one characteristic of the strong presidency during this new era.

When President Lee entered negotiations with the United Arab Emirates (UAE) for rights to develop oil resources there, he replaced the traditional first agent, the Ministry of Knowledge Economy (MKE), with the Presidential Committee of Future and Vision (PCFV) at the start of negotiations. Though the committee has a duty to provide advice on energy issues to the president,[13] excluding MKE from the negotiations seemed quite a surprise to

[8] June-Hoon Bae, "A Study on the Reorganization of the Presidential Office: The Case of President Myung-Bak Lee & Quotes Early Presidential Office," *Journal of Secretarial Studies* 18, no. 2 (2009): 48.

[9] Originally, "ace" means an elite pitcher in baseball. In Korean government, ace means a top-notch bureaucrat or an elite bureaucrat.

[10] Yong-joong Chu, "The Right of Personnel Management Returns to Minister?," *Chosun Ilbo*, June 24, 2009.

[11] Ho-Seob Kim, "Presidential Advisory Committees on National Policies: Actual Conditions, Unknown Performance and Future Tasks," *Korean Review of Orgarnizational Studies* 1, no. 2 (2004): 62–3. President Roh is known to follow Clinton's National Performance Review as a benchmark. Ibid., 59.

[12] Ibid., 64–5.

[13] The Committee's role is to provide advice on "making comprehensive national visions on the unity and safety of future society, demography, environment, education, culture, energy, food, water resource, health, IT, media and space development." Article 2–2 *Executive Order for Presidential Committee of Future and Vision*, Executive Order 20,652, (Dec. 29, 2008).

the ministry, which, until that time, had managed Korea's overseas oil development exclusively.[14] There were a few reasons for the PCFV's involvement. First, the government was concerned about the possibility that the contents of the deal would leak to foreign companies. The PCFV is a small independent presidential committee chaired by a member of the president's inner circle, so it was the right organization for this task.[15] Thus, President Lee ordered Kwak Sŭng-joon, the committee chair, to initiate the project secretly. In addition, Mr. Kwak was known to be acquainted with the Prince of Abu Dhabi from his studies at Vanderbilt University.[16] Furthermore, as stated in the official press release, *Chŏngwadae* regarded the deal not just as an oil deal, but as a deepening of its strategic partnership with the UAE, emboldened since Korea's successful bid to construct a nuclear power plant there. The oil deal was signed in return for Korea's cooperation in the UAE's future planning in areas such as economic development.[17] President Lee Myung-bak was committed to increase Korea's oil and gas self-sufficiency from 4 to 20% by the end of his presidential term, making this deal even more significant in this regard.[18] After finalizing the deal, he himself announced that this was "Korea's largest oil development deal," which would help increase Korea's oil and gas self-sufficiency ratio from 10.9 to 15%.[19]

However, it is important to know that there are some policies even a strong president cannot realize. One example is President Lee Myung-bak's failed

---

[14] Though the press release of the Committee mentioned that the Ministry of Knowledge Economy played a role in the negotiation, the accounts of journalists and government officials confirmed the exclusion of the MKE from the negotiation. Responding to the question from a reporter on MKE's being excluded, the MKE Minister indirectly affirmed this by saying "Why do you think that ŭipyŏng [volunteer militias] should not play a role and only kwan'gun [regular army] plays a role? If we can secure energy supply, it does not matter who does that. We should throw away the idea that only Ministry of Knowledge Economy works for securing resources." Also, what the MKE announced was different from what the Committee announced on the outcome of the negotiation, and the MKE had to correct its announcement later. This shows that the contents of MOU and HOT were not shared between the Committee and the MKE. See Sung-gyu Park, "MKE Announced without Knowing Details of Oil Deal," *Newsis*, Mar. 13, 2011; Chang Soon-won, "Choi Choong-Gyong Said Chung Woon-Chan Not to Talk More About Benefit-Sharing," *E-Daily*, Mar. 16, 2011.

[15] Interview with Lee Jin-myung, a reporter at *Chŏngwadae* Press Corps, conducted on Apr. 4, 2011.

[16] Ibid.

[17] At the MOU signing event, the MOU for future growth engine cooperation was also signed by the heads of Korea's PCFV and UAE's EAA (Executive Affairs Authority). The MOU states that system semi-conductors, IT contents, and finance will be the areas of the first cooperation. Office of the President, "Korea Opens the Door to UAE Abu Dhabi Oil Field in 37 Years," *Press Release*, Mar. 13, 2011.

[18] 20% was regarded as the minimum buffer for energy security. Practically, quantitative target of 20% for oil and gas is the level Japan is known to have achieved. While President Lee does not follow Japan's private sector-oriented institution, Japan has still served as a measuring stick of overseas oil development policy.

[19] Office of the President, "President Lee Wins International Prize, Oil Deal," *Press Release*, Mar. 15, 2011.

attempt to reverse the planned relocation of thirteen governmental ministries and agencies to Sejong City, located 120 km South from Seoul. The President wanted to maintain the efficiency of the central government, and thus he opposed the relocation plan. The government submitted a revised bill draft to the National Assembly in order to replace the original relocation plan with a new plan to make an education and science city, but the bill draft was rejected in the National Assembly with a vote of 105 in favor and 164 opposed in June 2010.[20] This result took place in spite of the fact that the President's party had the majority seats.

Ministerial Veto Players: The MKE and the MOSF

When the strengthening of a state-owned oil company was the issue in the early twenty-first century, Korea's MKE was the agenda setter, and the Ministry of Strategy and Finance (MOSF) was a ministerial veto player as the budget controller. Korea's ministerial structure on state-owned oil company in the early twenty-first century is similar to that of the 1970s.

Ministry of Knowledge Economy (MKE)

The MKE was a continuity of MOTIE (the Ministry of Trade, Industry, and Energy), which consolidated the Ministry of Trade and Industry (MTI) and Ministry of Energy and Resources (MER) in the 1993 government restructuring. While the trade policy function was transferred to the Ministry of Foreign Affairs and Trade in 1998's restructuring, 2008's government restructuring, which was completed before President Lee's inauguration of February 25, 2008, reinforced the power of the ministry.[21] The R&D policy function, internet technology (IT) policy function, and Korea Post were absorbed from two abolished ministries, the Ministry of Science and Technology and the Ministry of Information and Communication. Also, the administration of special economic zones was transferred from the Ministry of Finance and Economy.[22] This drastic change, which led to the MKE being touted

[20] See Jung-ho Yoon, "Revised Bill for Sejong City Rejected in the National Assembly: 105 vs. 164," *Chosun Ilbo*, June 30, 2010.

[21] The Transition Committee announced its draft for government restructuring on Jan. 16, 2008, and the final compromise with the majority party was reached on Feb. 20, 2008.

[22] Ministry of Knowldge Economy, "History: Ministry of Knowledge Economy," http://www.mke.go.kr/intro/intro/history07.jsp.

"another dinosaur ministry," was attributed to the ministry's human network in the presidential transition committee.[23]

The ministry's energy policy function was strengthened to support the president's energy diplomacy following the oil prices hike in the early twenty-first century. MKE maintained its jurisdiction over the state-owned energy companies, such as KNOC and KOGAS, and overseas oil and gas development associated with the president's resources diplomacy became MKE's main agenda in 2008.

Ministry of Strategy and Finance (MOSF)

Among Korea's policymaking and implementing institutions, the MOSF is the most powerful. In the pursuit of downsizing the government in early 2008, the MOSF was formed by merging the Ministry of Strategy and Budget (MSB, the successor of the EPB) and the Ministry of Finance and Economy, making it a powerful economic agency with 91 divisions.[24] The ministry's jurisdiction included policy planning, budget, and taxation policy, all essential powers to control other economic ministries.[25] While the status of the deputy prime minister was abolished as a part of the government restructuring, other ministries felt that the MOSF had become more powerful than the MSB. Also, MOSF's jurisdiction included supervising and giving guidelines to Korea's 298 state-owned companies, a function which was strengthened during the previous administration.[26] Therefore, the MOSF is still an effective veto player that ministries have to consult for managing state-owned companies and their budget allocations.

Legislative Institutional Veto Player: Unicameral National Assembly

With the 1987 Constitution, the National Assembly gained more power than before. Unlike the *Yushin* Constitution, the 1987 Constitution was drafted by

[23] Dong-suk Park, "Thank You, Minister Youn Jin-Sik!," *E-Daily*, Jan. 16, 2008.

[24] Ministry of Strategy and Finance, "Ministry of Startegy and Finance," ed. Ministry of Startegy and Finance (Kyunggido: Ministry of Startegy and Finance, 2011), 9.

[25] Korea's downsizing of the government meant a decrease in the number of ministries without decreasing the work of the government; therefore, the appearance of big ministries like the MOSF was an inevitable phenomenon.

[26] Dong-suk Oak and Young-jin Ham, "Institutional Framework of Ministries for Fiscal Policy in Korea," *Korean Review of Orgarnizational Studies* 6, no. 1 (2009): 212–13; Chong-kuk Chu, Youn, Kŭn-young and Shin, Ho-kyŏng, "MOSF Will Be Dinosaur Ministry?," *Chosun Ilbo*, Jan. 16, 2008.

eight members of the National Assembly from the ruling party and opposition parties, not by the president.[27] The power of the president to dissolve the National Assembly was removed, and the National Assembly gained the power to inspect and investigate state affairs. Furthermore, the limit on the length of the National Assembly sessions was removed and the power to check the president's emergency order was strengthened.[28] The power of the National Assembly to check the government's spending was also strengthened. In 2003, ministerial funds that had been regarded as off-budget were included as subjects to be inspected by the Assembly. In the same year, the National Assembly Budget Office (NABO), whose function is similar to that of the US Congressional Budget Office (CBO), was established to support the legislature's budget inspection.[29]

Despite its strengthened role, the National Assembly has not been an effective veto player. Unlike the United States and Japan, Korea has maintained, with the exception of 1960–1961, a unicameral legislative system.[30] In the National Assembly, the ruling party has also been the majority. If the ruling party did not occupy the majority of seats in the Assembly, the ruling party made efforts to gain the majority through other political measures. When the opposition parties gained more seats than the ruling party in the 13th National Assembly election in April 1988,[31] the then-President Roh Taewoo overcame political difficulties by creating the Democratic Liberty Party (DLP), which consolidated the ruling party, the Democratic Justice Party (DJP), with two opposition parties, the Democratic Party and the Republican Party, in January 1990.[32] When this ruling party lost its majority in the 15th election in 1996, the then-President Kim Young-sam changed the outcome by asking opposition and independent members to join the ruling party.[33] President Kim Dae-jung also utilized the same tactic by persuading 36 opposition members to switch to the ruling party.[34]

---

[27] Heo, *Hankuk Hŏnbŏmnon [Korean Constitution]*, 121.

[28] Ibid., 125.

[29] Bok Moon-Soo, "Analysis on the Fiscal Relationship between Government and National Assembly with Respect to Fiscal Democracy in Korea," *Korean Governance Review* 14, no. 3 (2007): 71, 73.

[30] Heo, *Hankuk Hŏnbŏmnon [Korean Constitution]*, 907. Though the 1952 constitution and 1960 constitution called for a bicameral system, bicameral elections were implemented only in 1960.

[31] The result of the election is as follows: 125 seats for Democratic Justice Party (the ruling party), 70 for Peace Democratic Party, 59 for Democratic Party, 35 for Republican Party, and 10 for others. Ibid., 121–22.

[32] Ibid., 125.

[33] Ibid., 127–28. The result of the election is as follows: 139 for New Korea Party (the ruling party), 79 for New Politics People's Congress, 50 for Liberal Democratic Federation, 15 for Unified Democratic Party, and 16 independent members. After the political tactic, the number of the ruling party increased from 139 to 165.

[34] Ibid., 129.

Table 1 Changed composition favorable to the president in the National Assembly

Party	17th National Assembly (May 30, 2004–May 29, 2008)	18th National Assembly (May 31, 2008–May 29, 2012)
Grand National Party	121 (40.4%)	153 (51.2%)
Future and Hope Solidarity	n/a	13 (4.3%)
Open Uri Party	152 (50.8%)	n/a
Democratic Party	9 (3%)	81 (27.1%)
Democratic Labor Party	10 (3.4%)	5 (1.7%)
Liberal Democratic Federation	4 (1.3%)	n/a
Liberal Advancement Party	n/a	18 (6%)
Creative Korea Party	n/a	3 (1%)
Others	3 (1%)	26 (8.7%)
Total members (Total constituencies)	299 (243 local constituencies)	299 (245 local constituencies)

Source Korea National Assembly Secretariat, *Taehanmin'gukkukhoe Yuksimnyŏnsa [60-Year History of the National Assembly of the Republic of Korea]*, 415,475
Note The data are based on the result of the general elections; therefore this does not reflect the slight change in numbers after the elections. The number in parenthesis is the portion of the seats occupied by the party.

Subsequent presidents have dealt with a possible divided government in different ways. President Roh Moo-hyun's party, the Open Uri Party, earned a favorable result in the 17th National Assembly election in April 2004, thanks to the political backfire after his impeachment by the National Assembly, where the opposition parties had been the majority previously.[35] Because the ruling party, the Grand National Party, was not the majority at the beginning of President Lee Myung-bak's Administration, he delayed important issues, such as the supplementary budget and privatization of state-owned companies, after the 18th general election on April 9, 2008, two months after his inauguration. When his party won 153 seats among 299 seats in the assembly,[36] President Lee could evade the difficulties of a divided government Table 1.

[35] Ibid., 133. The result was as follows. 152 for Open Woori Party (the ruling party), 121 for Grand National Party, 10 for Democratic Labor Party, 9 for Democratic Party, 4 for Liberal Democratic Federation, and one for People Unified Party.

[36] The National Assembly of the Republic of Korea, "Assembly Introduce: The Seats and Votes of Respective Parties," http://www.assembly.go.kr/renew10/asm/ifn/yat_04.jsp. The result of 18th general election was: 153 for Grand National Party, 81 for Integrated Democratic Party, 18 for Liberal Advancement Party, 13 for Pro-Park Coalition, 5 for Democratic Labor Party, 3 for Creative Korea Party, and 26 independent members.

In addition, the president's power in the ruling party, which is normally the majority party in the National Assembly, made the party cohesive. The president had always been chairman of the ruling party until November 2001, when President Kim Dae-jung quit his post as Chairman of the Democratic Party.[37] While President Lee has not held any party positions since his inauguration, he has been known to be involved in party activities including nomination through close party members.[38] The single five-year term limit generates relatively early lame duck presidencies, constricting the ability of the executive branch to guarantee the reelection of assembly members. In addition, unlike Kim Young-sam and Kim Dae-jung, who maintained broad support, recent presidents did not have a strong political base. Despite the ruling party's majority in the assembly, Korean presidents can enjoy a congruent National Assembly only during the early phase of his or her presidency.[39]

1.2 Partisan Veto Players: The Majority Ruling Party

As of 2008, eight parties have National Assembly members. Only the Grand National Party (GNP) and Integrated Democratic Party (IDP), which have more than 20 party members in the Assembly, are negotiation bodies. There are six non-negotiation bodies: the Liberal Advancement Party, the Future & Hope Solidarity, the Democratic Labor Party, the Creative Korea Party, the People-Centered Federation, and the Progressive New Party.[40] In a majority-based decision-making system, only the GNP, the majority ruling party, played a role as a veto player among parties.

Ruling Party: Grand National Party (GNP)

The GNP was established by consolidating the New Korea Party, the then-ruling party, and part of the Democratic Party in November, 1997. After two consecutive failures in presidential elections in 1997 and 2002, the

---

[37] Han-yong Sung, "President's Power to Nominate Party Candidate is the Retreat of Party Democracy," *The Hankyoreh*, Dec. 23, 2007.

[38] Choi Haangsoon, "Collaboration between the President and the Ruling Party: Focused on Analysis of Affecting Factors in Korean Regimes," *Korean Public Management Review* 22, no. 4 (2008): 191.

[39] Interview with Lee Jin-myung, a reporter at *Chŏngwadae* Press Corps, conducted on Oct. 12, 2011.

[40] The National Assembly, "The Current Status of the National Assembly Members," http://www.assembly.go.kr/renew10/mem/mem/mem_search.jsp.

party succeeded in gaining power in the 2007 presidential election with its candidate Lee Myung-bak.[41]

The ruling party maintained a high level of cohesion in 2008, the first year of the 18th National Assembly. According to Jeon Jin-young's analysis, the ruling party's Agreement Index and Rice Index, which measure the party's coherence, are 0.976 and 0.986, respectively.[42] This is equivalent to the party congruence in other successful parliamentary systems, such as the UK.[43] This situation can be attributed to strong party leadership shown in nominations. In the 18th general election, both the GNP and the IDP organized the Nomination Examination Committee, which was composed of party members and non-party members selected by the party leadership. This system made it possible for the party leadership to strengthen its influence by marginalizing possible challengers in the party.[44] With the influence in nomination in the general election and in the assignment of party posts, the party leadership could shape discipline in party members' voting behavior in the National Assembly.[45]

While the Policy Committee of Korea's ruling party held a high-level Party-Government Consultative Meeting (PGCM) in the Lee Myung-bak Administration,[46] this body cannot be categorized as a veto player in the same way as Japan's LDP PARC prior to 2009. Korean ministries consult with the policy committee at the PGCM, but the consultation is subsequent to the ministries' policy coordination with *Chŏngwadae*, which in turn reflects the president's policy direction. As a result, it is inevitable that the contents of the PGCM are at most fine-tuning of policy details, not substantial overhaul.[47] The PGCM is held once a month with the aim of getting the ruling party to

[41] Grand National Party, "History," http://www.hannara.or.kr/ohannara/hannara/hannara_05.jsp; National Assembly of the Republic of Korea, "History," http://www.assembly.go.kr/renew10/asm/ifn/yat_04.jsp.

[42] The Rice Index, invented by Stuart Rice, is the ratio of the number of party members who supported the majority's stance and the number who supported the minority's stance. If all agree or disagree with the bill, the index is 1. If half of the members agree with the bill and half disagree, the index is 0. The Agreement Index, suggested by Hix, Noury, and Roland, added the case of abstention. If all members show the same stance, the index is 1. If the vote splits 1/3 each for approval, disapproval, and abstention, the index is 0. See Jeon Jin-Young, "Party Cohesion of the 18th National Assembly," *Korean Party Studies Review* 9, no. 2 (2010): 124–27.

[43] Ibid., 127–28. The author uses the data of plenary sessions from May 30, 2008, the beginning of the 18th Assembly, until Mar. 3, 2009 in Korea's National Assembly.

[44] Park Kyungmee, "Party Nomination and Organization in the 18th General Election: The Grand National Party and the Democratic Party," *Korean Party Studies Review* 7, no. 2 (2008): 47–49.

[45] Jeon, "Party Cohesion of the 18th National Assembly," 128.

[46] Choi, "Collaboration between the President and the Ruling Party: Focused on Analysis of Affecting Factors in Korean Regimes," 194.

[47] Interview with a Korean government official, conducted on Oct. 2, 2011.

cooperate.[48] Associated with the party's congruence with the president, this cohesiveness affected voting behavior on the issue of strengthening the role of SOEs, which could be different from party's orientation of "big market, small government."[49]

Opposition Parties

Integrated Democratic Party (IDP)

The IDP, which was established by consolidating the United New Democratic Party (UNDP) and the Democratic Party (DP) in February 2008, is an opposition party that is considerably more progressive than the GDP. In particular, the IDP explicitly supports the role of government. The IDP especially opposes the privatization of state-owned companies that are related to national infrastructure as a basis for industrial development. Instead, the party supports the enhancement of publicity and efficiency through innovation.[50] The revision of KOGAS Law was submitted in September 2009 and introduced competition in Korea's monopolized natural gas imports, but this was effectively offset by the IDP's submission of a contrasting bill.[51]

The party could not gain a majority in the 2008 general election; therefore it sometimes obstructed parliamentary procedure to overcome its numerical disadvantage. For example, the IUP attempted to stop the budget bill from passing by occupying the speaker's seat on December 8, 2010, but the bill was rushed through by the GNP members.[52] This kind of incident in the National Assembly has taken place repeatedly between the majority

---

48 Choi, "Collaboration between the President and the Ruling Party: Focused on Analysis of Affecting Factors in Korean Regimes," 194.

49 *The Platform of Grand National Party,* (Jan. 9, 2006). Article 2 (On Economy and Government) "We bring about a free market economy robust with the energy of the private sector by ensuring private property rights, allowing equal opportunities and freedom of choice... The vestiges of bureaucratic economy will be removed and unnecessary regulations abolished. Both Korean and foreign enterprises will be provided with an environment that allows them to do business with freedom under the principle of individual responsibility. The Korean market will be liberalized in step with globalization...".

50 *The Platform of Democratic Party* (Oct. 3, 2010). The name of Integrated Democratic Party was changed to the Democratic Party on July 6, 2008.

51 Je-ho Lee, "Clash on Direct Import of Natural Gas between Opposition Party and Government," *Naeil Shinmun,* Feb. 9, 2011. The bill of Democratic Party prohibits the import of natural gas for non-commercial use.

52 To see two parties' stances on the incident, see Democratic Party, "Indulgence for Grand National Party's Illegal Rushing through Does Not Make Sense,"*Press Release,* Dec. 15, 2010; Grand National Party, "Unprecedented Violence in Plenary Session by Opposition Party," *Press Release,* Dec. 8, 2010.

ruling party and the opposition parties in a majority-based decision-making procedure in the Assembly.[53]

Liberal Advancement Party (LAP)

The LAP is a conservative party with a regional concentration in the Ch'ungch'ŏng province. With three members of Creative Party, the LAP organized a negotiation body in the National Assembly in August 2008; however, the negotiation body was dismantled in August 2009 due to political infighting in the LAP.

Democratic Labor Party (DLP)

Established as a social democratic party in 2000, the DLP member became part of the National Assembly for the first time in the 2004 general election. After the party split in 2008, the influence of DLP became weaker, and the number of assembly members decreased from ten to five. The party has maintained a pro-labor stance, and it attempted to block the passage of a temporary worker protection bill, which was agreed to by the Open Woori Party and the GNP, in 2006.[54]

1.3 No Interest Group Veto Players

While a few interest groups are included in the ministries' committee system, this does not guarantee their role as veto players. The MKE has many study groups that listen to the industry and academy, but their access in the policymaking process is restricted.

As in Japan, the interests of Korea's upstream oil industry are represented by an individual industry association rather than peak associations, such as the FKI, equivalent to Japan's *Keidanren*. The FKI and the Korea International Trade Association (KITA) expressed the private sector's opinions by asking for more governmental support during President Roh's

[53] The reverse case is the passage of the revised Private School Law during 2004–2005. See Hong Jing and Lee Hyeon-Woo, "Different Types of Violent Conflict in Korean National Assembly: Consensual Conflict, Discordant Conflict, and Resistant Conflict," *Korean Party Studies Review* 10, no. 2 (2011): 238–43.

[54] Ibid., 243–45.

Administration, especially in 2004.[55] However, their role of representing oil development-related industries shrank when the Energy & Mineral Resources Development Association of Korea (EMRD) was established in 2008 during President Lee's Administration.

In March 2008, against the backdrop of increased emphasis on energy diplomacy, the role of presenting the opinions of the overseas oil development industry was centralized in the EMRD. Under the government's third overseas resources development basic plan of August 2007, this association was established after the consolidation of three organizations: the oil development team of the Korea Petroleum Association (KPA); the Mineral Resources Development Association (MRDA); and the Energy Industry Association for Overseas Project (EIAOP).[56] Joining the EMRD were not only private companies, such as 20 oil developers and 50 mineral resources developers, but also SOEs, such as KNOC and the Korea Development Bank (KDB).[57]

While the role and organization of Korea's EMRD is similar to that of Japan Oil, Gas, and Metals National Corporation (JOGMEC), the EMRD is not a public agency. Instead, it is a membership association composed of oil and mineral resources developers.[58] In addition, compared to JOGMEC's 496 employees, the EMRD is much smaller with only 23 staff members and one standing vice chairman.[59] However, in September 2008, the EMRD

---

[55] Nam-sok Uhm, "FKI Asked the Support for Overseas Resources Development," *Yonhap News*, June 23, 2004. Soo-jung Ha, "Trade Industry Requested Ruling Party to Support Overseas Resource Development," *E-Daily*, Aug. 13, 2004. The GTCs, such as Daewoo International and Samsung Corporation, are membership companies of the KITA wanted to resume overseas resource development actively then. GTCs utilized the KITA as the platform upon which to express the private sector's opinions for overseas resources development.

[56] Ministry of Commerce Industry and Energy, "Oil and Gas Self-Sufficiency Ratio Will Be 28% in 10 Years: The Third Overseas Resources Development Plan Established," *Press Release*, Aug. 7, 2007.

[57] Jin-ho Lee, "Government and Business Speed up Overseas Resources Development," *etnews*, Feb. 14, 2008. While the government official favored this centralized system, some private sector people were skeptical about the establishment of new association. One criticism was about appointing Jung, Gyu-chang, former Chief Secretary to the Minister of Ministry of Energy and Resources, as the Vice Chairman, head among standing employees. See Yang-soo Park, "EMRD Became Common Eyesore. In This Week Launched. Without New Functions, It Would Be Redundant," *Munhwa Ilbo*, Mar. 10, 2008.

[58] JOGMEC was established on Feb. 29, 2004 by integrating the functions of the former Japan National Oil Corporation (JNOC), which was in charge of securing a stable supply of oil and natural gas, and the former Metal Mining Agency of Japan (MMAJ), which was in charge of ensuring a stable supply of nonferrous metal and mineral resources and implementing mine pollution control measures.

[59] Energy & Mineral Resources Development Association of Korea, "Organization," http://www.emrd.or.kr/intro/organ.jsp. Japan Oil Gas and Metals National Corporation, "About Us," http://www.jogmec.go.jp/english/aboutus/index.html.

expanded its functions from representing the industry to new public functions including the function to delegate governmental loans to the private oil and gas developers, which had previously been the function of the KNOC.[60]

2 Strengthening of the Korea National Oil Corporation (KNOC)

The discussion of reinvigorating KNOC began against the background of a sharp increase in oil prices. The president ordered a massive government investment to strengthen KNOC. The required budget was discussed among *Chŏngwadae* and the ministries, and the issue was debated in the National Assembly.

2.1 Background: Oil Price Hike and Korea's Overseas Oil Development

Oil Price Hike

In 2004, oil prices surpassed $30 per barrel, raising concerns over energy security in Korea. When the possibility of $100-per-barrel oil gained attention from the media, President Roh Moo-Hyun, the predecessor to President Lee Myung-bak, swiftly took the initiative and established the National Energy Advisory Committee to implement energy diplomacy. In order to show performance in energy diplomacy in the short period, the expansion of the role of state-owned oil company was inevitable.

The possibility of oil prices reaching $100 per barrel surprised a group of political and business leaders participating in the Korea Leader's Forum held on July 15, 2004. The oil prices were fluctuating around $40, which was drawing more and more attention. In response, Kim Tae-yoo, a professor of resources engineering at Seoul National University and former science and technology advisor to President Roh Moo-hyun, talked about the need to prepare for oil prices as high as $100 per barrel.[61] His comment made a number of members of the National Assembly attending the forum think about energy security seriously.[62] In turn, a few major party members in the

---

[60] Ministry of Knowldge Economy, "Improving Operation System to Enhance Transparency of Loan Delegation Committee on Overseas Resources Development," *Press Release*, Sept. 8, 2008.

[61] Sohn Hyun-duk, "Prepare for Era of $100 of Oil Price," *Maeil Economic Daily*, July 18, 2004.

[62] Interview with a Korean energy expert, conducted on Aug. 11, 2010.

National Assembly published an energy security policy report on October 3, 2004, which suggested the establishment of a quasi-governmental exploration subsidiary under KNOC and the reinforcement of a governmental organization for energy administration.[63]

Unlike Japan's prime minister, Korea's president can initiate concrete actions even before a relevant law or national plan is established.[64] President Roh began to emphasize energy diplomacy during his visit to foreign countries in September 2004. In November 2004, the President established the National Energy Advisory Committee, and made and implemented concrete plans for energy security. The committee, composed of energy-related people from the government, business, and academia, was convened without legal reference, because the concern over rising oil prices was too urgent to wait for the establishment of the Basic Energy Law, where the National Energy Committee was to be outlined.[65] Though the plans discussed in the advisory committee did not have legal binding power, the president's discretionary power was enough to implement the policy, barring legal obstacles. Those concrete plans were the President's initiative, requiring the president to show prompt performance in overseas oil development. In turn, the need to make quick achievement justifies the initiative of the public sector, which can take risks once the president authorizes projects national interests. The private sector, on the other hand, tends to make decisions based on short-term profit maximization. Accompanying CEOs of SOEs, President Roh visited foreign countries and signed MOUs with foreign heads to secure oil and gas contracts. In addition, the policy focus of the Ministry of Commerce, Industry, and Energy (MOCIE) was headed in a more energy-oriented direction, and the MOCIE was restructured for that direction thusly. While there was no increase in personnel in the ministry, the MOCIE newly established bureau-level headquarters for energy resources development in 2006.[66]

---

63 Kim Shin, "New Year's Interview: Open Uri Party Member Kim Tae-Nyon," *G&E TIMES*, Jan. 11, 2006.

64 This is a unique characteristic of the Korean presidential system, so it is not easily understood by foreigners. For example, in a white paper published by Japan's Ministry of Economy and Trade in 2010, it does not mention National Energy Advisory Committee but National Energy Committee. See Ministry of Economy Trade and Industry, *Enerugi Hakusho 2010 [Energy White Paper 2010]* (Tokyo: Ministry of Economy Trade and Industry, 2011), 62–5.

65 Ministry of Knowledge Economy, "President Held National Energy Advisory Meeting," *Press Release*, Nov. 3, 2004.

66 Ministry of Commerce Industry and Energy, "The Biggest Organizational Restructuring since the Consolidation between the Ministry of Trade and Industry and the Ministry of Energy and Resources," *Press Release*, June 9, 2006, 230–43.

PEDCO in Korea's Overseas Oil Development

Korea's overseas oil development had not reached a satisfactory level by 2007, when President Lee Myung-bak was elected. The oil self-sufficiency target set in 1993 was 10% by 2001, but in spite of this delayed target year, the achieved ratio was only 4.2% by 2007.

While Korea established the PEDCO (the predecessor of KNOC) in 1979, its activities were focused more on oil stockpiling than overseas oil development. Instead, Korea's private sector had become actively involved in overseas oil development since the successes in oil exploration in the Indonesian West Madura offshore field in 1984 and Yemen's Marib onshore field in 1984.[67] However, there had rarely been big successes, with the exception of the Vietnamese 15–1 giant oil field, and the failure to extend the contract of Marib oil field of Yemen in 2005 lowered the Korean oil self-sufficiency ratio. In addition, financial difficulties after the Asian Economic Crisis decreased the private sector's investment in overseas oil development.

Prior to the Asian Economic Crisis, PEDCO had focused on strategic petroleum reserves (SPR) rather than overseas oil development. The new political situation in the early 1980s and the lack of E&P manpower obstructed PEDCO's evolution toward an E&P company. Thus, since its inception in 1979, PEDCO's priority was 60 days of SPR capacity, so the budget of PEDCO was invested much more heavily in SPR than in E&P, as illustrated in Fig. 1.[68]

In the Chun Doo-hwan Administration in the 1980s, which pursued a more private sector-oriented economy,[69] PEDCO's role was scaled down to supporting private companies with technology and governmental loans, much like the JNOC.[70] PEDCO also lost the opportunity to be a vertically integrated national oil company in 1980 due to the political decision by the 1979 coup leaders. When Gulf decided to end its partnership with the Korean government in the management of *Yukong*, Korea's state-owned

---

[67] Korea Petroleum Development Corporation, *Han'guksŏgyukaepalgongsa Sibonyŏnsa, 1979–1994 [15-Year History of Korea Petroleum Development Corporation, 1979–1994]*.

[68] "The Import of Developed Coal Will Be Focused, Energy Minister Jang Reported. SPR Will Also Be Expanded," *Maeil Economic Daily*, Feb. 5, 1979. The Korean government achieved a 60-day of oil stockpiling in 1987. See Korea Petroleum Development Corporation, *Han'guksŏgyukaepalgongsa Sibonyŏnsa, 1979–1994 [15-Year History of Korea Petroleum Development Corporation, 1979–1994]*, 345.

[69] President Chun Doo-hwan made the case for a private sector-oriented economy often in the early 1980s, such as annual presidential addresses on Jan. 12, 1981, Jan. 22, 1982, and on Jan. 18, 1983.

[70] Nam-hun Heo, "Yuchŏn'gaepal Chiwondaech'aek [Report to the President: Measures to Support Oil Development]," ed. Ministry of Energy and Resources (Kyunggido: Presidential Archive, 1982), 18.

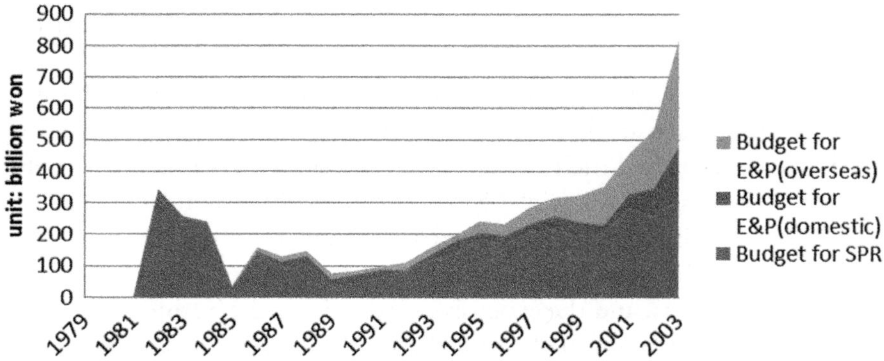

Fig. 1 PEDCO's budget allocation from 1979 until 2003 (*Source* Korea National Oil Corporation, *Han'guksŏkyukongsa Isibonyŏnsa 1979–2004 [25 Year-History of PEDCO, 1979–2004]*, [Kyunggido: Korea National Oil Corporation, 2005], 516–522)

refiner, how to deal with *Yukong* became the market issue. General Chun's office decided to privatize *Yukong*, a decision that was at odds with the MER, which wanted *Yukong* to be merged with PEDCO.[71] After some competition among *chaebol*s, especially between Samsung and Sunkyung, *Yukong* was sold to Sunkyung.[72]

One final reason PEDCO could not be active in overseas oil development was its lack of manpower. To remedy this situation, PEDCO hired Samuel B. Frazier, who had worked in oil exploration for Gulf for 29 years, as an advisor,[73]and sent its young workers to KODECO's West Madura offshore sites in the 1980s. As part of PEDCO's effort to increase its human capital, it hired engineers with Ph.D. and absorbed some former KODECO workers in West Madura.[74]

While KNOC had not been active in oil E&P in the 1980s and 1990s, it should be noted that the Asian Financial Crisis had a mixed impact on KNOC's future. On the one hand, the financial crisis introduced neo-liberalism to Korea on a full scale. This was accompanied by the IMF's pressure for privatization, which added more momentum to the domestic

---

[71] Ministry of Commerce Industry and Energy, ed. *Namkikosipŭn Iyakidŭl: Yŏktae Sanggong Tong-japuchanggwan Aesaeichip [Story That I Want to Hand Down: Essay Collection of All Successive Ministers of the Ministry of Trade and Industry and the Ministry of Energy and Resources]* (Kyunggido: Ministry of Commerce, Industry and Energy, 1999), 65–69, 103–21. Also interview with the then-Ministry of Energy and Resources officials on Oct. 14, 2011.

[72] For details, see SK corporate culture office, *SK 50nyŏn: P'aekiwa Chisŏngŭi Yŏjŏng [SK Group's 50 Year: The Journey of Aspiration and Intelligence]* (Seoul: SK corporate culture office, 2006), 176–86.

[73] "Personals," *Oil & Gas Journal*, Oct. 4, 1982, 148.

[74] Interview with Chang Kyung-chŏn, then-government official of the Ministry of Trade and Industry, conducted on Oct. 14, 2011.

effort than was normally present in Korean politics. The KNOC had no choice but to streamline its organization.

On the other hand, the government had to depend on the KNOC to achieve its target self-sufficiency ratio because many private companies had abandoned or decreased their overseas oil development activities. In the aftermath of 1997 Asian Financial Crisis, *chaebols*, Korea's main investors in overseas oil development, were forced to follow the 200 percent limit on debt/asset ratio that the government set as a measure to improve Korea's economic conditions, and the expensive oil development business became too burdensome. The KNOC had to fill the resulting vacuum, and the Korean government decided to repay all loans for KNOC's overseas oil development.[75] However, the support was provided at the level of ministry, not president, and it was not until the early twenty-first century that the KNOC was sufficiently supported because oil prices in the 1990s were too low to generate significant concern for the president.[76] Even during the Roh Moo-hyun Administration, the focus on exploration activities did not produce visible results in the oil self-sufficiency ratio in the short term.[77] That is why Korea's oil self-sufficiency ratio was still approximately 4.2% by 2007, when Lee Myung-bak was elected as Korea's President.

2.2 Discussion of the Reinvigoration of a State-Owned Oil Company

Order of the President-Elect

Energy security was such an important issue to President-elect Lee Myung-bak that he specifically mentioned the need to expand the KNOC on January 6, 2008.[78] Oil prices had just risen to $100 per barrel four days

[75] Policy Planning Committee, *Ch'amyŏjŏngbu Chŏngch'aek Pokosŏ 2–16: Haeoejawŏn Hwakpo Kyŏngjaeng [Roh Moo-Hyun Administration White Paper 2–16: The Competition to Secure Foreign Resources Supply]* (Seoul: Policy Planning Committee, 2008), 6.

[76] The MER was abolished in the 1993 government restructuring, and it was merged with Ministry of Trade and Industry (MTI) to form Ministry of Trade, Industry, and Energy (MOTIE). After the abolishment of the MER, the division for energy policy shrank substantially. The energy divisions were recovered in the early twenty-first century during the Roh Moo-hyun Administration.

[77] During Roh Moo-hyun Administration, the performance of energy diplomacy was expressed by reservoir size rather than production size, used to calculate the oil self-sufficiency ratio.

[78] Interview with a Korean government official of Ministry of Knowledge Economy, conducted on Oct. 21, 2011.

earlier, surprising the newly elected president.[79] In light of this development, lowering the cost of living by as much as 30%— one of the President-elect's ten campaign pledges—became unrealistic, as did the pledge of seven percent economic growth.[80] Also, the general public, normally indifferent to energy issues, began to realize the impact high oil prices had in terms of increasing gasoline prices.[81] Because oil was so crucial in the industrialized world, the impact of higher oil prices was stronger for low-income people than for high-income people,[82] creating negative political implications for the president. In this situation, he needed to craft a policy response to the volatile energy market.[83] The 17th Presidential Transition Committee stated energy diplomacy and a package-deal including both oil deal and construction contracts as two of 43 core agendas under the fifth target, "*Sŏngsukhan Segyekukka*" (Matured World-Class Nation).[84] In order to signify his commitment, the President-elect met N. Barzani, then Iraqi Kurdistan Premier on February 14, 2008, to show his willingness to engage in energy diplomacy.[85]

---

[79] On Jan. 2, 2008, the price for West Texas Intermediate (WTI), a benchmark for oil pricing, at New York Mercantile Exchange (NYMEX) recorded $100. It has been three years since the price of oil was $50 in Sept. 2004.

[80] See Grand National Party, *Illyukukka, Hŭimanggongdongch'ae Taehanminkuk [No.1 Nation, Community of Hope, Republic of Korea]*, ed. Grand National Party (Seoul: Bookmark, 2007), 31–2, 76–7. See also Soon-wook Park, Kim, Hŭng-soo, "Skyrocketing Oil Price Hurts Korean Economy?," *Chosun Ilbo*, Jan. 4, 2008. According to Korea's Samsung Economic Research Institute, a ten-percent rise of oil prices resulted in a 0.35 percent-decrease in economic growth.

[81] Interview with Lee Jin-myung, a reporter at *Chŏngwadae* Press Corps, conducted on Apr. 4, 2011. The gasoline price (standard one) increased from 1,410.72 won in Jan. 2007 to 1,652.25 won in Jan. 2008 and 1,906.80 won in June 2008. See Korea National Oil Corporation, "Opinet," (2011).

[82] Korea's National Assembly Budget Office, "Ichŏnp'allyŏndo Cheirhwae Ch'ukakyŏngjŏngyesananmit Kikŭmunyongkyehwaekpyŏnkyŏngan Punsŏk [the Analysis on the First Supplementary Budget and the Change of the Management of Fund in 2008]," (Seoul: Korea's National Assembly Budget Office, 2008), 87.

[83] Interview with Lee Jin-myung, a reporter at *Chŏngwadae* Press Corps, conducted on Apr. 4, 2011.

[84] The 17th Presidential Transition Committee, *Che Sipch'iltae Taet'ongnyŏngjik Insuwiwŏnhoe Paeksŏ: Sŏnggong Kŭriko Nanum [the White Paper of the 17th Presidential Transition Committee: Success and Sharing]*, 2 vols., vol. 1 (Seoul: The 17th President Transition Committee, 2008), 47.

[85] This meeting was controversial, because the Iraqi central government had warned that the contract between Kurdistan regional government and Korean companies was illegal, and had requested canceling it since 2007. There was an attempt to change President-elect Lee's mind by replacing the contract on Kurdistan oil field with the right to develop the Halfaya oil field, which KNOC used to negotiate for in 1997, but in vain, because the proposal could not be communicated to the President-elect. Later, the unsatisfactory outcome of KNOC's exploration in Kurdistan raised skepticism about President Lee's resource diplomacy in the National Assembly in 2011. See Jeong-hong Jeon, "President Lee Met Barzani to Discuss Oil Development in Kurdistan," *Maeil Economic Daily*, Feb. 13, 2008. "Committee on Knowledge Economy Reproved Insolvent Overseas Resource Development," *Yonhap News*, Sept. 26, 2011. Knowledge of the failed attempt is from an interview with a Korean energy expert, conducted on Oct. 8, 2011.

This new policy direction continued to materialize in the new administration. At a meeting with the MKE on March 17, 2008, President Lee gave an order to increase the size of KNOC five-fold. In order for the MKE to improve the oil and gas self-sufficiency ratio from 4.2% as of 2007 to 18.1% as of 2012, President Lee asked for a detailed plan of how this could be accomplished.[86] All compatible measures, such as governmental institutions, finance, and the education system, were aligned to achieve this target. Official development assistance (ODA) was also included as a potential measure to support overseas oil development.[87] The MKE also stated that a mid- and long-term development plan for state-owned resource development companies, made in the previous administration, would be implemented.[88]

Drafting

As oil prices continued to rise rapidly, the new administration began to seriously consider expanding the KNOC in April 2008. Using a study of this potential expansion from the previous administration, the MKE prepared a few options for implementing the President's order. Considering the difficulties of amalgamating the KNOC and KOGAS, one option was to establish a holding company, under which the KNOC and the KOGAS would exist as subsidiaries. The second option was to expand KNOC itself by capital injection and M&A.[89] The former option was modeled after "Temasek Holdings,"

[86] E-Daily, "President Asked Three Tasks," *Chosun Ilbo*, Mar. 17, 2008. The self-sufficiency ratio is the ratio between the oil developed by Korean companies and the imported oil.

[87] Though MKE's presentation provoked opposition from Korean NGO's, the reality has been different. ODA (Official Development Assistance) is not under the jurisdiction of the MKE, which is just one ministry to express its opinion. Regarding Korea's ODA, the Ministry of Foreign Affairs and Trade (MOFAT) deals with grants, and the Ministry of Strategy and Finance (MOSF) deals with concessional loans. According to a government official's account, the MOFAT prefers humanitarian assistance, such as women and children's healthcare, and the MOSF considers many categories. Therefore, oil and gas development, which is comparatively less competent than IT technology or green growth technology in Korea, does not easily attract attention. The Top-10 receivers, which account for 43% of Korea's ODA in 2008, show that Korea focuses on broad strategic partnership with neighboring countries rather than resources. (The Top-10 countries are Vietnam, Cambodia, China, Philippines, Indonesia, Mongolia, Sri Lanka, Jordan, Angola, and the Dominican Republic). In addition, Korea's membership in the OECD DAC (Development Assistance Committee) since January 2010 makes it difficult to tie national interests in with ODA. Interview with a Korean government official, conducted on June 16, 2011.

[88] Ministry of Knowldge Economy, "Ich'ŏnpallyŏndo Chisikkyŏngjepu Ŏmmupokocharyo: Chisik·Hyŏksinjutohyŏng Sanŏpkangguk Kŏnsŏl [the Ministry of Knowledge Economy Annual Work-Plan of 2008: The Construction of Innovation-Leading Industrial Power]," ed. Ministry of Knowldge Economy (Kyunggido: Ministry of Knowldge Economy, 2008), 26–9.

[89] Hak-ryul Lee, "Government Studies Various Options for Expansion of KNOC," *Money Today*, May 2, 2008.

a Singaporean state-owned investment holding company,[90] and was considered attractive because it did not require more capital to expand and make good use of the comparative advantages of KNOC's E&P knowhow and KOGAS's buying power. This would also evade the problem with the private shareholders of KOGAS and the possibility of clashes with labor unions in both companies. The latter option required a large amount of money,[91] but it provided the KNOC with an opportunity to increase its production capacity from 56 thousand b/d as of 2007 to 300 thousand b/d by 2010.

One interesting point to note here is that Japan's overseas oil development model was no longer a benchmark for Korea's policymaking. In the National Assembly delegation and the research reports, ENI and TOTAL, European major oil companies that began as state-owned oil companies, were regarded as models to follow. Korea would not pursue Japan's model as it was seen as a failure.[92] The only thing the Korean government watched carefully in Japan was the expansion of INPEX in the early twenty-first century, because this expansion was regarded as one quick solution to catch up with more advanced oil companies.[93]

Consulting with Veto Players

The MKE discussed its draft with *Chŏngwadae* and the Ministry of Strategy and Finance (MOSF), both effective veto players for this issue.[94] In *Chŏngwadae*, the issue was dealt with by the Office of Senior Secretary for Policy Planning, which also addressed the reform and privatization of state-owned companies. Originally, the MKE proposed to establish a holding company for KNOC and KOGAS, but *Chŏngwadae* did not want to contradict its political agenda of privatization and restraining the size of the public

[90] Sae-hyung Kim, "There is No Principle That Bureaucrat Has to Retire" *E-Daily*, Apr. 22, 2008.

[91] One idea was to utilize 620 billion won from the oil price stabilization fund. See Lee, "Government Studies Various Options for Expansion of KNOC."

[92] Comment by MKE Vice-Minister Lee Je-hoon in the Korea National Assembly, Standing Committee on Knowledge Economy, *6th Committee Meeting*, Interpellation, Sept. 17, 2008; Korea National Oil Corporation, *Han'gukhyŏng Sŏgyumaeichŏ Yuksŏngŭl Wihan Han'guksŏgyukongsa Hyŏksinpangan [the Innovation Plan of Korea National Oil Corporation to Promote Korean-Style Oil Major]* (Kyunggido: Korea National Oil Corporation, 2006), 51–3; MKE internal material written in 2008.

In China, state-owned oil companies, such as CNPC, SINOPEC and CNOOC, dominate overseas oil development. However, due to China's different conditions and environment, the study group regarded China's case purely for reference.

[93] A Korean consultant visited INPEX to learn about the expansion of INPEX when KNOC studied its long-term strategy. Interview with a Japanese industry person in Japan, conducted on July 30, 2011.

[94] "Lee Youn-Ho, "Part of Utility Will Be Privatized," *Yonhap News*, May 9, 2008.

sector. Rather than creating another state-owned holding company in addition to existing two state-owned companies, the MKE opted to expand the KNOC by providing money for M&A.[95]

Here, there was a gap between the MKE and the MOSF on how to supply money to the KNOC. The MOSF agreed with the President's orders, but the MKE proposed a different idea. This issue was discussed in front of President Lee at the budget allocation meeting on April 25, 2008. While the MKE wanted to realize KNOC's expansion by 2010 with legally guaranteed financing, the MOSF's idea was to delay the target year to 2012 and use fund under the control of the MKE.[96] The final compromise was closest to the MOSF's preference.

The procedure for policymaking reveals which institution is effective veto player and which is not. Due to the sensitivity of this issue, the plan was briefly explained to the ruling party at the Party-Government Coordination Meeting, and thus there was no room for the ruling party to intervene. The opposition parties did not hear the plan officially until it was released to the mass media.[97]

One more thing that has to be mentioned is that the private sector did not have an opportunity to share their opinions during the discussion on KNOC's expansion plan. The MKE consulted with the KNOC, but other private companies or industry associations, such as the EMRD, were excluded because of time constraints. Even advisory committees, such as the National Energy Committee, were not included in this planning. According to an MKE official, this plan was made in a secretive and speedy way in order to mitigate the side-effects of this massive financial support plan on the market.[98]

Plan Announcement

On June 13, 2008, the MKE announced the plan to transform the KNOC into a Korean oil major, which included substantial government funding for the company. In fact, this plan followed the basic framework of an extensive study conducted during the previous administration. There were some differences between the two blueprints, including the target year in which 300,000 b/d for KNOC's daily production would be reached. In the previous plan, it

[95] Interview with a government official, conducted on July 7, 2011.

[96] Interview with a Korean government official, conducted on Sept. 15, 2011.

[97] Interview with a Korea's government official, conducted on July 7, 2011.

[98] Interview with an MKE official who is versed on the planning, conducted on Jan. 14, 2012.

was 2013, but the newly-set target year was 2012. In addition, the focus shifted from the exploration of new fields to buying production fields as well as M&A of foreign E&P companies.[99] This new plan had greater potential to quickly achieve its objectives for the oil and gas self-sufficiency ratio by the target year and in increasing KNOC's capacity in terms of total oil production and the number of engineers. For this rapid expansion of the KNOC, the government would inject a total of 4.1 trillion won (about US$4billion) of capital into the KNOC by 2012, and the KNOC would borrow 15 trillion won (approximately US$15 billion) from financial markets.[100] The government's fund came not from its regular budget, but from a special energy fund, so the MOSF did not have to oppose the MKE's plan.

The KNOC in President Lee's SOE Privatization Plan

During President Lee's Administration, there were two competing agendas: achieving energy security by strengthening SOEs and the desire to privatize SOEs. When President Lee was elected, state-owned companies were expected to be privatized based on one of his campaign pledges. According to this pledge, the privatization was scheduled to follow the government restructuring. If a certain state-owned company competes with the private sector in the market and/or is diverged from its original purpose, the SOE was to be targeted for privatization.[101] As described in *The Economist*, this privatization was a hallmark of Lee's Administration.[102] Though his Presidential Transition Committee avoided finalizing the privatization plan, which had already been a controversial issue in Korean society, the promise of the President-elect had made was real to the workers of state-owned companies.

The President's attitude simultaneously attracted the support of conservatives and warded off laborers. The New Right Confederation, which had "united for change of regime," announced its support for presidential candidate Lee Myung-bak one month before the presidential election, believing

---

[99] Korea National Oil Corporation, *Han'gukhyŏng Sŏgyumaeichŏ Yuksŏngŭl Wihan Han'guksŏgyukongsa Hyŏksinpangan [the Innovation Plan of Korea National Oil Corporation to Promote Korean-Style Oil Major]*, 183–84.

[100] Ministry of Commerce Industry and Energy, "SK Energy Will Launch Oil Production at Brazil Bmc-8 Field from the End of July," *Press Release*, July 30, 2007.

[101] Grand National Party, *Illyukukka, Hŭimanggongdongch'ae Taehanminkuk [No.1 Nation, Community of Hope, Republic of Korea]*, 216–17.

[102] "The Bulldozer: A President-Elect Starts to Roll," *The Economist*, Jan. 17, 2008. "Privatization in Korea: Against the Tide," *The Economist*, Oct. 2, 2008.

him to be the right person to prevent excessive tax burdens and implement privatization of state-owned companies.[103] On the other hand, the Korean Confederation of Trade Unions (KCTU) established an organization to obstruct the privatization of the public sector,[104] and the union of an SOE was known to support the publication of a book that reinforced the legitimacy of the SOE.[105]

However, the KNOC had been exempted from privatization from the beginning due to the practical need for resource diplomacy. This became clearer when the privatization program was finalized, as it included special considerations to exempt the energy sector.

While some believe that the KNOC was not privatized as planned due to the loss of political capital after the 2008 US beef protest in Korea,[106] the energy market situation in 2008 is more responsible. Mad Cow protests, which began on May 2, 2008 and had continued for more than two months, provided a good opportunity for opponents of the Administration, like the KCTU, to speak out against privatization.[107] Though the overall reform effort for SOEs continued to include the energy sector, the level of reform for state-owned energy companies slackened in the finalized plan. In June 2008, the public's fear of utility price hikes increased, potentially adding to popular dissatisfaction with the administration. At this time, the government was already in a difficult place due to widespread demonstrations against the decision to resume US beef imports following the Mad Cow Disease scare of the previous few years. Therefore, the government retreated from its hardline stance by excluding electricity, gas, tap water, and medical insurance from privatization.[108] In the first plan of advancement of state-owned companies in August 2008, the KNOC was categorized as a functional modification,

[103] "New Right Confederation Announced the Support for Lee Myung-Bak," *Yonhap News*, Nov. 28, 2007.

[104] Korea Confederations of Trade Unions, "The Launching of Struggle Headquater to Stop Marketization and Privatization, and to Strengthen Social Publicness," *Press Release*, Mar. 10, 2008.

[105] One example is Kim Sang-gon and Kim Yun-Ja et al., *Kukka Aenŏji Chŏngch'aekkwa Hankugŭi Ch'ŏnyŏn'gasŭsanŭp Yŏn'gu [Research on National Energy Policy and Korea's Natural Gas Industry]* (Seoul: Nohkiyŏn, 2008).

[106] "Examine Reformation of State-Owned Company: Don't Make 'Bright Beginning, Dull Finish', Fix It," *E-Daily*, July 16, 2010.

[107] Korea Confederations of Trade Unions, "KCTU's Demonstration to Stop Marketization and Privatization," *Press Release*, Mar. 24, 2008.

[108] Mad Cow Mass Demonstrations took place due to the agreement between the US and Korean government on the import of US beef. However, unlike a prevalent idea that the demonstration was a leftist movement, Cho Ki-suk's analysis, which is based on surveys, shows that the demonstration was caused by people's dissatisfaction with the undemocratic procedures of decision making. See Kisuk Cho, "The Ideological Orientation of 2008 Candlelight Vigil Participants: Anti-American, Pro-North Korean Left or Anti-Neoliberalism?," *Korean Political Science Review* 43, no. 3 (2009): 142–43.

which is to restructure the company, in the first privatization plan in August 2008.[109] Furthermore, the government hoped that increased investment and financing from capital markets would strengthen its capacity for overseas oil development and streamline non-core functions, such as the strategic petroleum reserve (SPR).[110]

The most important reason for KNOC's exclusion from privatization was to support the president's resource diplomacy and his goals for energy security, the latter of which can be quantitatively estimated by Korea's self-sufficiency ratio. While this agenda could be seen as a repetition of the previous president's agenda,[111] though many items of that agenda had been abandoned, the new administration continued to emphasize energy issues during and after the Presidential Transition Committee. For the new president, who was responsible for every important issue in Korea, it was in his best interest to prepare for oil price spikes near his inauguration.[112] The energy security policies in the previous administration were thus selectively aligned to the policy setting in the new administration, an example of which is the President-elect's involvement in oil development in Iraqi Kurdistan.[113] His meeting with the Kurdistan Premier prior to his inauguration was the first concrete energy diplomacy action, and the resulting contract was implemented mainly by the KNOC. While the taskforce team planned the project with a goal to attract investors from the Presidential Transition Committee, the KNOC followed up by signing MOUs in February 2008, and then

---

[109] The categories were privatization, promoting more competition, merging, modifying functional structures, and streamlining.

[110] Ministry of Strategy and Finance, "Advancement of State-Owned Company Launched Including Consolidation of Housing Corporation and Land Corproation," *Press Release*, Aug. 11, 2008.

[111] President Roh Moo-hyun, Lee Myung-bak's predecessor, launched energy diplomacy policies for the first time as the presidential agenda. His assertive approach can be attributed not only to rising oil prices in the early twenty-first century, but also to other factors. For example, the white paper of the Policy Planning Committee states that President Roh's resource diplomacy was stimulated by Chinese President Hu's active role in China's quest for overseas oil. A government official involved in resource diplomacy alluded to political motivations as well. President Roh needed momentum to change the political geography and boost the morale of *Chŏngwadae* because he just escaped the opposition party's attempt to impeach him, thanks to the rejection by the Constitutional Court of Korea of May 14, 2004. Policy Planning Committee, *Ch'amyŏjŏngbu Chŏngch'aek Pokosŏ 2–16: Haeoejawŏn Hwakpo Kyŏngjaeng [Roh Moo-Hyun Administration White Paper 2–16: The Competition to Secure Foreign Resources Supply]*, 12; Interview with Korean government official involved in the early phase of President Roh's resource diplomacy, conducted on Mar. 31, 2011.

[112] Interview with Lee Jin-myung, a reporter at *Chŏngwadae* Press Corps, conducted on Apr. 4, 2011.

[113] During the Roh Moo-hyun Administration, the KNOC made a contract for the Bazian oil field in the Iraqi Kurdistan area in 2007. Korea National Oil Corporation, "KNOC Made PSC Contract for Exploration of Iraqi Bazian Field," *Press Release*, Nov. 12, 2007; Policy Planning Committee, *Ch'amyŏjŏngbu Chŏngch'aek Pokosŏ 2–16: Haeoejawŏn Hwakpo Kyŏngjaeng [Roh Moo-Hyun Administration White Paper 2–16: The Competition to Secure Foreign Resources Supply]*.

signing a contract in June and September of 2008 with the Iraqi Kurdistan Regional Government.[114]

Passage of a Supplementary Budget in the National Assembly

The National Assembly finally agreed to appropriate a supplementary budget in September 2008. While there was some debate in the Standing Committee on Knowledge Economy and Special Committee on Budget Settlement, attention was drawn more to the usage of the budget for political distributive purposes than the appropriate role of the public sector. The effect of the opposition parties' argument was limited in the National Assembly, where the ruling party had gained the majority seats in the recent general election in May 2008.

The opposition party members argued that financial support for peoples' livelihoods, especially for farmers and independent enterprises, should be considered before the KNOC. From their perspective, support for the KNOC could come from the regular budget, not this supplementary one.[115] Funding for the KNOC was also an ideal target for attacking such a sizeable item in the supplementary budget, as the draft would have given the opposition party an opportunity to replace it with their preferred items. In addition, though the National Assembly Budget Office (NABO) agreed that the supplementary budget fit the goals of the National Energy Plan,[116] the analysis report on the supplementary budget for the deliberation in the committee raised suspicions about why the supplementary budget, which is supposed to be allocated for urgent needs, should be distributed for this moderate issue. While the sharp increase in oil prices did fit "the critical

---

[114] Keun-Wook Paik, a senior research fellow at the Oxford Institute for Energy studies (OIES) at the University of Oxford, stated that this was President Lee's first major blunder. The decision completely ignored the importance of energy diplomacy based on a central-to-central relationship and consequently opened the door to pay a very dear price for E&P projects in Iraqi Kurdistan area in the coming years. Interview with Keun-wook Paik, conducted on Oct. 7, 2012. The 17th Presidential Transition Committee, *Che Sipch'iltae Taet'ongnyŏngjik Insuwiwŏnhoe Paeksŏ: Sŏnggong Kŭriko Nanum [the White Paper of the 17th Presidential Transition Committee: Success and Sharing]*, 2 vols., vol. 2 (Seoul: The 17th President Transition Committee, 2008), 102–03; Korea National Oil Corporation, "Korean Consortium Agreed on Package Deal-Style Resource Development with Iraqi Kurdistan Regional Government," *Press Release*, Feb. 14, 2008; Korea National Oil Corporation, "KNOC Signed Contracts with Iraqi Kurdistan Regional Government for 8 Oil Fields," *Press Release*, June 25, 2008; Korea National Oil Corporation, "KNOC Agreed on Implemetation Plan of Oil-SOC Deal with Iraqi Kurdistan Regional Government," *Press Release*, Sept. 25, 2008.

[115] A comment by Choi In-ki in Korea National Assembly, Special Committee on Budget Settlement, *The First Supplementary Budget Draft of 2008*, Interpellation, Sept. 9, 2008.

[116] Korea's National Assembly Budget Office, "Ichŏnp'allyŏndo Cheirhwae Ch'ukakyŏngjŏngyesananmit Kikŭmunyongkyehwaekpyŏnkyŏngan Punsŏk [the Analysis on the First Supplementary Budget and the Change of the Management of Fund in 2008]," 178.

change of the situation," the report doubted the need to urgently add the supplementary budget to the already-allocated regular budget for 2008.[117] In addition, there was the issue of whether KNOC's focus on M&A was an appropriate task for state-owned oil enterprises, which are supposed to take on riskier ventures, such as exploration projects.[118]

Finally, a total 865 billion won, substantially less than the original budget demand of 1.214 trillion won, was agreed to be allocated to KNOC's expansion in 2008.[119] While the ruling party's attempt to force the supplementary budget through failed on September 11, 2008, the ruling party and opposition parties passed the budget six days later on September 17th.[120]

3 Active in M&A and Increasing Self-Sufficiency Ratio

Since early 2009, the KNOC has purchased foreign oil companies and production assets to expand and achieve the targeted oil self-sufficiency ratio, which has been used as an indicator of energy security in Korea. The KNOC purchased a half share of Petrotech, a Peruvian oil company, along with its production assets in February 2009—the first acquisition of a foreign oil company by a Korean company.[121] Soon after in October 2009, the KNOC merged with Harvest Energy, a Canadian oil company, which was praised for allowing the KNOC to expand into the refinery business and enhancing

117 Tae-soo Kwon, "Cheirhwae Ch'ukakyŏngjŏngyesananmit Chŏllyŏksanŏpkipankikŭm Unyongkye-hwaekpyŏnkyŏngan [the Inspection Report on the First Supplementary Budget and the Changed Proposal for the Management of Utility Industry Foundation Fund]," ed. The Committee on Knowledge Economy (Seoul: Korea's National Assembly, 2008), 7–9.

118 Comment by Lee Kang-rae at Korea National Assembly, Standing Committee on Knowledge Economy, *5th Committee Meeting*, Interpellation, Sept. 10, 2008. And comments by Woo Che-Ch'ang and Kim Ki-hyun in the *6th Committee Meeting*.

119 Korea's National Assembly, "Ich'ŏnp'allyŏndo Cheirhoe Ch'ukakyŏngjŏngyesan [the First Supplementary Budget in 2008]," (Seoul: Korea's National Assembly, 2008), 354.

120 Yonhap News, "Parties Agreed on Supplementary Budget Rapidly," *Chosun Ilbo*, Sept. 17, 2008. This quick reversal of the situation was attributed to the opposition party's concern about "politics of numbers." Hong Joon-pyo, the ruling party's representative in the National Assembly, was criticized for his failure to force the budget proposal through, and he was in danger of being ousted. For the opposition parties, he was better than other possible successors who might prefer utilizing the numerical superiority of the ruling party in the National Assembly. Therefore, the opposition parties weakened the political attack on the ruling party by agreeing to pass the budget proposal after a short negotiation. See Yonhap News, "Democratic Party Saved Hong Joon-Pyo?," *Chosun Ilbo*, Sept. 17, 2008.

121 Angela Moon, "Korea National Oil Acquires Half of Peru Petro-Tech," *Reuters*, Feb. 6 2009.

Korea's self-sufficiency from 6.3 to 8.1%.[122] The KNOC also took over Sumbe, a Kazakhstani oil company, in December 2009 and overtook DANA Petroleum, a British oil company, in 2010 through a hostile M&A, which drew attention from the local media.[123]

In order to secure positive results from the merger, the KNOC preferred closed negotiations or M&A rather than those that were open-bidding. This approach is likely related to the failure in winning the bid for Addax Petroleum, a Swiss oil company, to the aggressive betting of Sinopec, one of three Chinese state-owned major oil companies, in June 2009. This loss came as unpleasant news to the Korean government, which had been focusing on improving its oil self-sufficiency ratio through M&A. The incident also fomented Sino-phobia in KNOC and in the Korean resource diplomacy sector.[124] Therefore, the Korean government decided to target companies that China would not be interested or companies averse to Chinese M&A.[125] Following this strategy, the KNOC negotiated with Canada's Harvest Energy exclusively and bought the company with US$3.9 billion.[126]

KNOC's aggressive purchase of oil assets and companies was paralleled by an inevitable increase in debt and a weakening of exploration activities. KNOC's debt increased sharply by 18 trillion won (approximately $17 billion) from 2008 until June 2012 due to massive loans to subsidize its M&A.[127] While such massive debt may have been seen as inappropriate for a state-owned company in a different political context, it was excused as a necessary evil in the pursuit of energy security. The situation was similar

122 Cho Mee-young and Michael Erman, "South Korea Wins Biggest Energy Deal with Canada's Harvest," *Reuters*, Oct. 22 2009; Min-Hui Jeon, "300 Thoudand B/D of Daily Production by 2012, 'World Top-50s'," *Korea Energy News*, Jan. 11, 2010; Ministry of Knowledge Economy, "KNOC Succeeded in Overtaking Harvest," *Press Release*, Oct. 22, 2009; Ministry of Strategy and Finance, "2009 Evaluation Report for Management Performance of State-Owned Companies (SOC)" (Kyunggido: Ministry of Strategy and Finance, 2010), 21.

On the other hand, whether the KNOC paid too much became a controversy later. See Sang-jin Han, "Issue: Overcharged for M&A? Low Quality Mine? With Low Prices, Many Assets Would Not Be Profitable," *Shin Dong-A*, Aug. 2011, 82–91.

123 Terry Macalister, "DANA Board Relents in KNOC Battle," *Guardian*, Sept. 24, 2010.

124 See Cha-ryong Ku, "China's Hunting for Company is Our Concern," *Dong-A Ilbo*, July 6, 2009; Yong-bum Park, "In Order to Avoid the Second Failure of 'Addax'," *Maeil Economic Daily*, July 15, 2009; Jin-sok Lee, "Black Hole China, 'Adverse Effect on Resource Supply,' HRI Report," *Korea Economic Daily*, July 22, 2009.

125 Si-hoon Ryu, "Approach to North American Company That Has Negative Feeling Towards China Works," *Korea Economic Daily*, Oct. 22, 2009.

126 One interesting point was that KNOC began to make use of investment banks to precede negotiation and plan financing schemes. For example, Bank of America Merrill Lynch has joined KNOC's M&A activities for the bidding of Addax Petroleum as one of financing institution and for the M&A of Harvest Energy and DANA Petroleum as the superintendent financial advisor.

127 Se-jin Chung, "40 Trillion Won of Debt, Repressed Charge Hike of State-Owned Energy Companies for 3 Years," *Dong-A Ilbo*, Aug. 29, 2011.

to KNOC's unsatisfactory exploration performance. In December 2010, the Vice Minister of the MKE mentioned that while M&A has been successful, the company's performance in terms of exploration had been unsatisfactory.[128] During interpellation in the Standing Committee on Knowledge Economy in 2011 and 2012, the National Assembly members, including the ruling party members, also accused the KNOC of accruing enormous debt without satisfactory exploration results.[129]

4 One Argument: Level of Economic Development?

While Korea's economy was still behind Japan in terms of GDP in 2008, when the Korean government expanded the size of its state-owned oil companies, Korea was by any account an industrialized country.[130] Considering Korea's per capita GDP was $15,446 in terms of constant value as of 2000, Gershenkron's thesis that the level of economic development determines the method of government intervention does not fit this case.[131] For an industrialized country, the expansion of state-owned oil companies is not a normal policy choice. The question persists, then, why Korea expanded activity within a state-owned company when it was already economically developed. For the most part, political feasibility was considered in determining how to expand Korea's oil development industry. However, this logic was disguised by the deterioration of the private sector's capacity for overseas oil development following the Asian Economic Crisis and some misunderstanding of facts on the private sectors' capacity and foreign companies.[132]

4.1 Overseas Oil Development Before 2004

By the end of 2003, the performance of Korean companies had not been satisfactory in overseas oil development, thus oil developers did not have enough

---

[128] Young-taek Kim, "2010 Overseas Resource Development Symposium," *Tomato News*, Dec. 2, 2010.

[129] Mahn-gyu Cho, "KNOC's Success in Exploration is Only One since 2008," *Energy & Environment News*, Sept. 26, 2011; "Blind Investment Produced Increase of Debt by 18 Trillion Won," *Newsis*, Oct. 12 2012.

[130] World Bank data show that both Korea's economic size and the GDP per capita are 15th in the world, as of 2008.

[131] Source: World Bank World Development Index (WDI).

[132] As examined later, once the policy direction was fixed, the logic justifying KNOC's stronger role in 2004 and after was provided. Sometimes, the materials do not reflect the reality accurately.

Table 2 Korean Companies' overseas oil development as of the end of 2003

# of on-going project			# of finished project	Total
Exploration	Development	Production		
22	9	24	67	122

Source Pyŏng-do Han, Kim, Tae-nyŏn, Lee, Kwang-jae and Seo, Kab-won, *Enŏchi Munje, Saeroun Daeannŭl Ch'ajasŏ [Energy Problem, Pursuing New Alternatives]*, Policy Material for the Regular Session in 2004 (Seoul: Legislature Research Center, 2004), 17

capacity to compete with foreign oil companies. After two oil development successes in Indonesia's West Madura offshore field and North Yemen's Marib field in the early 1980s,[133] the total number of production fields by the end of 2003 was only 24, among which was KNOC's 2001 discovery of a giant oil reserve of the Southeast coast of Vietnam Table 2.

Most Korean companies could not become competitive enough to catch up with Western oil majors or other big national oil companies (NOCs), and this was also true for large general trading companies (GTCs). Among these enterprises, Hyundai Corp., Samsung Corp., and Daewoo International had been active in resource development and import, but only had minor achievements in this field.[134] Hyundai Corp. in particular was more successful than other GTCs in overseas oil development by finding a lucrative oil field in Marib, Yemen and beginning production there in 1985.[135] Oil development activity was not an original purpose of Korean GTCs. Instead, it was a consequence of the new oil supply system of Korea in the late 1970s and Korean GTCs' relative advantage in information gathering through a global network.[136] The second oil shock in the late 1970s drew policymakers'

[133] For details on these stories, see Ho Lee, *Nanŭn Asŭpaltŭ Kkallin Kirŭn Kachi Anŭnda [I Don't Go on the Asphalted Road.]* (Seoul: Ollim, 2007); Ray Fairchild, "Yemen Oil Hunt–A Gift from the Gods," in Allen G. Hatley, *The Oil Finders: A Collection of Stories About Exploration* (Utopia, Tex.: Centex Press, 1995).

These two symbolic successful projects took different paths. KODECO renewed its contract on West Madura Offshore oil field with 20% of equity participation in May 2011. On the other hand, Yemen Exploration & Production Co. the joint-venture of Hunt, Exxon, SK Energy, Hyundai Corp., and Samhwan Construction, could not extend the contract in Nov. 2005. See PERTAMINA, "PERTAMINA Operator Blok WMU," *Press Release*, May 9, 2011; "Contract Dispute Halts Marib Operations in Yemen," *Oil & Gas Journal*, Nov. 28, 2005.

[134] Cho, *Han'gukŭi Chonghapmuyŏksangsa: Ponjilgwa Chŏllyak [Korea's General Trading Company: Nature and Strategy]* 209.

[135] Shareholders were Hunt (38.505%), Exxon Mobil (36.995%), SK Energy (15.925%), Samhwan Construction (3.675%), Hyundai Corp. (2.45%), and KNOC (2.45%).

[136] Cho, *Han'gukŭi Chonghapmuyŏksangsa: Ponjilgwa Chŏllyak [Korea's General Trading Company: Nature and Strategy]*, 68.

attention to Korea's insecure oil supply, so the government expanded the eligibility of oil imports from PEDCO and refiners to Korean GTC's in 1980.[137] Young elite employees, half of which had majored in engineering and the other half in social science, were selected to implement oil projects secretly in Korean trading houses.[138] Interestingly, there is a possibility that the desire of businesses in the 1970s and 80s contributed to their active involvement in overseas oil development. Historically, the merchant class has been underestimated due to Korea's Confucian culture, where the bureaucrat stands at the top of the social hierarchy. Thus, oil development, which was seen in Korea as a valuable public good, offered an incentive to gain further respect in society.[139]

Among Korean refiners, those that had not affiliated with foreign oil companies have been involved in overseas oil development. In particular, the SK Group had demonstrated its eagerness in oil ventures to streamline its value chain from downstream to upstream. This effort came after declaring its synergizing goal of "*Sŏmyuaesŏ Sŏkyukkachi*" (from textile to petroleum) since its merger with *Yukong*, the former state-owned refiner, in 1980. However, other refiners had been inactive in oil ventures, joining a total of only four projects by 2003.[140] Refiners could get a stable oil supply from their own oil major partners. In addition, refiners, which were mainly a member company of the *chaebol*, had been assigned only to downstream work, while oil development had been a task for GTC's, another member company in the *chaebol* Table 3.

Furthermore, Korean refiners' profits had been stable, and consequently, they had less interest in participating in overseas oil development than in expanding refinery capacity.[141] This tendency was changed due to rising oil prices in the early twenty-first century. The upstream activity by GS-Caltex in Cambodia in 2003 is one example, when GS-Caltex pursued its own oil development for its refinery after its separation from the LG Group in 2002.[142]

137 "Eligibility of Oil Import License Expanded to the Private Companies," *Maeil Economic Daily*, May 29, 1980.

138 Sŭng-gu Lee, "The Oil Men, the Guard of Petroleum War Frontier," *Kyung-hyang Shinmun*, Feb. 23, 1981.

139 Cho, *Han'gukŭi Chonghapmuyŏksangsa: Chaetowa Kwalli [Korea's General Trading Company: Institution and Management]*, 45.

140 Based on the material by Energy & Mineral Resources Development Association of Korea (EMRD).

141 It was not until June 1993 that the refinery market was open to new-comers.

142 Interview with a Korean industry insider, conducted on Apr. 14, 2011.

Table 3 Division of oil business in *chaebol* before the Asian Economic Crisis

	Downstream	Upstream
SK	SK Corp[143]	SK Corp
Hyundai	Oil bank (partnership with Getty Oil)[144]	Hyundai Corp
LG	LG-Caltex (partnership with Caltex since 1966)	LG international
Hanwha	Kyung-In Energy (partnership with Union oil since 1968)[145]	Hanwha Corp
Samsung	n/a[146]	Samsung Corp
Ssangyong	Ssang-Yong refinery (partnership with Saudi-Aramco since 1991)[147]	n/a

Sources Homepage of Korea Petroleum Association (http://www.petroleum.or.kr/) and Energy & Mineral Resources Development Association of Korea (http://www.emrd.or.kr/intro/intro.jsp)

In addition to state-owned major oil companies and *chaebols*, there have also been independent oil developers, such as KODECO Energy. The unique success story of KODECO is similar to Japan's Arabian Oil. Choi Gye-wol, CEO of KODECO, who had established a network with power elites in Indonesia, signed a production sharing agreement (PSA) with PERTAMINA, the Indonesian state-owned oil company for the West Madura offshore field in May 1981.[148] This venture is recorded as Korea's first oil discovery in a foreign country and the first operatorship by a Korean company, a rare occurrence in Korea's E&P history. Also, the West Madura project was utilized for on-the-job training for Korean engineers who had much knowledge, but

[143] *Yukong* (the former entity of SK Energy) was established in partnership with Gulf Oil in 1962 as the first refiner of Korea. The share of Gulf had been 50%, but Gulf sold it to the Korean government in 1980. Yukong was sold to Sunkyung (the former entity of SK Group) without foreign company's partnership.

[144] Originally, Kŭktong Petroleum (the former entity of Oil Bank) had the partnership with Royal Dutch Shell from 1968 until 1978. In 1984, Getty Oil Company invested in Kŭktong Petroleum as a partner.

[145] Kyung-In Energy (or Hanwha energy) was sold to Hyundai Oil Bank in Sept. 1999 and resold to SK Corp. in 2005.

[146] Samsung Group has not owned a refiner, but a petrochemical company called Samsung Petrochemical since 1974.

[147] Ssang-yong Refinery was segmented from Ssang-yong Group in Dec. 1999, and the name of the corporation was changed to S-Oil, whose "S" means Saudi Arabia.

[148] Mr. Choi's biography and industry person's account shows that the connection began with his contribution to Indonesia's unification with Irian Jaya (West Papua). See Lee, *Nanŭn Asŭpaltŭ Kkallin Kirŭn Kachi Anŭnda [I Don't Go on the Asphalted Road.]*, 87–105. Interview with a Korean industry person, conducted on Apr. 14, 2011.

little field experience.[149] On the other hand, this project shows the still-emergent state of Korea's overseas oil development at that time. For this finding, American petroleum engineers were employed as the main workers, while Korean staff was sent from the Korea Resources Corporation, the state-owned mining company. Offshore oil rigs were also borrowed from the Japan Drilling Corporation.[150]

This first *Taegeuk* oil project also implicitly shows how political interests in the presidential system can intervene in oil development projects.[151] Just as the Japanese-owned AOC's oil development in the Neutral Zone between Kuwait and Saudi Arabia was a "star of hope" in postwar Japan,[152] this first Korean oil venture attracted much attention from the political world, especially from President Chun Doo-hwan for his propaganda of "*Sae Sidae*" (New Era) in 1980.[153] Considering PEDCO's low capacity in E&P at that time, General Chun had to depend on an entrepreneur capable in managing an overseas project. Mr. Choi's account states that even before General Chun's presidential inauguration in September 1980, General Chun bypassed the Korean government and asked Mr. Choi to launch an oil development project in Indonesia. The Korean government offered a special loan called a success repayable loan for the first time in 1981. PEDCO played a role not as a partner but as a supervisor of KODECO's venture, because PEDCO was assigned to the role of managing this success repayable loan, which had already been introduced in Japan since the establishment of the JPDC to support the private sector's risky oil E&P business.[154]

149 This incubating function was regarded significantly by the government. See Heo, "Yuchŏn'gaepal Chiwondaech'aek [Report to the President: Measures to Support Oil Development]," 6, 11. See also Lee, *Nanŭn Asŭpaltŭ Kkallin Kirŭn Kachi Anŭnda [I Don't Go on the Asphalted Road.]*, 198–99.

150 Interview with a Korean industry person, conducted on Apr. 14, 2011.

151 *Taegeuk* oil means the oil developed by Korean companies in a foreign country. *Taegeuk* originally means the center of Korean national flag, and the author invents the term to compare with *Hinomaru* oil, the oil developed by Japanese companies in a foreign country.

152 Interview with a Japanese industry person on June 23, 2011.

153 Lee, *Nanŭn Asŭpaltŭ Kkallin Kirŭn Kachi Anŭnda [I Don't Go on the Asphalted Road.]*, 194. Chun Doo-hwan gained power through coup on Dec. 12, 1979, and became the Korean President on Sept. 1, 1980. It seems that President Chun was eager to differentiate his administration from the Park Chung-hee Administration for the 1960s and 1970s.

154 Success repayable loans were introduced officially in December 1983 by government decree. For KODECO's West Madura project, the council of economic ministers decided on this special loan offer in 1981. While the loans from the JPDC were always success repayable loans, the loan from the Korean government was composed of success repayable loans and low-interest loans. See PEDCO, *Han'guksŏkyukaepalgongsa Sibonyŏnsa 1979–1994 [15 Year-History of PEDCO, 1979–1994]*, 224–31.

Table 4 Fluctuation of the number of Oil E&P engineers near the Financial Crisis

	1997	2000	2003	2005
KNOC	90	81	92	112
Private Companies	106	35	46	67

Source Korea National Oil Corporation, *Han'gukhyŏng Sŏgyumaeichŏ Yuksŏngŭl Wihan Han'guksŏgyukongsa Hyŏksinpangan [the Innovation Plan of Korea National Oil Corporation to Promote Korean-Style Oil Major]*, 116

4.2 The Private Sector's Capacity and the Asian Financial Crisis

While the Asian Financial Crisis in 1997 seriously impaired the capacity of private companies in oil development, it is unclear whether KNOC's situation fully justified the reinforcement it received. After the crisis, Korean companies abolished oil development divisions or sold oil field equities in order to adapt to the standard debt/asset ratio of 200% by the end of 1999, a condition of the economic rescues measure forced upon the government. Fifteen companies removed their E&P teams, and the total number of companies involved in oil E&P decreased from 26 in 1997 to 20 in 2003.[155] Considering that 30 of Korea's largest industrial groups' average debt/asset ratio was 387% in 1996,[156] the private sector had no choice but to reduce investments and sell assets for a quick adjustment. As seen in Table 4, the extent of the decrease in engineers in the private sector was more significant than that in the KNOC.

However, the E&P capacity of the KNOC was not high enough to legitimize government support, and the private sector still had a certain E&P capacity. In 2003, SK Energy's oil reserve was bigger than KNOC's, but SK Energy's oil production was 23 thousand barrels per day, almost equal to KNOC's daily production of 24 thousand barrels.[157] Similarly, in 2004, the total daily production of the private sector was 68 thousand barrels per

155 Korea National Oil Corporation, *Han'gukhyŏng Sŏgyumaeichŏ Yuksŏngŭl Wihan Han'guksŏgyukongsa Hyŏksinpangan [the Innovation Plan of Korea National Oil Corporation to Promote Korean-Style Oil Major]*, 116.

156 "Debt Ratio," *Han'gukkyŏngje Magazine*, Apr. 1998.

157 Pyŏng-do Han, Kim, Tae-nyŏn, Lee, Kwang-jae and Seo, Kab-won, *Enŏchi Munje, Saeroun Daeannŭl Ch'ajasŏ [Energy Problem, Pursuing New Alternatives]*, Policy Material for the Regular Session in 2004 (Seoul: Legislature Research Center, 2004), 18.

day, which was much larger than KNOC's daily production of 42 thousand barrels.[158] In addition, the KNOC had not been visible in the world energy market. The Energy Intelligence Top 100, a reliable benchmark for comparing the world's top oil and gas companies, did not list the KNOC in the world's top 100 companies as of 2002, and in 2009, it was ranked 77th after SK Energy, ranked at 71st.[159]

Political consideration primarily drove the government's decision to expand KNOC's role. First, there was a concern over whether the government could implement a policy that looked favorable to a specific *chaebol*. If the government supported specific private interests, rather than public interests, it would face criticism from people general, the media, and other public interests. When options to strengthen Korea's upstream are considered, special favors to specific *chaebol* could result in controversy that would deteriorate political capital in Korean politics. Also, considering that the private companies' decision-making is based more on monetary benefit than on securing a steady oil supply, the effectiveness of the support from the private sector in enhancing the equity oil ratio is dubious for policymakers.[160] This concern was proved valid when SK Energy sold out its shares in a Brazilian offshore oil field in 2011. For this project, SK Energy received a US$27 million success repayable loan, the aim of which was to enhance Korea's oil security in terms of its oil self-sufficiency ratio. The news that BMC-8 would enter the production phase resulted in the government's rosy expectation that BMC-8 would enhance Korea's oil sufficiency ratio by 0.8%.[161] However, when SK Energy sold out all shares of this production field along with two other offshore exploration fields in Brazil for US$2.4 billion in July 2011, these expectations went up in smoke.[162] For the government, which supported a US$7.7 million share of the success repayable loan for SK's Brazil project, SK's sale of its share of the successfully explored oil field was disappointing.[163]

158 Korea National Oil Corporation, *Han'gukhyŏng Sŏgyumaeichŏ Yuksŏngŭl Wihan Han'guksŏgyukongsa Hyŏksinpangan [the Innovation Plan of Korea National Oil Corporation to Promote Korean-Style Oil Major]*, 55.

159 Energy Intelligence, "The Energy Intelligence Top 100: Ranking the World's Oil Companies," *Petroleum Intelligence Weekly* 2003; Energy Intelligence, "The Energy Intelligence Top 100: Ranking the World's Oil Companies," *Energy Intelligence* 2010. The world's top 100 oil companies are rated based on the size of reserve and production, refinery capacity, sales, net income, total assets, and the number of employees.

160 Han, *Enŏchi Munje, Saeroun Daeannŭl Ch'ajasŏ [Energy Problem, Pursuing New Alternatives]*, 23.

161 Ministry of Commerce Industry and Energy, "SK Energy Will Launch Oil Production at Brazil Bmc-8 Field from the End of July."

162 Ministry of Knowledge Economy, "SK Innovation's Selling Shares of 3 Fields in Brazil Approved," *Press Release*, July 13, 2011.

163 In the same year, SK Energy also sold its North Sea oil assets to First Oil. It is known that the price was virtually one tenth of the value among industry people. Interview with Keun-wook Paik,

Another reason for expanding the KNOC is that SOEs are easy to mobilize in an effort to enhance the performance of energy diplomacy in the short term. Unlike private companies, which are normally cautious about risky ventures, the KNOC could be brought to oil-rich countries to support energy diplomacy.[164] This took place during the Roh Moo-hyun Administration's energy diplomacy, in efforts such as the West Kamchatka offshore project in Russia.[165]

This tilt toward public enterprise was reinforced with the foreign examples studied in policy choice. In the reports, Total, ENI, and Repsol, which began as SOEs in France, Italy, and Spain, respectively, are favorably quoted as company models that Korea should follow, because they achieved more than 50% of an equity oil ratio in each country.[166] In contrast, Japan is mentioned as a failure, because Japan's oil self-sufficiency ratio was estimated at around 9.8% in 2003. In that regard, experts have concluded that the privatization needs to take place after the KNOC achieves a certain level of capacity, as seen in the cases of former European state-owned oil companies.[167]

5 Summary and Conclusion

The small number of veto players contributed to the expansion of the KNOC, which is a decisive outcome compared to the government's moderate

conducted on Oct. 7, 2012; See First Oil, "Signed SPA for Celtic Oil Acquisition," *Press Release*, April 2011; First Oil, "Kraken 9/02b-5z Tests at 4,550 Barrels of Oil Per Day," *Press Release*, September 2011.

164 The Vice Chairman of SK Energy, Korea's biggest private oil company, accompanied the Prime Minister's energy diplomacy tour to Central Asia in May 2008. Though SK had already received an offer from Turkmenistan for natural gas development, this was not formalized during the energy diplomacy tour. From a simple cost/benefit perspective, it is likely that SK energy viewed gas development in Turkmenistan as unfeasible.

165 The West Kamchatka offshore oil exploration project was agreed among members of the Korean consortium, composed of KNOC and Korean private companies, and Rosneft, a Russian state-owned oil company, after President Roh visited Moscow in Sept. 2004. However, the license for the project was expired in August 2008 with the reason that the Korean consortium did not conduct the required number of exploratory drilling outings in 2007. The next Korean President Lee Myung-bak, during his visit to Moscow in September 2008, asked Prime Minister Putin a favor on Korean consortium's expired contract on the West Kamchatka offshore project, no positive change was taken place. In August 2010, the Korean consortium decided to withdraw from the project with the loss of U$252.84 mil.

166 In Japan, Total and ENI were exemplified as models of oil developers in the report by Arisawa's European Oil Policy Study Tour Group in1961.

167 Han, *Enŏchi Munje, Saeroun Daeannŭl Ch'ajasŏ [Energy Problem, Pursuing New Alternatives]*, 30–3; Korea National Oil Corporation, *Han'gukhyŏng Sŏgyumaeichŏ Yuksŏngŭl Wihan Han'guksŏgyukongsa Hyŏksinpangan [the Innovation Plan of Korea National Oil Corporation to Promote Korean-Style Oil Major]*, 43–53. In 2003, Japan's equity oil ratio was 14.9%. However, this may not have altered the conclusion of policymakers.

support. In Korea's policymaking on the reinvigoration of a state-owned oil company in the early twenty-first century, the number of veto players was six: the President (*Chŏngwadae*), the MKE, the MOSF, the GNP, the National Assembly, and the Standing Committee on Knowledge Economy. Compared to Japan, no interest groups had an opportunity to become veto players in policymaking, even in the early twenty-first century, because the government made the decision secretly and excluded interest groups.

Moreover, the number of effective veto players becomes even smaller when the absorption rule is applied. Korea's top-down decision-making system and its decision-making norm affect the distance among veto players. The preference of the GNP, the ruling party, is included in that of the president. Still, the majority-based decision-making norm, where the preferences of the standing committee and the National Assembly were included in that of the president through majority members from the ruling party, reduced the number of effective veto players, in this case to three.

While the MOSF placed a few constraints on how to finance KNOC's expansion, the President's eagerness to reinvigorate the KNOC since winning the election made the plan possible without much challenge. The supplementary budget was reviewed easily under favorable conditions in the National Assembly after the ruling party won the majority of the Assembly in the general election of April 2008. Even though Korea was already industrialized, a state-owned oil company, not a private one, was supported. The reason for this has more to do with political considerations than with the capacity of the KNOC. While there was a capable private company available, the possible political dispute over special favors to a specific *chaebol* made the private-oriented option unfeasible. The President's practical choice at a time when oil was priced at $100 per barrel in the early twenty-first century, as well as the convenience of utilizing an SOE in implementing energy diplomacy, could explain the conservative support for the active role of the KNOC. This outcome also reaffirms the relevance of veto player theory, which the outcome passed through a small number of veto players is likely to be more public-regarding than private-regarding.

One thing that needs to be mentioned is how Japan addressed the same situation in 2008. Unlike Korea, the Japanese state did not attempt to establish SOEs, but increased its support for the overseas oil ventures by private companies. In April 2007, JOGMEC increased its extent of equity investment and loan guarantee to 75% of total investment in the project. In the guideline for resources acquisition (*shigenkakuho shishin*) approved by the Cabinet in March 2008, the government's role is to support the private sector's resource development by improving diplomatic relations with oil producing countries and making good use of ODA.[168]

[168] Japan Petroleum Development Association, *Wagakuni Sekiyu·Tennengasukaihatsu No Genjō to Kadai [the Situation and Issues of Our Country's Oil and Natural Gas Development]* (Tokyo: Japan Petroleum Development Association, 2009), 123–27.

6

Conclusion

This research has examined state capacity/incapacity in terms of achieving intended policy change to solve problems. The book uses Tsebelis's veto player theory with a few modifications, such as clarifying the range of veto players in vertical and horizontal dimensions. The application of the theory requires detailed information, such as the situation of status quo, the preferences of veto players, and the distance among veto players, all of which justifies a case-based qualitative study. In this case study, the establishment or reinvigoration of a state-owned oil company for energy security in Japan and Korea is the dependent variable, and the number of veto players, especially those that are effective, is the independent variable. This comparative analysis examines why the Japanese state's attempt to establish a state-owned oil company was neither achieved in the 1960s nor in the early twenty-first century, whereas the Korean state's efforts to establish and strengthen a state-owned oil company was achieved in the 1970s and in the early twenty-first century.

The finding of this study is that while the presence of many veto players in Japan's decision-making system permitted only resolute change from the status quo, fewer veto players in Korea's decision-making system meant more decisive change was possible. This answer goes beyond the statist argument and revisionist approach, which have dominated the direction of scholarly research in East Asian Studies. This conclusion chapter recaps the significant findings of this empirical analysis in the preceding chapters and discusses the

S.-i. Oh, *Overseas Energy Investment of Korea and Japan*, https://doi.org/10.1007/978-981-99-0285-9_6

implications of this research both for the world of scholarship and the world of praxis. Related questions for future research will also be raised at the end of this chapter.

1 Main Findings

As seen in Chapter 1, Japan's private-oriented overseas oil development is in a sharp contrast to Korea's public-oriented one. This phenomenon boils down to one core question: Despite the similarities in socio-economic situation, why could Japan not establish a state-owned oil company in the 1960s and the early twenty-first century, while Korea did so effectively in the 1970s and the early twenty-first century? What explains the differences in the state's role in overseas oil development between Japan and Korea? If state capacity is defined as a state's power to make and implement policies to resolve problems, this puzzle can be reshaped into: What accounts for the differences in capacity in these two "strong" states?

This question is not easily answered in the existing literature. Statists would expect Japan and Korea to converge in the structure of overseas oil development because the two countries have many similarities, including the fact that they are both developmental states, where effective bureaucracies contributed to the economic development of each country. Scholars, such as Peter Evans, would have predicted that the Korean state would have played a role as a supporter for the private sector's overseas oil development in the early twenty-first century. This prediction would be shared by revisionists who emphasize the role of the private sector in policymaking.

To better explain the puzzle, this study hypothesizes that political structure matters because the number of effective veto players determines the range of agreement on change in decision-making. In order to account for the significant determinant of establishing or strengthening SOEs for overseas oil development, this research qualitatively examined how the number of effective veto players shaped the different outcomes. This case study thus examines the status quo, the policy preferences of veto players, interactions among them, and interest group politics. The overseas oil development of Japan and Korea are selected to test the hypothesis not only because their economic systems have been regarded as similar enough to be comparable, but because both countries attempted to found SOEs for overseas oil production to overcome the scarcity of domestic oil resources. The analysis based on political institutions provided us with a better understanding of why an SOE is established or strengthened in a certain country.

The cases in this research provide strong empirical support for the inversely proportional relationship between the number of effective veto players and the establishment/strengthening of SOEs. As shown in Table 1, Japan had many more veto players than Korea both in the developmental and post-development eras, and failed to achieve the state's intention to establish a state-owned oil company. After applying the absorption rule by incorporating congruence among the veto players and cohesion of respective veto players, the number of effective veto players in Japan is still more than that of Korea.

As shown in Chapters 2 and 4, the interaction among many veto players in Japan, such as MITI/METI, MOF, the prime minister, two important councils in the LDP, the bicameral Diet, and interest groups, resulted in a failure to launch a state-owned oil company. In the 1960s, Japan's attempt to launch an ENI-style state-owned oil company to promote Japan's autonomous oil supply was not realized. The result of the debate from 1961 to 1967 was the establishment of the JPDC, a state financier for the private sector's overseas oil ventures. This outcome is resolute compared to the status quo of non-existent SOEs for overseas oil development. The resolute result is mainly due to different preferences among many veto players, including those between MITI and the MOF, between MITI and interest groups, and between the PARC and the Executive Council of the ruling party, and even among interest groups. In addition, consensus-based decision-making norms after the incident of the US-Japan Mutual Security Treaty during the Kishi Administration added to the number of effective veto players by making the bicameral Diet non-negligible in Japan's decision-making system.

In the early twenty-first century, the initiative to establish a Japanese state-owned oil company was a countermeasure to the abolishment of the

Table 1 Number of veto players and policy outcomes in Japan and Korea

	Development era		Post-development era	
	1960s in Japan	1970s in Korea	Early twenty-first century in Japan	2008 in Korea
# of veto players	12	7	14	6
# of effective veto players	8	3	5	2
Outcome of policy-making process	Failure in establishment of SOE	Success in establishment of SOE	Failure in establishment of SOE	Success in reinvigoration of SOE

debt-ridden JNOC during the Koizumi Administration. METI (the successor ministry to MITI) proposed the establishment of a state-owned company to deal with important oil and gas assets of the JNOC. However, METI's attempt to establish a state-owned oil company failed, and all METI could do was to privatize INPEX, a subsidiary of the JNOC, to inherit its assets. This obstruction of the proposal was due to disagreement mainly among the Prime Minister, the ministries, and the councils of the ruling party in Japan's un-Westminster parliamentary system. Furthermore, a coalition government increased the number of veto players by the number of coalition partners with the LDP.

On the other hand, Chapters 3 and 5 show unambiguous contrast to Japan. The small number of veto players in Korea's political system, such as the energy ministry, the finance ministry, and *Chŏngwadae*, contributed to the establishment and reinvigoration of Korea's state-owned oil producer. In the 1970s, Korea's MTI/MER succeeded in making a state-owned oil company called PEDCO (the former entity of KNOC) in 1979. While the MTI/MER faced opposition and indifference from the EPB due to the dire economic situation in the aftermath of the first oil shock in 1973, the MTI and its successor could realize their intention. The result was decisive one compared to the status quo of non-existent SOEs for overseas oil development. This result was mainly due to a lack of veto players in Korea's centralized decision-making system, where the president, MTI/MER, and the EPB were veto players. Unlike in Japan, the ruling party did not have power to review policy, and interest groups were too immature to express their voice. The majority-based decision-making system made the unicameral National Assembly only a formality.

Facing an oil price hike in 2008, the Korean government made efforts to increase the country's oil self-sufficiency ratio through overseas oil development and chose to directly supply massive investments to increase E&P assets and the capacity of the KNOC, Korea's state-owned oil company. The government succeeded in realizing enormous funding in the budget process of the unicameral legislature. This reinvigoration of the KNOC was possible due to a small number of veto players in Korea's policymaking process. Compared to the *Yushin* period in the 1970s, the number of veto players in the early twenty-first century has rarely changed because the 1987 Constitution focused on the replacement of the indirect presidential election with direct one, which does not accompany a substantial change of political structure. The majority-based decision-making system still hindered the National Assembly and the ruling party from playing effective roles as veto players, at least during the early period of the five-year term presidency. Also, interest groups were excluded from this discussion for prompt decision-making without the intervention of the private sector.

A Comprehensive understanding of this case study provides the reasons for the disparity between the number of veto players in Japan and Korea. One reason is the dissimilarity in political structure. While these two developmental states have same number of ministerial veto players, the policymaking system in Japan and Korea shows stark differences in four dimensions: the number of the legislative chambers, the role of the ruling party, the involvement of interest groups, and the possibility of a coalition government. Secondly, examination of the effective veto players by analyzing the distance among veto players and the cohesion of respective veto players shows notable dissimilarity in political norms. Comparing Japan and Korea's policymaking system, whether the system is consensus-based one or majority-based one determines how the preferences of veto players are absorbed. Also, whether the decision-making system is top-down or bottom-up affects the absorption of the preferences.

As seen above, the cases of Japan and Korea confirm the hypothesis that the number of effective veto players is inversely proportional to the likelihood of success in the establishment or reinvigoration of SOEs. A large number of veto players resulted in the failure of the initiative to establish a state-owned oil producer in Japan. On the contrary, fewer veto players made it possible to realize or strengthen an SOE.

2 Alternative Explanation

Many variables exist that could potentially explain the state's role in overseas oil development. However, their accounts offer incomplete explanations on why Japan did not establish SOE but Korea did. Here, two possible explanations are examined: a cultural explanation and a national characteristics explanation.

A cultural explanation had been regarded as one main factor in the political economy of Japan and Korea. Japan's uniqueness had been emphasized as determinants of phenomena that look different from the Western world. Also, Korean culture had been the explanation for the absence of development in the past time.

For this analysis, these explanations would suggest that the Japanese people's hatred of public companies and the Korean people's preference for them might produce the policy divergence in Japan and Korea. According to Kiyohiko Yoshitake, the dislike of state control was a prevalent sentiment in Japan's postwar era, and there was psychological reaction against a totally controlled economy in wartime. State intervention has been associated with the notion of inefficiency, poor quality, and meager supply of goods.[1] Some culturalists would argue that this culture was institutionalized in the Japanese political system, thereby confining Japan's SOE policy. However, this approach is limited in analyzing policy change in Japan and Korea. The dislike of state control is too superficial to explain why Japan could not establish a state-owned oil company. The culture does not explain why a state financier for Japan's overseas oil development has been permitted despite the fact that Japan had had relatively strong private financial institutions. Also, the culture may cause a misunderstanding that veto players' preferences are unified rather than diverse, which does not reflect the reality of conflict among veto players.

For Korea, pragmatism ultimately affected the establishment of SOEs. Leroy P. Jones points out that the choice of policy instruments to address the economic issue is not restrained by any ideological tendencies. According to his research, Koreans are willing to try any measures from public ownership, to free markets, to coercion.[2] However, the culture of pragmatism itself cannot fully explain why Korea chose an SOE as a main measure for its overseas oil development.

The national character studies cannot explain the divergence of policy outcomes well, either. From the viewpoint of *nihonjinron* (discussions about Japanese people), the Japanese limit their actions, do not claim rights, and always obey those with higher status, not because they have no other choice, but because it comes naturally to them.[3] If this were true, the state's intention would have been easily materialized. The national character studies can explain neither the outcome of the decision-making process nor the process itself. Someone might say Korean's *ppalli ppalli* (hurry up) tendency would make SOEs a better choice as an energy policy measure. However, as seen in Chapter 3, taking approximately five years to establish PEDCO in the 1970s cannot be explained by the Korean people's *ppalli ppalli*.

---

[1] Kiyohiko Yoshitake, *An Introduction to Public Enterprise in Japan* (Beverly Hills, CA: Sage Publications, 1973), 57.

[2] Jones and SaKong, *Government, Business, and Entrepreneurship in Economic Development: The Korean Case*.

[3] Karel van Wolferen, *The Enigma of Japanese Power: People and Politics in a Stateless Nation*, 1st Vintage Books ed. (New York: Vintage Books, 1990), 265.

3 Theoretical Implications of This Study

As seen in this research, veto player theory can explain why Japan and Korea diverged in terms of the government's role in overseas oil development. By including the legislature, the ruling parties, and interest groups, state capacity can be assessed more accurately in this study. Also, this approach provides a balanced evaluation of the role of decision-makers compared to the overemphasis of the private sector. Furthermore, this study shows how veto player theory can be more useful in analyzing the policymaking process.

Considering Japan and Korea transformed from East Asian "developmental states" to more market-rational nations, the statist argument needs to be fine-tuned. While scholars have traditionally focused on similarities between developmental states, veto player theory can explain the less-examined divergence in policy outcomes. This difference can be examined by incorporating the roles of the legislature, political parties, and interest groups. In other words, examining the political structure addresses the downside of statist arguments in Japan Studies and Korea Studies. Such an analysis of a broader political structure is more relevant in contemporary policymaking, where the role of politics cannot be ignored.

This study introduced an empirical analysis of Japan and Korea, two primary "developmental states" that diverged in their state's attempts to realize or reinvigorate state-owned oil companies. Considering the socio-economic similarities between these neighboring countries, we would expect that similar attempts in energy security policies, especially in terms of making state-owned oil companies for overseas oil development, would produce similar outcomes. Since it has been regarded that Korea learned or inherited key characteristics of the Japanese developmental state, such as meritocratic and autonomous bureaucracies, this expectation is not a strange one. However, this overemphasis on the role of a strong bureaucracy in economic development has led scholars to neglect the function of broader political structures and the private sector. Application of the veto player framework corrects this oversight by revealing the roles of different political structures in producing different policy outcomes for otherwise similar developmental states. This analysis also shows that the role of the legislature and the parties in policymaking has been more substantial than previously believed.

Compared to the statist argument, there is one thing worth mentioning about the ruling triad model, which had been influential in describing postwar Japanese political economy. This model explains that the bureaucracy, the LDP, and big business together formed a power elite group that controlled most policy making. While both the ruling triad model and veto player theory define the scope of participation in policymaking more broadly than the statist model does, the former easily assumes cooperation and harmony among the triumvirs and cannot reflect fragmentation and conflict among the participants in decision-making, which may be considered in veto player theory. In addition, the triad model would predict that MITI's initiative can get support from the LDP and big business and achieve any goal, which does not reflect reality.

For revisionist thesis, veto player theory provides the opportunity to analyze decision-making systems in a more balanced way. Overemphasis on the role of the private sector by the revisionists, such as Samuels, produces the conclusion that policy outcome depends on what the private sector desires. This study examines recognized interest groups with public sector veto players, making it possible to consider the preferences of relevant actors in the political economy. Compared to Samuels's "reciprocal consent," which blurs what will be the final outcome, veto player theory predicts the direction of the final outcome compared to the status quo.

This study's attempt to operationalize veto player theory improves the theory's utility in academic research. In quantifying the number of veto players, this case study adds a few modifications to Tsebelis's veto player theory for its operationalization. In particular, this study considers the role of relevant ministries in the government, certain councils of the ruling party, and interest groups. In a horizontal dimension, in order to clarify which interest groups are veto players, I borrow the concept of insider/outsider, whose main difference is recognition by the government. In a vertical dimension, relevant ministries and committees in parties are considered if they fit the definition of veto player in the policymaking system and have internal cohesion in Japan and Korea. This fine-tuning contributes to the accuracy of its application and the prediction of policy change.

4 Practical Implications

The implications of this study are not limited to the realm of scholarship. This examination of the relation between political structure and policy change offers a new interpretation for the prevalent ideas of Japanese incrementalism and the Korean strong presidency. While this study does not prescribe an optimal number of veto players, it contains practical implications for the respective political economies of its case studies.

First of all, this study implies that Japan's gradualism in policymaking, if not immobilism, is due to its political structure rather than its culture.

The main policy changes need to be associated with law-making and the budget review process through decision-making procedures by the ruling party and the bicameral Diet. In addition, agenda setters, such as MITI, must consult with the finance ministry or other relevant ministries. While there are some measures for ministries to evade policy review, such as administrative guidance and ministerial ordinance, their effectiveness is constrained when legitimizing the use of budget and establishing an SOE. For example, Japanese budget politics explained in Campbell's *Contemporary Budget Politics* can be reinterpreted to show that the involvement of more veto players could make the size of the agreeable budget change smaller.

As seen in Chapters 2 and 4, the involvement of many veto players in decision-making processes reduces the range of agreement and lowers the possibility of policy change from the status quo. In turn, agenda setters become inactive in policymaking that requires the review of the legislature. In this situation, it is difficult to find a set of policy options that meet the diverse preferences of many veto players. In other words, having many veto players in the ruling party and the Diet lowers the possibility of achieving originally intended changes. In that vein, if decisiveness in policymaking is desired, the number of veto player needs to be decreased by removing the review of the ruling party or making the legislature unicameral.[4]

This research also provides an alternative to the prevalent idea that parliamentarism is the only solution to keep Korea's "imperial presidency" in check.[5] In Korea, the relevant ministry sets the agenda under the guidance of the president, and thus the president is the de facto agenda setter. With a small number of veto players, it is relatively easy for the president to promptly create decisive changes from the status quo, which tends to be intensified by small window of opportunity offered by Korea's five-year one-term presidency. Seeing the negative aspects of this decisive policy change as the inevitable outcome of presidentialism, some politicians and scholars offer a parliamentary system as a solution. However, as seen in Chapters 3 and 5,

[4] In the seminar at the Japan Center of Dongseo University on May 31, 2012, Mr. Yoshifumi Wakamiya, Chairman of the Editorial Board of *Asahi Shinbun*, argued that Japan's impotence is caused by its bicameral parliament. See Young-hwan Oh, "Seosomun Forum: Korean Politics Is Really Following Japan?" *Korea JoongAng Daily*, June 11, 2012.

[5] Some scholars raise the opposite concern. Cho Yoon-je worries about the fragmentation of power structure and promotes a more centralized decision-making structure. I call this "neo-developmental state argument." Emphasizing the efficiency of decision-making, this argument raises the need to insulate the state's economic decision making from the outside pressure, including that from the legislature. See Yoon-je Cho, *Hankugŭi Kwŏllyŏkkuchowa Kyŏngjechongch'aek: Saeroun Chŏngch'i, Kyŏngjeŭi T'ŭrŭl Ch'ajasŏ [State Governance Structure and Economic Policies in Korea: Pursuing for New Politics and Economy]* (Paju: Hanul, 2009).

one of the main attributes of Korea's decisive policy change in the strong presidency is the outcome of the Korean Constitution, in which there is a mix of presidentialism and parliamentarism, rather than characteristics of presidentialism exclusively. It is worth noting that a fragmented system does not necessarily result from parliamentary system. Westminster and single-party-dominant systems tend to show power concentration, while coalition systems tend to diffuse power, as seen in the separation of power in the US presidential system.[6] If the prototype of a parliamentary system is the Westminster system, the reality that the prime minister has strong power is contrary to the expectation that the prime minister will be a weak leader and that power will be fragmented. The remedy to the "imperial presidency" is to increase the number of veto players by creating a bicameral legislature or making the policy review power of the ruling party substantial.

5 Minor but Interesting Findings

While this study focuses primarily on the effect of political structure in policymaking, other factors may play an intervening role in the political reality. Therefore, the study implicitly examined the role of two factors often treated as key determinants related to the establishment of SOEs: the political attitude of leaders/the ruling parties and the level of economic development. I found that similarities between the political attitudes of Japan and Korea exclude the relevance of ideology in analyzing the establishment of SOEs. In the case of the level of economic development, its applicability is constrained to the early phase of economic development, such as the 1960s and the 1970s. This does not necessarily imply a refutation of Gerschenkron's thesis, but it suggests that veto player theory answers more persuasively the question of why Japan did not establish SOEs for overseas oil development, while Korea did.

This study also found that certain shocks provide momentum to accelerate changes, and sometimes even create them. In the 1960s, a domestic political scandal known as the Black Mist Incident reduced the finance ministry's veto power in the process of creating the JPDC, a state financier for Japan's overseas oil development. In the 1970s, Japan's Diet delayed ratification of the Korea–Japan Joint Development of Continental Shelf Agreement for almost three and a half years after both governments signed it in January 1974. The

[6] R. Kent Weaver and Bert A. Rockman, *Do Institutions Matter? Government Capabilities in the United States and Abroad* (Washington, DC: The Brookings Institution, 1993), 450.

diplomatic tension caused by this stalemate revived the initiative to found a Korean state-owned oil producer.

It is worth noting that external energy shocks did not produce the same response regarding SOEs in Korea and Japan. In Korea, the first oil shock in the 1970s and the oil price spike in the early twenty-first century created the need to establish or reinvigorate Korea's state-owned oil company. However, the same was not true in Japan. There, these shocks made the country's private-oriented institutional setting more proactive. This can be reinterpreted by looking at the association between the number of veto players and crisis response—countries with a small number of veto players tend to respond to the crisis more assertively than those with many veto players.

This research revealed a few interesting findings that undermine prevailing notions. For example, Japan's *Keidanren* has been considered the most politically powerful economic organization in Japan's private sector, but the case of the 1960s shows that *Keidanren*'s opinions were not always reflected in policymaking. In fact, in the process of crafting the Petroleum Industry Law (PIL-II) in the early 1960s, some of *Keidanren*'s demands were rejected by MITI and the Diet.

This case study also draws out some unexpected facts regarding the capabilities of strong leaders. In the 1970s, Korea's powerful president Park Chung-hee asked the energy ministry to establish a state-owned oil company, but it took approximately five years to realize this due to budget constraints. This shows that policy outcomes in nations with powerful leaders need to be examined through the veto player framework rather than through a one-dimensional examination of leadership style, which might expect a quick policy outcomes as well as the realization of the policy itself. Prime Minister Koizumi's case also provides a different perspective. In the case of Koizumi's effort to remove the JNOC, Japan's state financier for overseas oil development, he was unsuccessful due to complex discussion among many veto players. This shows that policy outcomes need to be analyzed in the context of political structures rather than through the personal factors of individual leaders.

6 Questions for Further Study

Thus far, this research has shown that the number of effective veto players in the policymaking processes of Japan and Korea influenced how each country realized, or did not realize state-owned oil companies. Pursuing the integration of Japan Studies and Korea Studies into mainstream comparative politics in the West, this research on the politics of SOE policy in overseas oil development suggests a few important directions for future research.

This application of veto player theory to the establishment and strengthening of state-owned oil company can be extended to the privatization and operation of SOEs. This case study partially mentions a similar issue in Chapter 4 by dealing with Japan's attempt to abolish its state financier for overseas oil development and its gradual change. The research on the effect of political structure upon the operation and privatization of SOEs can explain how the issue of efficiency is addressed in politics.

In addition, while veto player theory has not been well researched in Japan Studies and Korea Studies, the theory can be applied to other issues like the relocation of foreign military bases in host countries. Such a study would add a new perspective to explain what factors affect the prospects for successful military base management in different host countries.[7] This can be connected to a broader inquiry on the nature and origins of policy differences between Japan and Korea.

In utilizing veto player theory, the dependent variable can be more than policy outcomes. Another promising direction is the affiliation between the number of veto players and the intensity of advocacy by the civil society, which has rarely been studied in veto player theory. Veto players are important actors in decision-making and also offer channels for those outside the government to voice opinions, and thus the number of veto players corresponds to the number of chances civil society organizations have in order to affect the final outcome. In that situation, it is possible that such actors would attempt to push harder for their agenda if there were a smaller number of veto players. For example, Korea has fewer veto players than Japan, as well as a higher intensity of advocacy. Analyses of this phenomenon have so far ascribed the intense political struggle in Korean politics to an inchoate political culture. However, if there is a negative relationship between the number of veto players and the intensity of advocacy, this relationship can explain the political situation in both countries more systematically and provide possible solutions.

[7] One example is Kent Calder's five hypotheses. For details, see Kent E. Calder, *Embattled Garrisons: Comparative Base Politics and American Globalism* (Princeton, NJ: Princeton University Press, 2007).

7

Epilogue

Previous chapters have examined the evolution of the structure of overseas oil ventures by Japan and Korea until the first decade of the twenty-first century. Then, it is only fitting that the story of the following decade from 2010 is told to overcome the time gap. In Japan, energy security policy has been consistent, as is expected in a decision-making structure with a large number of veto players. The role of INPEX in Japan's energy security has been stronger. The outcome of this consistency is an increasing trend in terms of Japan's oil and natural gas self-sufficiency ratio, as seen in Fig. 1.

On the other hand, Korea's energy security policy has fluctuated due to political leaders' changing levels of interest or indifference through different administrations. As a result, while KNOC or KOGAS was not abolished, their roles in overseas oil and natural gas development have been drastically downsized. The inconsistency in Korea's energy security policy produced a different result from that of Japan in terms of the numerical target of oil and natural gas self-sufficiency.

S.-i. Oh, *Overseas Energy Investment of Korea and Japan*,
https://doi.org/10.1007/978-981-99-0285-9_7

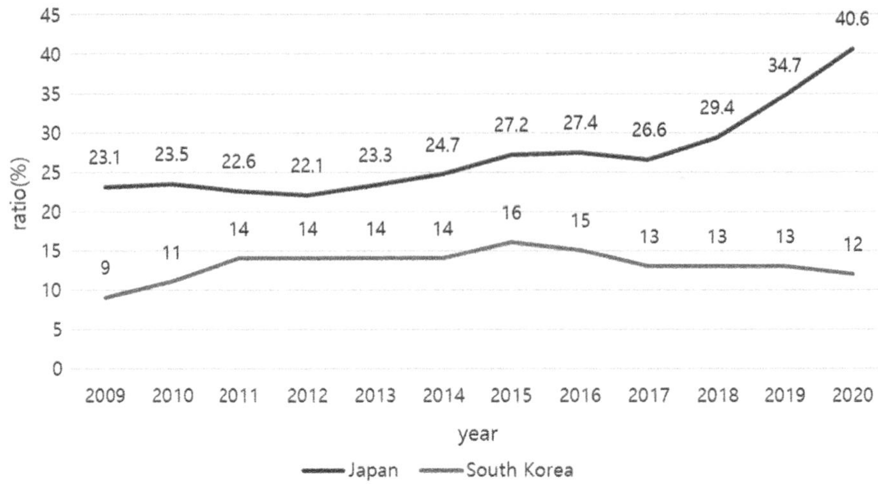

Fig. 1 Oil and natural gas self-sufficiency ratio from 2009 until 2020 (*Sources* For Japan, Japan's Ministry of Economy, Trade and Industry. "METI Published Our Country's Oil and Natural Gas Self-Sufficiency Ratio as of 2020," *Press Release*, Nov. 24, 2021. https://www.meti.go.jp/press/2021/11/20211124001/20211124001.html [Accessed Nov. 11, 2022]. For Korea, Statistics Korea, "Status of Overseas Resources Development," https://www.index.go.kr/potal/main/EachDtlPageDetail.do?idx_cd=1167 [Accessed Nov. 11, 2022])

1 Japan: Consistent Energy Security Policy

1.1 Policy Direction

Japan has maintained its overseas oil and natural gas development policy under the government's target of resource acquisition in two ways: independent development and diversification of supply sources. In that regard, in 2017, the Japanese government quadrupled its investment compared to 2012 in order to take advantage of low oil prices as the opportunity for Japan to expand its overseas oil and natural gas reserves.[1] Its concern also affected the increase in investments because the low level of investment, which normally takes place under low oil prices, would cause the energy security risk to grow.

The increase in fossil fuel imports since the 2011 Tohoku Earthquake added to the importance of energy security in Japan. The Fukushima nuclear power plant shutdown in March 2011 caused by the earthquake and tsunami created an urgent need for additional electricity supply by thermal power

[1] Dong-won Sung, "A Comparative Study on the Participation of Policy Financial Institutions in Project Finance for Overseas Resource Development: The Case of Korea and Japan" (Ph.D. Thesis, Korea University, 2021).

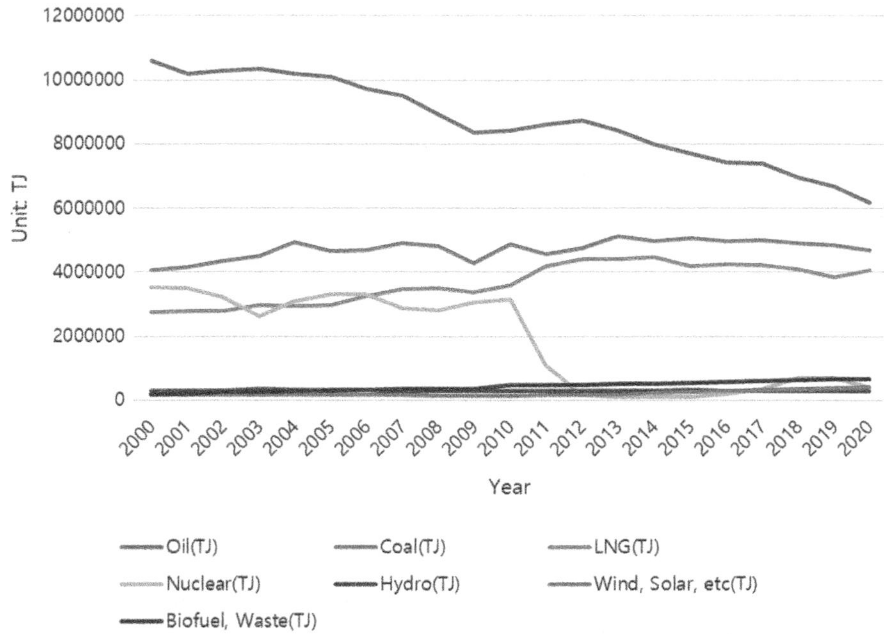

Fig. 2 Japan's total energy supply by source, 2000–2020 (*Source* IEA, "Country Profile: Japan", https://www.iea.org/countries/japan [Accessed Dec. 4, 2022])

stations.[2] In particular, the LNG thermal power ratio reached nearly 50%.[3] The level of fossil fuel in electricity generation increased from 65% in 2010 to 85% in 2012, and the level of Japan's self-sufficiency in primary energy supply decreased from 20% in 2010 to 8% in 2012 (Figs. 2 and 3).[4]

The increase in LNG imports became one of the main causes of Japan's trade deficit from 2011 through 2015, and thus Japanese government emphasized a stable and economical energy supply more than before.[5] Japan's LNG imports sharply increased from 91.8 bcm in 2010 to 108.3 bcm, 113.1 bcm, 114.2 bcm, and 115.9 bcm in 2011, 2012, 2013, and 2014, respectively.[6] Considerable increase in LNG imports by Japan, the world's biggest LNG

[2] Jane Nakano, "Japan to Invest in the Latest Russian LNG Project," *Center for Strategic and International Studies*, July 12, 2019, https://www.csis.org/analysis/japan-invest-latest-russian-lng-project (Accessed Nov. 22, 2022).

[3] Ministry of Economy, Trade and Industry, "The Road to Securing LNG," *METI Journal*, May Issue, 2014, https://www.meti.go.jp/english/publications/pdf/journal2014_05b.pdf.

[4] Ministry of Foreign Affairs of Japan, *Diplomatic Bluebook 2018* (Tokyo: Ministry of Foreign Affairs of Japan, 2018).

[5] Ministry of Foreign Affairs of Japan, *Japan's Enernrgy Diplomacy* (Tokyo, Ministry of Foreign Affairs of Japan, 2019).

[6] IEA, Japan Natural Gas Security Policy, https://www.iea.org/articles/japan-natural-gas-security-policy (Accessed on Dec. 4, 2022).

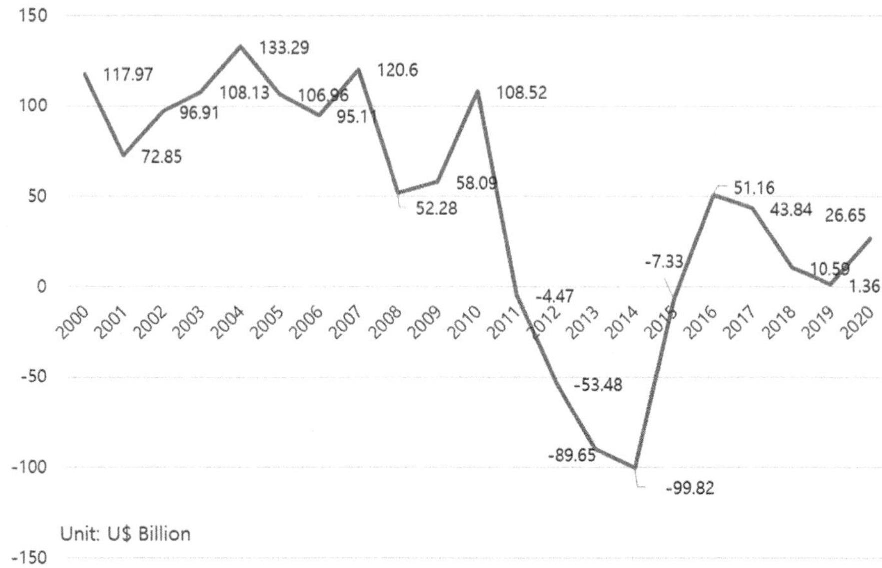

Fig. 3 Japan's net trade goods, 2000–2020 (*Source* World Bank Open Data, https://data.worldbank.org/indicator/BN.GSR.MRCH.CD?end=2021&locations=JP&start=1999&view=chart [Accessed Dec. 4, 2022])

importer at that time, affected the world's LNG market, and thus Japan's LNG import price rose sharply.[7] Japan's LNG importing price per million BTU rose from U$11 in 2010 to U$14.7 in 2011, U$16.7 in 2012, U$16.0 in 2013, and U$16.3 in 2014.[8] Now, reducing costs of LNG procurement has become an imminent mission for the Ministry of Economy, Trade and Industry and the Agency for Natural Resources and Energy.[9]

1.2 Energy Diplomacy

The Japanese government was active in energy diplomacy in the 2010s. In addition, political leaders like Abe Shinzo have been enthusiastic about energy diplomacy. In 2013, then-Prime Minister Abe Shinzo visited resource-rich countries, such as the U.S., Russia, Saudi Arabia, the UAE, Qatar, Canada, and Mozambique. Thanks to such energy diplomacy, Japan succeeded in the

[7] In 2021, China imported 78.93 million tons of LNG, which surpassed Japan's LNG import of 74.32 million tons. See Yuka Obayashi, "Japan's 2021 LNG imports slip, gives up world's top buyer spot to China," *Reuters*, Jan. 20, 2022.

[8] IEA, "LNG Importing Prices in Selected Countries, 2010–2019," https://www.iea.org/data-and-statistics/charts/lng-import-prices-in-selected-countries-2010-2019 (Accessed on Dec. 5, 2022).

[9] Ministry of Economy, Trade and Industry, "The Road to Securing LNG".

acquisition of export approvals for U.S. LNG to Japan on May 17, 2013. It was a rare case for non-FTA countries like Japan and was helpful for Japan in overcoming the trade deficit caused by the sharp increase in LNG imports in the aftermath of the closing of the Fukushima nuclear power plant.[10] Also, Japan's energy strategies, including New International Resource Strategy, have consistently emphasized the importance of energy diplomacy.[11]

Furthermore, Japan's energy diplomacy diversified energy import sources by evading the political difficulties that could have manifested. Considering Japan is closely allied with the U.S., Japan's achievement in energy diplomacy with Russia, which has abundant energy resources, was discernable. The Arctic LNG-2 project in Yamal, one of the world's largest natural gas reserves, had gained the Japanese government's attention, which pursues an economically competitive LNG supply. Following the summit between Putin and Abe in May 2016 and in June 2019, JBIC's 200 million euro loan to support engineering, procurement, and construction by Japanese companies in the Yamal project and JOGMEC's U$3 billion investment to get a 10% share of the project with Mitsui & Co. were decided respectively.[12] Also, in July 2020, the first Yamal LNG cargo was shipped to Japan's Ohgishima LNG terminal through the Northern Sea Route.[13] As a result of balancing in energy diplomacy, the US is the fourth largest LNG supplier to Japan with a 10% portion, and Russia is the fifth with 9% as of 2021.[14]

1.3 Subsidy and Loan

JOGMEC has been involved in 30 projects in the way of capital investment and 11 projects in the way of liability guarantee in 2017.[15] In particular,

---

[10] Office of Legacy Management, Department of Energy, "Timeline of Events: 2013," https://www.energy.gov/lm/timeline-events-2013 (Accessed Nov. 22, 2022).

[11] Ministry of Economy, Trade and Industry, "New International Resource Strategy," Ministry of Economy, Trade and Industry, March 2020.

[12] Reuters Staff, "Japan's JBIC to Provide Loan to Novatek-led Yamal LNG Project," *Reuters*, Nov. 24, 2016; Mitsui & Co. and JOGMEC, "Final Investment Decision for Arctic LNG 2 Project in Russia," *Press Release*, Sept. 5, 2019; Jane Nakano, "Japan to Invest in the Latest Russian LNG Project".

[13] Mirza Duran, "Novatek Delivers 1st Yamal LNG Cargo to Japan via NSR," *LNG Prime*, July 24, 2020.

[14] Larry Persily, "Japan Eyes Support for US LNG Project Expansions to Shore Up Its Energy Needs," *Alaska Journal of Commerce*, May 17, 2022.

Even under the situation of Russia-Ukraine War, Japanese government maintained its interest in Russia. See Kyodo News, "Japan to Keep Interest in Russia's Sakhalin 2 Gas Project," *Kyodo News*, Aug. 20, 2022.

[15] JOGMEC, "Financial Assistance to Japanese Companies," https://www.jogmec.go.jp/english/oil/oilgas_10_000007.html (Accessed Sept. 11, 2022).

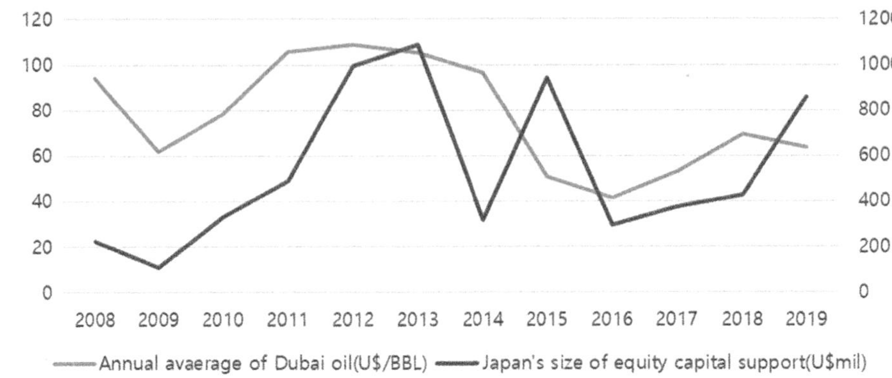

Fig. 4 Oil price and JOGMEC's annual equity capital (*Source* Sung, "A Comparative Study on the Participation of Policy Financial Institutions in Project Finance for Overseas Resource Development." 140)

capital investment to special purpose companies played a role as success repayable loan, which JNOC had previously executed. While JOGMEC capital investments were limited to exploration of the field and buying oil and natural gas field until 2016, the revision of JOGMEC Law in November 2016 expanded its area of investment to the development of the field and M&A.[16] In addition, New International Resource Strategy, formulated in 2020, included JOGMEC's more active role in providing support for risk coverage to Japanese companies' upstream oil and natural gas development, in particular for natural gas. In that regard, JOGMEC's support area included the liquefaction business.[17]

JOGMEC's equity investment has increased from U$223 mil in 2008 to U$857 mil in 2019. The amount of equity investment spiked at the time of energy emergencies, such as the years of 2012 and 2013, resulting from the 2011 shutdown of the Fukushima nuclear power plant (Fig. 4).

1.4 Strengthened Role of INPEX

INPEX has played a major role in Japan's equity oil and natural gas. Its production is 584.3 thousand BOE/D as of 2021, which is about half of Japan's equity oil and gas.[18] INPEX succeeded in developing a big natural

---

[16] Reuter Staff, "Japan Passes Law to Allow JOGMEC to Invest in Foreign Oil, Gas Firms," *Reuters*, Nov. 11, 2016. https://www.reuters.com/article/us-japan-jogmec-idUSKBN1360UC (Accessed Aug. 15, 2022).

[17] Ministry of Economy, Trade and Industry, "New International Resource Strategy".

[18] INPEX Corporation, *Annual Report 2021: A New Wind for Energy* (Tokyo: INPEX, 2021).

gas reserve in Ichthys, an Australian offshore LNG project, where it produces 8.9 million tons of LNG per year, 1.65 million tons of LPG a year and 100 thousand barrels of condensate a day.[19] Considering the development cost of Ichthys is about U$40 billion, the project was not so economically successful.[20] However, for Japan, the project was meaningful because this was Japan's first LNG project where a Japanese company play a major role as the operator. Also, the LNG production volume is equivalent to more than 10% of Japan's LNG import and 70% of LNG produced at Ichthys is planned to ship to Japan. INPEX began production of LNG in July 2018, and shipped the LNG to Japan in October 2018.[21]

One noticeable project of INPEX is the Abadi LNG project in Indonesia, which is also known as the Masela LNG project and a similarly integrated LNG project with Ichthys. INPEX discovered natural gas in 2000, and its long-delayed deal with the Indonesian government on the $20 billion-worth Abadi LNG was finalized in June 2019. The deal was approved by the Indonesian government in July that year after they changed the originally proposed plan of floating LNG (FLNG) to a plan based on onshore LNG processing facility. Considering the cost of the project is expected to be U$15 billion, Abadi is expected to be an economically successful project, which will produce 10.5 million tons of LNG a year including pipeline gas for Indonesia's domestic use and 35 thousand barrels of condensate a day. The production sharing contract (PSC) was extended by 27 years from 2028 to 2055. INPEX, which has 65% of the share, is preparing for the development with Shell, which has 35%.[22]

INPEX's role includes the exploration and drilling in the domestic onshore and offshore as a measure of energy security. The onshore field producing natural gas is the Minami-Nagaoka field. For its offshore project, INPEX

---

19 INPEX Corporation, "Ichthys Project: Outline," https://www.inpex.co.jp/english/ichthys/index.html?__utma=47397631.138530629.1655342165.1669535386.1670394784.4&__utmb=47397631.1.10.1670394784&__utmc=47397631&__utmx=-&__utmz=47397631.1670394784.4.3.utmcsr=inpex.co.jp|utmccn=(referral)|utmcmd=referral|utmcct=/&__utmv=-&__utmk=88167134.

According to INPEX Annual Report 2020, INPEX invested 66.245%, and TOTAL, Taiwan's CPC, Tokyo Gas, Osaka Gas, Kansai Electric, JERA and East Gas has 26, 2.625, 1.575, 1.2, 1.2, 0.735, and 0.42% respectively.

20 Interview with industry specialist on Dec. 12, 2022.

21 INPEX Corporation, "INPEX-operated Ichthys LNG Project Commences LNG Shipment," *Press Release*, Oct. 23, 2018.

22 INPEX Corporation, "Southeast Asia: Main Projects: Abadi LNG Project" https://www.inpex.co.jp/english/business/asia.html (Accessed Nov. 13, 2022); NS Energy, "Abadi Project," https://www.nsenergybusiness.com/projects/abadi-lng-project/ (Accessed Nov. 13, 2022); INPEX Corporation, "INPEX Receives Approval of Revised Plan of Development for Abadi LNG Project, Masela Block, Indonesia," *Press Release*, July 16, 2019; Bloomberg, "INPEX Inks $20 Billion Deal on Abadi LNG Project, the Largest Investment Ever in Indonesia," *Japan Times*, June 17, 2019.

executes exploration 130 km offshore from Shimane prefecture of Japan, which is quite close to Korea's East Sea gas field.[23]

2 Korea: Shrinking Role of KNOC and KOGAS

2.1 Policy Direction

Korean government reviewed and changed its overseas oil and gas development policy since the end of the Lee Myung-bak Administration. As Hyun-don Shin, one Korean energy security expert, described, Korea faced a new "decade of abandoned energy resource security." The Korean government's investment in oil and natural gas development plummeted as a response to the criticism on the previous administration's activities because low oil prices made the value of invested oil and natural gas assets less lucrative. The pessimistic mood of "keeping your head down!" became prevalent.[24] In Korea, there were no incidents to signal alarms for energy security, like Japan's closing of a nuclear power plant. Low oil prices only indulged the people's indifference to energy security. Meanwhile, the role of KNOC and KOGAS, which had previously been active in oil and natural gas development, decreased sharply.

2.2 Not Energy Diplomacy but Inspection and Investigation

In Korea, energy security could not gain its leaders' attention due to political conflict or low oil prices, and thus energy diplomacy was not prioritized by Korea's leaders. Unlike Japan, where low oil prices were considered an opportunities for overseas energy development, low oil price trends removed the drive for energy diplomacy in Korea for the following Park Geun-hye Administration, which was from February 2013 through March 2017, and Moon Jae-in Administration, which was from May 2017 through May 2022. Under the evaluation based on lowering oil prices, the financial situation of public

[23] INPEX Corporation, "INPEX to Commence Exploratory Offshore Drilling Offshore Shimane, Yamaguchi Prefectures in Japan," *Press Release*, Jan. 17, 2022; Jae-hang Lee, "Kim Kyung-man, Member of National Assembly said We Should Make More Efforts for Exploration of Oil Near Korean Peninsula," *Straight News*, Oct. 20, 2022.

[24] Hyun-don Shin, "National Energy Security Has Been Abandoned for a Decade," *Gas News*, July 15, 2022.

energy corporations looked worse, and thus the investment in the previous administration was considered one "failure."[25]

During the Park Geun-hye Administration, there were few foreign visits related to energy diplomacy. In contrast, during the Park Administration, the Board of Audit and Inspection (BAI) inspected continuously, and its reports were publicized in June 2013, January 2015, and November 2015. The report in 2013 concluded that "reckless overseas resources development caused inefficient investments and bad financial status."[26] With the BAI's asking public energy corporations to sell non-core assets and non-profitable assets and to reduce their size of investments, in October 2013, the Ministry of Trade, Industry and Energy (MOTIE) made the restructuring plan of energy public corporations like KNOC and KOGAS, where 25 among 190 projects were to be sold.[27] In 2015, prosecutors investigated into the Lee Administration's energy diplomacy, and the success repayable loan was one main target.[28] Even the National Assembly established the special investigation committee on overseas resources development in December 2014. In the 14th National Energy Committee of 2016, it was decided that KNOC would not make new investments anymore, and its role would be limited to supporting private companies' overseas oil and gas development.[29]

The next administration put more restrictions on KNOC's overseas activities. During the Moon Jae-in Administration, energy diplomacy, which had been regarded as President Lee's legacy, was stigmatized as deep-rooted corruption, *Jŏkpe* in Korean, and energy diplomacy for overseas oil and natural gas development was almost abandoned.[30] While the Park Administration rendered the bad situation of public energy corporations to have come from foreign factors, like low oil prices, the Moon Administration regarded

[25] Eui-jin Jung, "Government is Busy in Selling Overseas Mines. It Sold 4 Mines with Loss of 80 Billion Won," *Korea Economic Daily*, Jan. 17, 2022.

[26] Board of Audit and Investigation, *Investigation Result Report: Status of Financial and Business Management of Public Corporation* (Seoul: Board of Audit and Investigation, 2013).

[27] Mi-young Kim, "Kim Je-nam, Member of National Assembly, Said Park Administration Executed Massive Restructuring," *e-Today*, Apr. 9, 2015.

[28] Timothy Kim, "Park Geun-hye, What Does She Want through Prosecutors' Investigation on Resource Diplomacy," *Business Post*, Apr. 9, 2015.

[29] Ministry of Trade, Industry and Energy, "Remove Distressed Asset and Upgrade Capability," *Press Release*, June 29, 2016.

[30] Joon-bum Jeon, "Acquisition and Profit… Australian Gas Field Lee Administration Invested Became Good Asset," *Chosun Biz*, Aug. 24, 2022; Chan-gu Kang, "Yoon Administration Sell Resource Public Corporation, either? Lost Decade in Overseas Resource Development," *Weekly Chosun*, Oct. 12, 2022; Eun-gyul Koh, "Moon Administration Invested Half. New Administration's Financing is Important," *Newsis*, Apr. 7, 2022.

On the other hand, when oil prices go high and political situation changes like in 2022, energy diplomacy was reevaluated as a good policy.

that it came from domestic factors, pointing to the absence of decent management and control.[31] In November 2017, the MOTIE set up a task force, mainly consisting of private-sector people, to take thorough measures to address the problems identified in the development of overseas resources and the task force concluded that public corporations needed to minimize debts and sell non-beneficial projects.[32] In May 2018, the MOTIE asked the Prosecutors' Office to investigate two energy projects, KNOC's Harvest project and KOGAS's West Cut Bank project.[33] In addition, the BAI inspected and publicized the report on KNOC's management in May 2019.

There were two points during the Moon Administration. While the Zero-carbonization policy of European countries considered energy security together, the Korean one did not, and thus even overseas development of natural gas, which had been regarded as a "bridge fuel" between coal and renewable energy, did not get support.[34] Also, one unusual try needs to be mentioned during the Moon Administration. One Korean politician in the ruling party at that time, who had a close relationship with Russia, attempted to check the possibility of Korea's participation in the Yamal project. However, the trauma from the failure of overseas oil and natural gas projects in Russia, such as the one in Kamchatka, and the high level of debt of KOGAS halted the progress.[35]

2.3 Subsidy and Loan

The absence of a long-term strategy created the vicious circle of Korea's inveterate investment pattern, which means buying at a high price and selling at a low price. The government's financial support also expanded during the time of high oil prices and shrank during the opposite. This becomes quite visible when comparing the budget allotment of Korea and Japan. While Japan's

---

[31] People's Solidarity for Participatory Democracy (PSPD). *PSPD Issue Report: First 2 years of Moon Jae-in Administration, How Much is Achieved in Removing Deep-rooted Corruption*, https://www.peoplepower21.org/pspd_press?mod=document&uid=1642054 (Accessed Oct. 2, 2022).

[32] Sung-pil Jeon, "Moon Jae-in's Resource Diplomacy Begin. What is different from the Lee Administration," *Kookmin Ilbo*, May 12, 2020.

[33] Ministry of Trade, Industry and Energy, "MOTIE asked Prosecutors' Office to Investigate on Overseas Resource Development."

[34] Se-young Oh, "Moon Administration Had Neglected Overseas Resource Development, But Now Changed Stance on Resource Security," *Energy Economy*, Dec. 14, 2021. There is controversy on whether natural gas is bridge fuel. On positive opinion, see MarketWatch, "Biden's Climate Envoy Kerry Says Natural Gas Could Be a 'Bridge Fuel,' But Emissions Capture is Key," *Spectra*, Mar. 14, 2022.

[35] Interview with a Korean expert on Russia on June 18, 2022, and In-hwan Kim, "KOGAS Has Never Implemented Exploration Since 2013," *The Korea Industry Daily*, Oct. 22, 2020.

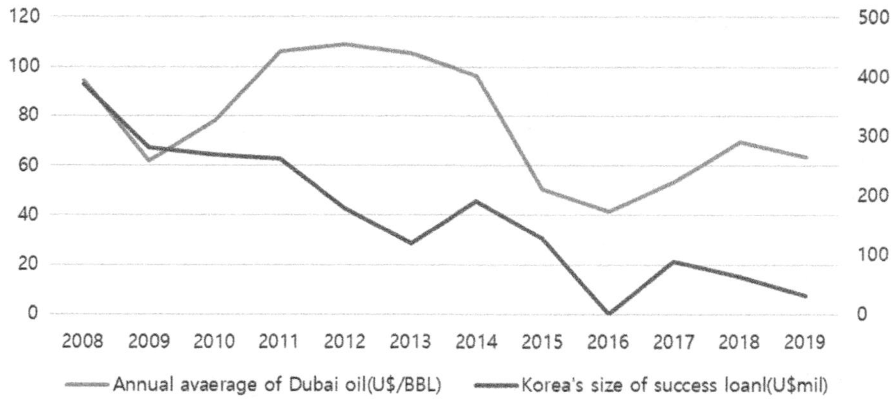

Fig. 5 Oil price and Korea's size of success repayable loan (*Source* Sung, "A Comparative Study on the Participation of Policy Financial Institutions in Project Finance for Overseas Resource Development." 135)

budget to support its overseas oil and gas development increased by 57%, from 48 billion yen in 2015 to 74 billion yen in 2016, the Korean government abolished budget support for success repayable loans itself in 2016.[36] This tendency was also reflected in the government's funding for KNOC (Fig. 5).

2.4 Role of Public Energy Corporation

While KNOC and KOGAS explain about 55% of overseas oil and natural gas development of Korea in 2020, that did not mean KNOC or KOGAS were still active in the sector.[37] In a depressing situation, KNOC only sold its assets and reduced the size of its organization.[38] KOGAS has not executed overseas exploration of natural gas reserve since 2013.[39]

After the Lee Administration, KNOC faced enormous criticism that it had paid too much to acquire oil field assets. For this matter, there are different interpretations among experts. Hyun-don Shin said, "Lee pressured

[36] Jae-joon You, "Overseas Resource Development as Mid and Long-term Policy Should Be Implemented Consistently: Budget for Success Repayable Loan was Removed in the Next Year's Budget," *Gas News*, Nov. 21, 2015.

[37] According to an industry expert, in the year of 2020, while KNOC and KOGAS produced 164 BOE/D and 74 BOE/D respectively, Korean private companies did 196 BOE/D in total. Interview on June 18, 2020.

[38] Soo-yong Jeon, "Heavily-Indebted KNOC Will Sell Overseas Assets and Restructure. Managers of 3rd Rank Will be Reduced by 10%," *Chosun Biz*, Mar. 12, 2019.

[39] Jung-hwan Lee, "Overseas Resource Development, Which Has Been Neglected for a Decade, Should Fall Into Place," *Gas News*, May 6, 2022.

the public corporation into quickly buying new oil fields before his term came to an end, so companies skipped the proper screening process."[40] On the other hand, some experts mentioned KNOC was not the only company to pay too much for their acquisitions: companies with decades of experience, such as Rio Tinto, also made devastating faults, and KNOC's projects, such as oil Block 15-1 in Vietnam, appear to be performing well. In addition, SNCF, French railways, along with Air France and GDF-Suez, of which the French government is a major shareholder, are mentioned as examples of favoring short-term objectives to long-term ones by making high dividends paid in 2015.[41]

KNOC had tried to evade criticism by increasing press releases on its achievements, such as a series of oil discoveries at Hawler field in the Iraqi Kurdistan area in 2013 and 2014, and in the UAE in 2014. However, the criticism still grew.[42] Accusations against KNOC became prominent, when the oil price plummeted in late 2014, which made KNOC's assets look devalued and its sales decreased with reduced oil production in the oil field where KNOC had owned shares.[43]

Under this pressure, KNOC had to sell its non-profitable assets at a lower price than at the buying time and reduce its overseas investment. In November 2014, KNOC sold North Atlantic Refining Limited (NARL), a subsidiary of Harvest Operations, at the price of C$97.3 million, which is almost 1/10 of KNOC's original payment.[44] The sold assets included Sabia Peru and Anchor in the U.S.[45] For the investment, as seen in Fig. 3, since the change of the presidency in the early 2010s, KNOC's overseas investment had shrunk to 1/10 of its highest level, and the level of investment has yet to

[40] Ji-yoon Kim, "Energy Resources Drive Cost Korea Billions," *Korea JoongAng Daily*, Nov. 12, 2014.

[41] Margaret Armstrong et al., "How Resource-Poor Countries in Asia are Securing Stable Long-term Reserves: Comparing Japan's and South Korea's approaches," *Resources Policy* 47 (2016): 58–60.

[42] Korea National Oil Corporation, "Oil Discovery in the Field of Hawler in Iraqi Kurdistan, *Press Release*, Mar. 27, 2013; Korea National Oil Corporation, "KNOC Discovered Oil Again in the Field of Hawler," *Press Release*, Nov. 11, 2013; Korea National Oil Corporation, "260 Million Barrel of Commercial Discovery in the Field of Hawler," *Press Release*, Apr. 2, 2014; Korea National Oil Corporation, "More than 100 Million Barrel of Oil was Discovered in UAE, Premier League of Oil Industry," *Press Release*, Dec. 26, 2014.

[43] Na-gyung Lee, "Due to Covid-19, Eleven Public Corporations, including Inchon International Airport Corporation and Korea Racing Authority, Went Into Red," *Energy Economic News*, May 2, 2021.

[44] Kim, "Energy Resources Drive Cost Korea Billions."

[45] Chul-hoon Kim, "KNOC Sold All Shares of Peruvian Oil Company, Which is a Good Signal for Improving Financial Structure," *Global Economic*, Jan. 21, 2021; Joon-hyung Lee, "KNOC Sold Anchor Oil Field, for Which It Had Invested 1.2 Trillion Won, at a Clearance Price," *Asia Economy*, Sept. 6, 2022.

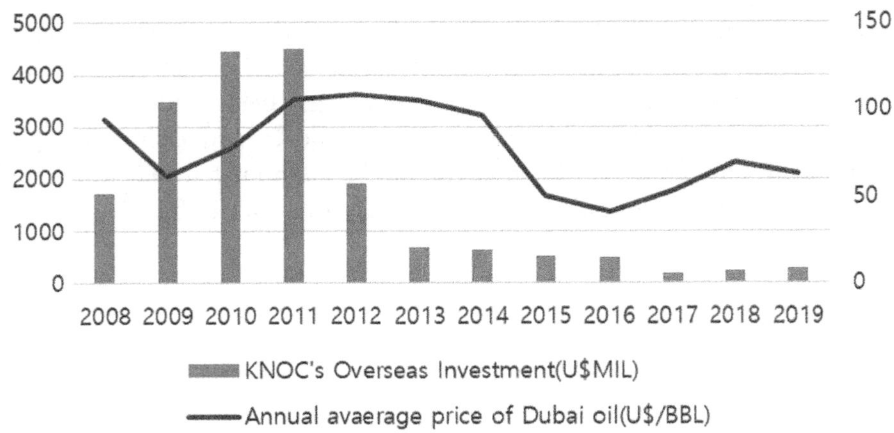

Fig. 6 Oil price and KNOC's overseas investment (*Source* Jung, "KNOC was weakened due to the Administration's Greed. Resource Development Ignored")

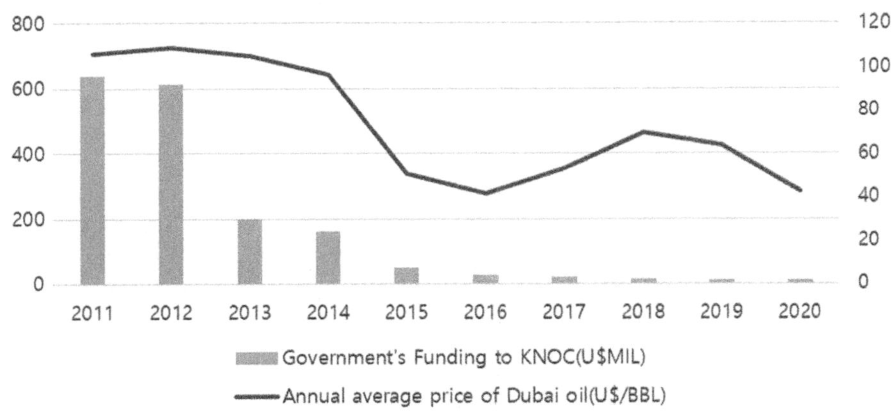

Fig. 7 Oil price and Korean government's investment in KNOC (*Source* Ibid.)

recover despite the change of oil prices due to the government's indifference to overseas energy development (Figs. 6 and 7).[46]

As of 2020, the KNOC operated 32 projects, which is about 30% of that of 2011. Its daily production also shrank to 164 thousand b/d, which is 75% of that of 2011, and 1.25 billion barrels of reserve.[47] This change was quite the opposite of its "Great KNOC 3020," the action plan to become a major oil producer with a daily production of 300 thousand b/d and reserves of 2

[46] Sang-pil Jung, "KNOC Was Weakened Due to the Administration's Greed. Resource Development Ignored," *Energy Platform News*, Aug. 2, 2021.

[47] Jung, "KNOC Was Weakened Due to the Administration's Greed. Resource Development Ignored".

billion barrels by 2012.[48] Considering the fact that Korea lost its status as an oil and natural gas producing country at the end of 2021, when Korea's only domestic oil and natural gas field, the East Sea gas field, was depleted, the public conversations suggested that KNOC should execute domestic explorations more actively.[49] For this matter, MOTIE announced that it invested about U$10 mil for domestic discovery and exploration in 2022.[50]

[48] Korea National Oil Corporation, "Overview of Oil Development Business," http://www.knoc.co.kr/ENG/sub01/sub01_3.jsp.

[49] Byung-chul Lee, "Flare of East Sea Gas Field Disappeared. Exploration is Halted in Korea," *Dong-a Ilbo*, Apr. 29, 2022. KNOC discovered the East Sea gas field in 1998, and had produced 48 million boe of natural gas from 2004 till the end of 2021. The field is planned to be utilized as a space for carbon capture and storage (CCS).

[50] Ministry of Trade, Industry and Energy, "Government is Supporting Discovery and Development of Oil and Natural Gas in 3 Areas Including the Continental Shelf in the East Sea by Investing about US$10 mil in KNOC in 2022. Explanation on the Seoul Economic Daily's report that No Budget Had East Sea Natural Gas Development Stopped," *Press Release*, Feb. 4, 2022. Related article is Woo-bo Kim, "No Budget Had East Sea Natural Gas Development Stopped," *Seoul Economic Daily*, Feb. 4, 2022.

Appendix A: State-Business Arrangement in Overseas Oil Development

The main hypothesis of this research is that the Japanese and Korean states in those countries' overseas oil development are closely linked to differences in their political institutions. As a point of departure for this case study, this appendix chapter explains the specific differences between the institutional settings of overseas oil development in Japan and Korea. When we examine which sector initiates overseas oil ventures, the difference between Japan and Korea is quite distinctive: Japan is more private sector-oriented, and Korea is more public sector-oriented. This, in turn, also shows that political institutions affect whether policies become public-regarding or private-regarding.

This appendix chapter illustrates the different institutional arrangements in Japan and Korea for overseas oil development. Throughout this chapter, "the state" will refer to a set of governmental institutions including committees and SOEs, while "business" will refer to private companies and relevant associations.

Institutional Setting at the State Level

While Japan's METI and its energy agency, the Agency for Natural Resources and Energy (ANRE, *Shigenenerugichō*), are tasked with policymaking for overseas oil development, its Korean counterpart, the MKE, does not have an energy agency.[1] In certain cases, presidential councils, such as the Presidential

[1] The second vice minister is responsible for the energy business in the MKE.

S.-i. Oh, *Overseas Energy Investment of Korea and Japan*,
https://doi.org/10.1007/978-981-99-0285-9

Council for Future and Vision (PCFV), replace the MKE as the president's agent for oil development deals.

Ministry: Agenda Setter or First Agent to Head of State

Japan's METI and the ANRE

The ANRE under METI deals with Japan's energy policies including overseas oil and gas development.[2] In response to the new international oil market situation of the 1970s, the ANRE was established by Nakasone Yasuhiro, the then-MITI Minister, and Morozumi Yoshihiko, the then-MITI Administrative Vice-Minister, in July 1973 as a part of MITI's most substantial organizational restructuring since its inception in May 1949.[3] The establishment of the agency can be seen as a revival of old energy agencies, such as the Coal Agency (*sekitanchō*), existing under the MCI from December 1945 until the establishment of MITI in 1949, and the Agency of Natural Resources (*Shigenchō*), existing from the establishment of MITI in 1949 until August 1952.[4]

In the ANRE, overseas oil development is addressed by the Department of Natural Resources and Fuel, which is known to be more powerful than the agency's other two departments, the Department of Energy Conservation and Renewable Energy and the Department of Electricity and Gas Industry.[5] The agency utilizes the oil and gas self-sufficiency ratio, which is the ratio between the oil and gas developed by Japanese companies and total oil and gas imports, for its quantitative policy target. While the numeric target was abolished around 2000, it was revived in Japan's comprehensive energy strategy of 2006, where the government set a target of 40% by 2030 (Table A.1).

In the policymaking process, issues pertaining to overseas oil development are the purview of METI advisory committees (*shingikai*). The Comprehensive Resources and Energy Review Committee and Petroleum Subcommittee are maintained under the METI minister. The former is composed of 25 private sector members, including ten professors, and the latter is composed

[2] See Articles 4, 16 and 17 in *METI Establishment Law*.

[3] The establishment of the ANRE took place just before the first oil shock of October 1973. Ministry of International Trade and Industry, *Tsūshōsangyō Seisakushi [The History of International Trade and Industry Policy]*, 510–21.

[4] Ministry of International Trade and Industry, *Tsūshōsangyōsesakushi [The History of International Trade and Industry Policy]*, 204–46.

[5] Interview with a METI official, conducted on Aug. 19, 2010.

Table A.1 METI's self-sufficiency ratio target and results

Planned year	Target	Result
1967	30% of oil by 1985	10.7% by 1985
1978	1.5 MMBD by 1990	0.45 MMBD by 1990
1983	1.2 MMBD by 1995	0.69 MMBD by 1995
1993	1.2 MMBD by the beginning of twenty-first century	0.58 MMBD
2000	Numeric target dissolved	N/A
2006	40% of oil by 2030	N/A

Source Masanari Koike, Gento Mogi, and Waleed H. Albedaiwi, "Overseas Oil-Development Policy of Resource-Poor Countries: A Case Study from Japan," *Energy Policy* 36, no. 5 (2008): 1765

of 32 members, including 14 high officials of industrial associations.[6] In terms of policy, there has been little change in overseas oil development policy since the DPJ gained power.[7]

Korea's Ministry of Knowledge Economy (MKE)

Since 2008, the MKE has been responsible for dealing with Korea's energy issues including its overseas oil development, despite not having an agency specifically assigned to this task.[8] However, as seen in the case of the Abu Dhabi oil deal, other agencies, such as the Presidential Committee of Future and Vision, have replaced the MKE's role of the first agent to the president.[9]

Leading up to the establishment of the MKE in 2008, there were a number of structural changes in the administration of Korea's oil development policy. In 1978, there was a decisive change in the organization of energy administration. With President Park Chung-hee's instruction, the MER (*Tongnyŏkchawonpu*) was established by severing the energy policy

[6] Ministry of Economy and Trade, "Information on Advisory Committee," http://www.enecho.meti.go.jp/info/committee/index.htm.

[7] Interview with a METI official, conducted on December 24, 2011.

[8] Article 32 in *Government Structure Law* (July 17, 1948). There is one noticeable difference between Japan and Korea in stipulating the function of ministries. While each ministry has its own establishment law in Japan, Korea's Government Structure Law stipulates the function of all ministries.

[9] During the Roh Moo-Hyun Administration, the Korean National Railroad was involved in the oil deal for the Sakhalin-8 onshore oil field, but this task was not officially assigned by law or presidential order. One of the purposes was to decrease the deficit of railway business through oil business. See Korea National Assembly, Standing Committee on Construction and Transportation, *Inspection of Korea Railway Corporation*, Interpellation, Oct. 5, 2005.

Table A.2 MKE's self-sufficiency ratio target and results

Planned year	Target	Result
2001	10% of oil and 30% of gas by 2010	10.8% of oil and gas by 2010
2004	10% of oil and gas by 2008 15% of oil and gas by 2013	5.7% of oil and gas by 2008
2007	20% of oil and gas by 2013 28% oil and gas by 2016	14% of oil and gas by 2013, 15% by 2016
2010	28% of oil and gas by 2013 30% of oil and gas by 2019	13% of oil and gas by 2019

Note Overseas Resources Development Plans have ten-year policy targets and have been announced every three years since 2001
Source Press release on the first, second, third and fourth Overseas Resources Development Plans, and other related MKE press releases, Statistics of Korea, "Status of Overseas Resources Development"

function from the MTI (*Sanggongpu*) in an effort to overcome the energy crisis.[10] Since the MER was abolished and merged to the MTI in 1993 by the then-President-elect Kim Young-Sam, the Ministry of Trade, Industry, and Energy (MOTIE) and its subsequent ministries, such as the Ministry of Commerce, Industry, and Energy (MOCIE) and the MKE, have dealt with overseas oil development.[11] In the MKE, the issue is dealt with by the Division of Resources Development Strategy, which was expanded from a section of a division in a 2006 organizational restructuring. Like Japan's METI, Korea's MKE also utilizes a self-sufficiency ratio, the ratio between the oil and gas developed by Korean companies and the total oil and gas import, as a policy target. As of 2012, the energy strategy target is 40% by 2030[12] (Table A.2).

In October 2010, the MKE replaced the president-led National Energy Committee, which reviews mid- and long-term policy, with the Energy Committee chaired by the MKE minister.[13] This committee became

[10] Interview with O Won-Chol, the Second Senior Presidential Secretary for the Economy to President Park Chung-hee, conducted on July 5, 2011. According to Mr. O, due to the significance of the energy administration at that time, Jang, Ye-Jun, the then Minister of Trade and Industry, was appointed as the first Minister of Energy and Resources by President Park.

[11] The last MER Minister was Jin Nyum, who later became Deputy Prime Minister.

[12] Ministry of Knowldge Economy, "Establish National Energy Basic Plan, Foundation of Green Growth," *Press Release*, Aug. 27, 2008. Japan and Korea use essentially the same term meaning to refer to the self-sufficiency ratio. The only difference is its pronunciation: *Jishukaihatsuryō* in Japanese and *Chachukaepallyul* in Korean.

[13] The high status of the National Energy Committee established during the Roh Moo-hyun Administration was transferred to the Presidential Committee for Green Growth, founded during the Lee Myung-bak Administration, due to the shift of presidential agendas.

similar to Japan's *shingikai* in terms of hierarchical location and role.[14] The committee is composed of seven government officials and 19 non-governmental members, out of which ten are professors and nine come from the private sector.[15]

Public Institution to Support Overseas Oil Development: JOGMEC

While Korea does not have a public institution for overseas oil development, Japan has an independent administrative public institution called JOGMEC (Japan Oil, Gas and Metals National Corporation) to encourage the private sector's oil ventures in foreign countries.[16] JOGMEC is not an oil producer but rather a financier to support private sector business. JOGMEC provides private companies with funds for risky investments through equity investment and loan guarantees without intervening in the management of projects. However, loans, another possible financial measure, was not provided anymore because this had been identified as the main cause of JNOC's enormous debt. JOGMEC also has the functions of conducting geological surveys, gathering and disseminating information, and managing natural resource stockpiles.

JOGMEC is closely connected to METI. METI provides JOGMEC with capital and guarantees JOGMEC's long-term loans and bond issues. These operations require METI's consent or approval, for which METI also needs to consult with the MOF.[17] As of April 2012, the President of JOGMEC is Hirobumi Kawano, former Chief of the ANRE, and a few METI officials are working for JOGMEC, such as the chief of the branch office in Washington, DC.[18]

---

[14] *Energy Law* (Mar. 3, 2006).

[15] Ministry of Knowldge Economy, "Energy Committee Launched, Showing Energy Vision," *Press Release*, Oct. 5, 2010.

[16] In Korea, the same function is implemented by an industry association called the EMRD, Energy and Mineral Resources Development Association of Korea.

[17] Articles 14, 16 and 21 in *Japan Oil, Gas and Metals National Corporation Law* (July 26, 2002).

[18] Japan Oil Gas and Metals National Corporation, "About Us."

Institutional Setting for Overseas Oil Development at the Firm Level

Japan and Korea show a stark difference in the implementation of overseas oil development. Whereas private companies invest in overseas oil ventures in Japan, state-owned companies invest in the major portion in Korea. An in-depth examination of this difference is connected to how companies in the two countries behave differently based on incentives and constraints.[19]

Japan's Private Companies

In Japan, there is no state-owned oil company, and only private companies invest in oil ventures. While INPEX Corporation is affected by a special-class share of METI, the decision-making is basically determined by the criteria of profit maximization. While the amount of investment is smaller than that of upstream-only companies, the investment of general trading companies (GTCs) and utility companies are to expanding for the purpose of gaining the opportunities of profit-earning in overseas oil development.

It is worth noting that the major producers in overseas oil development are upstream-only companies, such as INPEX, not GTCs (*sōgōshōsha*) in Japan. As seen in Fig. A.1, the contribution of upstream-only companies, such as INPEX and JAPEX, is almost half of Japan's overseas oil production.[20] On the other hand, contrary to widely held beliefs, GTCs explain only 18%, and vertically integrated oil companies only 32%[21] (Fig. A.1).

While INPEX has considered national energy security as one of its management criteria due to a special relationship with METI, profit-maximization is the most significant aim of all Japanese oil developers. This is different from Korea's state-owned oil and gas companies, which are required to be accountable to the citizenry through the National Assembly and relevant ministries rather than shareholders. Also, while companies are obligated to import developed oil during energy crises, in Japan this duty is based on companies' loan contracts with JOGMEC, unlike in Korea where it is based on the law.

---

[19] This does not mean that we can evaluate the superiority of public-oriented or private- oriented characteristics between Japan and Korea.

[20] The AOC (Arabian Oil Company) is categorized as vertically integrated company, because the AOC merged with Fuji Oil Company to become the AOC Group.

[21] The decreasing trend of cumulative investment by trading companies in total investment of overseas oil development also supports their weak role in overseas oil development: 15.8% by 1970, 9.9% by 1980, and 5.5% by 1990. Masanari Koike, "Japan's Overseas Oil Development and a Role of Technology," in *GraSPP Working Paper Series* (Tokyo: The University of Tokyo, 2008), 11.

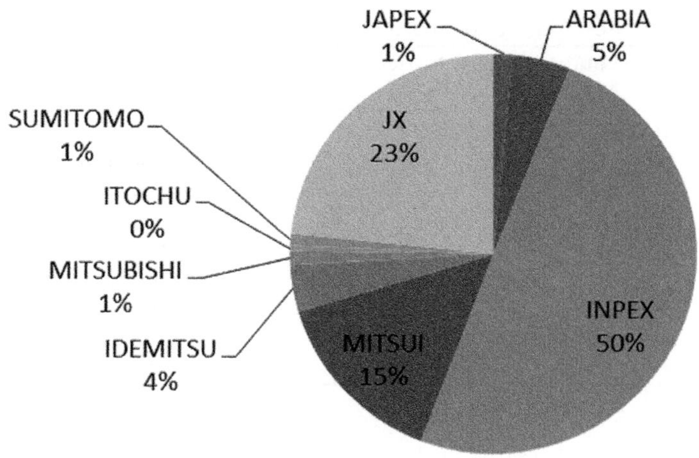

Fig. A.1 Japan's Overseas Oil Production in 2011 (*Source* Japan Petroleum Development Association, "Statistics: Production of Oil and Natural Gas," *Sekiyu Kaihatsu Jihō [Petroleum Development Magazine]*, May 2011, 68–71; Japan Petroleum Development Association, "Statistics: Production of Oil and Natural Gas," *Sekiyu Kaihatsu Jihō [Petroleum Development Magazine]*, May, 2012, 84–87)

Upstream-Only Companies

As mentioned above, the role of upstream-only companies, such as INPEX and JAPEX, is important in overseas oil development and, in particular, INPEX plays a leading role in Japan's upstream ventures. While some portions of shares of INPEX and JAPEX are owned by the government, both companies are basically private companies based on Commercial Law.

(a) **INPEX**

In the evaluation of the PIW (Petroleum Intelligence Weekly) in 2010, INPEX is ranked Japan's No. 1 oil developer and the world's 55th oil company.[22] It has approximately 1.3 billion BOE of proved reserves, and produces 423 thousand BOE per day, as of March 2011. Through aggressive exploration as an operator and asset acquisition, the production output of INPEX is expected to reach one million BOE/day by 2020.[23]

In the early twenty-first century, INPEX became privatized from a quasi-SOE. INPEX had been established as a project company of JAPEX, whose

[22] This ranking is based on the Petroleum Intelligence Weekly (PIW) of December 2010.

[23] INPEX Corporation, "Annual Report 2011: Firm Commitment to Sustainable Energy," 36–37.

Table A.3 Decrease of equity or share ownership for INPEX by the government

Time	Equity or share ownership by METI or JNOC (%)	Related event
October 1967	98	JPDC established
April 1970	50	JAPEX separated from JPDC
May 2004	53.96	INPEX merged with JODCO
March 2005	36.06	Initial public offering (IPO) at the Tokyo Stock Exchange on Nov 17, 2004
April 2006	29.35	INPEX merged with Teikoku
September 2010	18.94	Increased the stock for financing

Source Material of INPEX

major shareholder had been the JNOC. Therefore, INPEX used to be considered a kind of state-owned oil company; however, the reformation of Japan's oil upstream industry caused by the controversy of the debt-ridden JNOC since 1997 led to INPEX becoming a Japanese oil major (*wasei meja*) as a private company. METI transferred JNOC's assets, including Japan Development Corporation (JODCO) in the U.A.E., to INPEX (Table A.3).

INPEX also merged with Teikoku Oil. During this process, as shown in Table A.3, the portion of METI's share ownership decreased sharply to 18.9%, which was even smaller than METI's share ownership for Mitsui Oil Exploration Corporation (MOECO) at 20.03%.[24]

In the past, based on JNOC's major equity ownership, the president of INPEX and a few executive members came from MITI/METI, and this practice of *amakudari* has been maintained.[25] Also, former high government officials are helpful in communicating with the upper echelon of METI and financial institutions, such as JOGMEC and Japan Bank for International Cooperation (JBIC), from which INPEX is required to receive financial support.[26] While such characteristics distinguish INPEX and JAPEX from other companies, the practice of *amakudari* takes place in both private and public corporations, thus *amakudari* itself cannot provide evidence that

[24] Mitsui Oil Exploration Co. Ltd., "Shareholders," Mitsui Oil Exploration Co. Ltd., http://www.moeco.co.jp/en/about/stockholder.html.

[25] In Korea, the corresponding practice is called *nakhasan* (parachute), which refer to an institutionalized practice where outside persons use their connections with powerful people to occupy high-profile positions of state-owned companies or private companies.

[26] Interview with Keun-wook Paik, senior research fellow at Oxford Institute of Energy Studies, conducted on July 16, 2011.

INPEX is a state-owned company.[27] The special-class share, or golden share, does give METI veto power on certain issues,[28] but the main purpose of the golden share is to protect INPEX and its important oil field assets from a hostile takeover by foreign companies.[29] One noticeable point that distinguishes INPEX from typical SOEs in Japan is that governmental institutions, such as METI and the Board of Audit of Japan (BAJ), do not inspect the management of INPEX.[30]

(b) **JAPEX (Japan Petroleum Exploration Co. Ltd.)**

JAPEX has been an important player in the history of Japan's overseas oil development, but its annual production is only 39 thousand BOE/day, as of March 2011, which is less than one-tenth of INPEX.[31] That said, JAPEX is the only Japanese oil company joining the Iraqi oil field development after the Iraq War, which could be a source of momentum to enhance its performance.[32]

---

27 During the past 50 years, there have been 68 cases of *amakudari,* and now 13 retired high officials from METI hold high positions in private utility companies. See Kyodo News, "Utilities Got 68 Ex-Bureaucrats Via 'Amakudari'," *Japan Times*, May 4, 2011.

28 According to METI's guideline in April 2006 for the exercise of the golden share, METI is supposed to exercise veto power over the decisions that (1) are inconsistent with its goal of securing supply of energy to Japan as a flagship company, (2) adversely affect the goal of efficiently securing a stable supply of energy to Japan as a flagship company, and (3) may affect the voting power of the golden share holder. See INPEX Corporation, "Annual Report 2011: Firm Commitment to Sustainable Energy," 95–96.

29 Interview with a METI official, conducted on June 30, 2011, and an industry insider, conducted on June 23, 2011.

30 The Board of Audit of Japan audits 218 public corporations, more than half of whom have government funded capital. Some examples are Japan Finance Corporation, Japan Tobacco, Inc., and Tokyo Metro Co., Ltd. See Board of Audit of Japan, "Coverage of Audit," http://www.jbaudit.go.jp/english/effort/coverage.html#header.

31 INPEX Corporation, "Annual Report 2011: Firm Commitment to Sustainable Energy," 37; Japan Petroleum Exploration Co. Ltd., *Annual Report 2011* (Tokyo, 2011), 2. In terms of proved reserves, JAPEX has 72% of its oil and 95% of its gas in domestic fields, as of March 31, 2011. It seems to be attributed to JAPEX's tilt towards explorations in Japan's domestic area, which has less potential for oil and natural gas than does a foreign area. Also, its diversification of business areas, such as geothermal power generation and sand oil development after the oil shocks in the 1970s, could be reasons. Furthermore, continuous shifts in regional focus in its overseas ventures restricted JAPEX from accumulating geophysical information related to oil development in certain areas. These are the reasons JAPEX could not have enough proved oil and gas reserve in foreign countries.

32 JAPEX had an opportunity to boost its overseas E&P business in Iraq in the second petroleum licensing bidding in December 2009. With PETRONAS, the Malaysian state-owned oil company, JAPEX won the bidding for the Garraf oil field with a remuneration fee of $1.49/BOE. While other Japanese companies, such as INPEX, focused on the Nasiriyah oil field through negotiation, JAPEX conducted in Iraq independently and got a successful result. See Petroleum and Contracts Licensing Directorate, "Iraq's Second Petroleum Licensing Round: Garraf Contract Area-Bidding Results," ed. Republic of Iraq Ministry of Oil (Baghdad, 2009).

When the Japanese government established JAPEX for domestic oil development in 1955, JAPEX was a state-owned company.[33] After three years of co-existence with the JPDC (the predecessor to the JNOC), it restarted as an independent oil venture in 1970 with 50% equity ownership by the JNOC. During the restructuring of Japan's upstream in the early twenty-first century, JAPEX was listed on the Tokyo Stock Exchange in December 2003, and the share portion of METI decreased to 34% of the total as of March 31, 2010.[34] Although the presidents of JAPEX have been former METI officials,[35] METI has a memorandum with JAPEX, which does not significantly restrict the company's activities[36] (Table A.4).

General Trading Companies (GTCs or Sōgōshosha*)*

As the prices of natural resources sharply increased in the early twenty-first century, so too did the potential for GTC's to earn profits from such materials. For example, natural resources accounted for 55.8% of Mitsubishi Corporation's profit and 63.4% of Mitsui & Co's profits in 2006.[37] In particular, oil and gas development currently accounts for 10–15% of profit for Japan's trading companies.[38]

In Japan, GTCs involvement in overseas oil development has a long history, and the investment has been affected by the business environment

---

33 However, JAPEX also participated in overseas ventures, such as Arabian Oil Company's Khafji field exploration and the NOSODECO (North Sumatra Oil Development Co., Ltd.) in Indonesia in the early 1960s. See Japan Petroleum Exploration Co. Ltd., *Uchiyu Hyakugojuokunenni Sakuita Hana: Sekiyushigenkaihatsu Kabushikikaisha Gojunensi [Flower in the 15 Billion Years of Space: 50-Year History of Japex]* (Tokyo: Japan Petroleum Exploration Co. Ltd., 2006), 88b–91b.

34 JAPEX'S Corporate Data 2010.

35 One interesting point is that the presidency of JAPEX has been considered more prestigious than that of INPEX from the perspective of MITI/METI officials, and thus administrative vice minister-level officials tended to go to JAPEX after retirement. Osamu Watanabe, the current President of JAPEX as of 2012, used to be MITI Administrative Vice Minister, equivalent to deputy secretary in the United States. On the other hand, Naoki Kuroda, Chairman of INPEX, used to be the Administrator of METI's Agency for Natural Resources and Energy, which is director-general level, and Toshiaki Kitamura, current President of INPEX used to be Vice Minister for International Affairs, which is equivalent to assistant secretary-level in the United States.

36 The memorandum says JAPEX needs to consult the following: changes to the Articles of JAPEX, changes in capital or issuance of bonds, settlement and disposition of earnings, transfer of a portion or all of operations or acquisition of operations, selection of director candidates and matters with a significant influence on assets or operational administration. See Japan Petroleum Exploration Co. Ltd., *Annual Report 2011*, 27.

37 Kubo Iwao and Kato Shigeru, *Zukai Shōsha Gyōkai Hendobuku [An Illustrative Guide to the Trading Companies]* (Tokyo: Tōyōkeizaishinbunsha, 2008), 23.

38 Interview with industry insider, conducted on June 30, 2011.

Table A.4 Decrease of equity or share ownership for JAPEX by the government

Time	Equity or share ownership by METI or JNOC (%)	Related event
December 1955	56.47	Old JAPEX was established based on JAPEX Law
October 1966	69.63	The government injected more capital to JAPEX
1967–1970	–	JAPEX was merged into the JPDC
September 1970	65.74	JAPEX was separated from the JPDC and new JAPEX was launched based on commercial law
December 2003	49.94	Initial public offering (IPO) at the Tokyo Stock Exchange
June 2007	34.00	METI sold 15.94% of stockholdings

Source Japan Petroleum Exploration Co. Ltd., *Sekiyushigenkaihatsu Kabushikikaisha Junensi [The 10-Year History of Japex]* (Tokyo: Japan Petroleum Exploration Co. Ltd., 1977), 10; Japan Petroleum Exploration Co. Ltd., *Uchiyu Hyakugojuokunenni Sakuita Hana: Sekiyushigenkaihatsu Kabushikikaisha Gojunensi [Flower in the 15 Billion Years of Space: 50-Year History of Japex]*, 242b-43b; Japan Petroleum Exploration Co. Ltd., *Annual Report 2011*, 78

facing trading companies. When manufacturers implemented trading functions by themselves in the 1960s, trading companies instead strengthened their function as an investor. In the 2000s, when the economic situation became worse, GTCs expanded into more risky business areas to increase profits. The method of investment by GTCs is normally equity participation or acquisition of assets, which have relatively shorter payback periods than exploration, and they tend not to have a regional focus. Unlike other types of companies, trading companies are also involved in the businesses closely associated with overseas oil and gas development, such as pipeline construction.

One thing that needs to be clarified is the contribution of GTCs in overseas oil development, for there is a common misperception that GTCs lead Japan's overseas oil development. As seen in Fig. A.1, the production and investment by Japan's trading companies in overseas oil development are smaller than that of upstream-only companies, and they don't have enough engineers to implement operatorship. Their investment strategy is "small investments in many projects," and they cooperate with Western major companies. Thus, businesses of famous GTCs with well-known international

Table A.5 Division of work in Mitsui & Co's overseas energy development

	Mitsui & Co. headquarter	MOECO
Regional division	• Whole business of Australia • 60% of participation in the Middle East and the US	• Whole business of South East Asia including Thailand • 40% of participation in the Middle East and the US[39]
Functional division	Mainly LNG E&P	• Mainly Oil E&P

Source Interview with Japanese industry person, conducted on June 29, 2011

companies could have a larger impact on public consciousness than that of upstream-only companies.

(a) **Mitsui & Co. Ltd.**

Mitsui & Co. has recently been the most active in overseas oil development among Japanese trading companies. In 2010, it produced 250 thousand BOE/day, including MOECO's 92 thousand BOE/day. Mitsui has Energy Unit I and II in the headquarters and Mitsui Oil Exploration Co. Ltd (MOECO) as an affiliated oil development company (Table A.5).

Mitsui & Co. has been involved in overseas oil development since AOC's Khafji development with 2.20% of equity participation.[40] The company participated in the establishment of MOECO, the oil developer of Mitsui keiretsu, with 16 other Mitsui Group companies in 1969. Mitsui & Co's oil and gas development has been active in the early twenty-first century, and its profit became the biggest among Japan's six largest general trading companies in 2006.[41] It prefers equity investment as a non-operator, though MOECO operates one onshore project in Thailand, as of 2011.[42] Interestingly, METI owns 20.03% of MOECO's shares as of January 2012.[43]

(b) **Mitsubishi Corporation**

[39] In the case of the oil spill accident in the Gulf of Mexico that took place on April 20, 2010, MOECO is to pay U$1.065 billion to BP, and BP will waive all invoiced claims against MOECO, except punitive damages or civil penalties. See Mitsui Oil Exploration Co. Ltd., "Settlement of Claims Arising from Gulf Oil Spill," *News,* May 20, 2011.

[40] Japan Petroleum Consultants. Ltd., *Japan Petroleum & Energy Yearbook 1978* (Tokyo: Japan Petroleum Consultants. Ltd., 1978), T-5.

[41] Kubo Iwao and Kato Shigeru, *Zukai Shōsha Gyōkai Hendobuku [An Illustrative Guide to the Trading Companies]*, 96.

[42] Interview with Japanese industry person on June 29, 2011.

[43] Mitsui Oil Exploration Co. Ltd., "Shareholders."

Mitsubishi Corporation, which is Japan's No. 1 trading company in terms of profit as of 2010, has been focusing on LNG development and trading through its ten project units in the twenty-unit Energy Business Group at the company's headquarters.[44] While the company's main product has traditionally been coal, Mitsubishi Corporation has also been the forerunner in the LNG business among Japan's GTCs. In 2010, the production level of oil and gas was 110 thousand BOE/day. The energy business has played an important role as a driver of the company's earnings, accounting for 20% of net income in FY2011.[45]

Mitsubishi Corporation joined a natural gas project in Brunei by investing 25% of its equity in the project in 1968 and was involved in the marketing and import of Alaska gas in 1969.[46] It also jointly participated in Sakhalin II in 1994 with Mitsui & Co. In the case of 2006, the sales of LNG amounted to ¥775 billion.[47] In the LNG business, Mitsubishi has established a good, long-standing relationship with KOGAS, Korea's state-owned gas company, through projects, such as Mitsubishi's coordination of LNG swap transactions between KOGAS and Japanese utilities.[48]

(c) **Itochu Corporation**

Itochu Corporation produces approximately 40 thousand BOE/day, and at the same time, it makes good use of oil and gas ventures to utilize its strength as a trading company. While Itochu Group established CIECO (C. Itoh Energy Development Co., Ltd.) as a subsidiary to develop overseas oil and gas in 1972, its full-scale energy development began in 1992 by acquiring the UK North Sea oil field. The Itochu Corporation acquired interests in the Azeri-Chirag-Gunashli (ACG) oil field development and Sakhalin-I project, and the company has acquired trade rights to create synergies through resources development and trade in both ventures.[49] Also, the company is a partner in

---

44 Mitsubishi Corporation, *Mitsubishi Corporation 2011* (Tokyo, 2011), 13.

45 Ibid., 24.

46 Mitsubishi is known to create Japan Customs-cleared Crude (JCC, so called Japanese Crude Cocktail), an index used for long-term LNG contracts in Japan, Taiwan, and Korea, where the LNG price was set from Alaska project. Interview with Japanese industry person, conducted on Jan. 18, 2012.

47 Kubo Iwao and Kato Shigeru, *Zukai Shōsha Gyōkai Hendobuku [An Illustrative Guide to the Trading Companies]*, 72–77.

48 These two companies also signed a Memorandum of Understanding (MOU) for strategic collaboration in their natural gas businesses on February 25, 2011. See Mitsubishi Corporation, "MC Signs MOU with Korea Gas Corporation in Natural Gas Business," *Press Release*, Feb. 25, 2011.

49 Itochu Oil Exploration Co. Ltd., *Itochu Corporation Annual Report 2010* (Tokyo: Itochu Oil Exploration Co. Ltd., 2010), 25.

the BTC (Baku-Tbilisi-Ceyhan) pipeline project, which transports oil from ACG oil field in the Caspian Sea region to the Mediterranean Sea.[50]

(d) **Sumitomo Corporation**

Sumitomo was Japan's first oil developer to enter the UK North Sea, where the company produces seven thousand BOE/day in nine blocks and was appointed as an operator, which is rare among Japanese trading companies.[51] Originally, Petro Summit had been the core business of Sumitomo's E&P operations in the UK, but it was sold to Idemitsu Kosan in November 2009.[52] For LNG production, Sumitomo established LNG Japan Corporation with Sojitz, and LNG Japan joined projects, including Tangguh in Indonesia, as an equity investor. In 2009, the company became Japan's first company to participate in a shale gas project by joining the Barnett shale field drilling project in Texas.[53]

(e) **Marubeni Corporation**

Marubeni also attempts to develop oil and gas to enhance its earnings base through Energy Division-II. The company holds interest in oil and gas fields in the US Gulf of Mexico, the UK North Sea, the Indian Ocean, Qatar, and Sakhalin.[54]

Vertically Integrated Companies

Unlike Western countries, oil companies integrating upstream and downstream have not been dominant in Japan. This is partly attributed to the evolution of Japan's oil industry, where downstream companies have shown a strong presence with stable oil supply by Western oil majors.

(a) **JX Nippon Oil and Gas Exploration Corporation**

---

[50] Itochu Oil Exploration Co. Ltd., "Itochu's Activities: Azerbaijan (Btc Pipeline)," Itochu Oil Exploration Co. Ltd., http://www.itochuoil.co.jp/e/world_index-e.html.

[51] Sumitomo Corporation, "Our Business: Mineral Resources, Energy, Chemical & Electronics Business Unit," http://www.sumitomocorp.co.jp/english/business_overview/resource_chemical/outline15b.html.

[52] Sumitomo Corporation, "Share Sale of Petro Summit Investment UK Ltd.," *News Release*, Nov. 13, 2009.

[53] Sumitomo Corporation, *Annual Report 2011* (Tokyo: Sumitomo Corporation, 2011), 21.

[54] Marubeni Corporation, "Our Business: Energy Division-2," www.marubeni.com/division/energy_2.html.

JX Holdings, Inc. was created through the M&A of Nippon Mining in 2010, which was subsequent to the merger between Nippon Oil and Mitsubishi Oil in 1999. JX Holdings is composed of JX Nippon Oil and Gas Exploration Corporation for oil and gas upstream, JX Nippon Oil and Energy Corporation for oil and gas downstream and JX Nippon Mining and Metals Corporation for metal business.[55] While JX Nippon Oil and Gas Exploration Corporation has a stable value chain with JX Nippon Oil and Energy Corporation, the refiner company of JX Holdings, the company's upstream business has its own independent aim of maximizing profit.

Its predecessors had origins in mining: Nippon Oil since 1888, and Nippon Mining since 1914. In addition, the oil venture is encouraged to make its own profit rather than to supply crude oil to its affiliated refiner. Currently, the company pursues its own exploration opportunities with about 140 engineers, which is different from oil ventures by Japan's trading houses.[56] With Malaysia, Vietnam, and the UK as its core areas, JX Nippon Oil and Gas Exploration Corporation produces 140 thousand BOE/day, as of March 2011.[57]

(b) **Idemitsu**

Idemitsu is strong in the petroleum products business, accounting for 81.5% of sales with 640 thousand b/d refining capacity and 4,100 service stations in Japan. It does not emphasize oil and gas E&P, producing only thirty thousand BOE/day, or less than 2% of net sales in FY 2010.[58] Though its first oil venture began with an offshore field in Thailand by acquiring interests from Amoco Thailand Petroleum Company (now BP), active E&P activities have been implemented in Norway, the UK, and South East Asia since the establishment of Idemitsu Oil & Gas Co., Ltd., Idemitsu's E&P arm, in 1989.[59]

[55] JX Holdings. Inc., *Jx: From Integration to Sustaining Power* (Tokyo: JX Holdings. Inc., 2010).

[56] Interview with Japanese industry person, conducted on July 1, 2011.

[57] JX Holdings. Inc., *Jx: From Integration to Sustaining Power*, 30–31. JX Holdings. Inc., *Annual Report 2011: Bold Strategies To...* (Tokyo: JX Holdings. Inc., 2011), 4.

[58] Idemitsu Oil & Gas Co. Ltd., "Message from the President," Idemitsu Oil & Gas Co., Ltd., http://www.iog-idemitsu.co.jp/eng/company/greeting.html; Idemitsu Kosan Co. Ltd., *Annual Report 2011* (Tokyo: Idemitsu Kosan Co. Ltd., 2011), 2, 10, 11.

[59] Idemitsu Oil & Gas Co. Ltd., "Project: Thailand," Idemitsu Oil & Gas Co. Ltd., http://www.iog-idemitsu.co.jp/eng/project/thailand/index.html; Idemitsu Kosan Co. Ltd., "Resource Development: Oil and Gas," http://www.idemitsu.com/products/resource/oil.html.

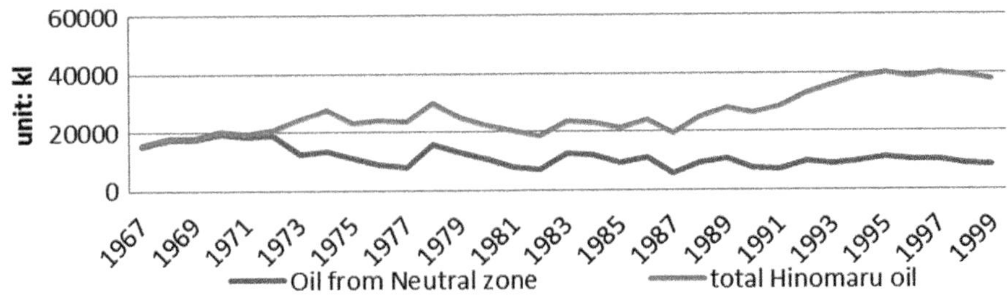

Fig. A.2 AOC's Contribution to *Hinomaru* Oil Imports (*Source* Japan National Oil Corporation, *Sekiyukōdan Nijūnensi [20-Year History of Japan National Oil Corporation]* 492–493; Japan National Oil Corporation, *Kono Jūnenno Mayumi: 1987nen-1996nen, Sōritzu San Jūnen Kinensi [This 10-Year of Footprint: 1987–1996, 30 Year-Anniversary Report]* (Tokyo: Japan National Oil Corporation, 1997), 210; Japan National Oil Corporation, *Sekiyukōdanshi (1997 <Heisei9> Nendo ~2004 <Heisei16> Nendo) [The History of JNOC (1997~2004)]*)

(c) **AOC Group**

Since its first success in Japan's overseas oil development history in 1961 in the Neutral Zone between Saudi Arabia and Kuwait, the oil developed by the AOC in the area had accounted for more than 90% of *Hinomaru* oil until 1972, after which its portion fell to about half of *Hinomaru* oil import.[60] The AOC had been the main producer of *Hinomaru* oil until it failed to extend a concession with Saudi Arabia for the Neutral Zone in 2000 (Fig. A.2).

However, the expiration of the concession agreements in February 2000 with Saudi Arabia and in January 2003 with Kuwait changed the situation completely. While the AOC continued to be engaged in petroleum operations under the Technical Services Agreement with Kuwait until January 2008, the company could not find an alternative to the Neutral Zone oil fields.[61] In response to this situation, the AOC merged with Fuji Oil, a local refiner established in 1964 with the AOC's investment, to make AOC Holding Company, and moved to the Norwegian continental shelf, the Gulf of Suez, and Egypt for oil ventures. However, its oil production fell to 37.2 thousand b/d in 2011, 99% of which is composed of crude oil produced in Kuwait under the 20-year crude oil sales and purchase agreement concluded with that country.[62]

60 In 1972, the pro-rata system, where MITI guided Japanese refiners to purchase AOC's oil based on market share, was finished.

61 Arabian Oil Company, "Agreements with Kuwait," Arabian Oil Company, http://www.aoc.co.jp/e/pro/agreement.html.

62 AOC Holdings, *Annual Report 2011* (Tokyo: AOC Holdings, 2011), 9–10.

Utility Companies

Japan's utility companies also have been involved in overseas oil and gas development. The purpose of this business is to create new opportunities for profit creation and to stabilize the energy supply through long-term contracts. Equity investment is the main investment method of these firms.

(a) **Tokyo Gas and Osaka Gas**

Among gas companies, Osaka Gas is more active than Tokyo Gas in overseas oil and gas development, although the sales volume of Osaka Gas is two-thirds of that of Tokyo Gas.[63] In addition, while Tokyo Gas's activity is limited to investment in natural gas development, Osaka Gas expands its investment not only for gas but also for oil projects, such as Idemitsu's Snorre oil development in the Norwegian North Sea.

For Tokyo Gas, the main purpose of upstream business is to develop a new profit source,[64] and as of 2011, its equity investments were focused in Australia.[65] On the other hand, Osaka Gas, the first Japanese gas utility company involved in overseas oil and gas projects, participated in nine upstream projects including the Snorre oil field, as well as a 1.25% stake in the Gorgon gas project. It also joined a shale gas project in Canada in 2011.[66] Osaka Gas's acquisition of equities in upstream business aims to guarantee a stable value chain by securing a long-term gas supply for downstream operations.[67]

(b) **Tokyo Electric, Kansai Electric, and Others**

Electric companies joined oil and gas ventures primarily to secure their fuel supplies. Tokyo Electric has been investing in overseas oil development since the AOC's venture in the Neutral Zone. It also has an 11.25% equity share in the Wheatstone LNG project. Chubu Electricity, located in the Nagoya area, has invested in a shale gas project in British Columbia, Canada by acquiring

63 Osaka Gas Group, *Osaka Gas Group Annual Report 2011* (Osaka: Osaka Gas Group, 2011), 3.

64 Interview with Japanese industry person, conducted on July 1, 2010.

65 Its investments in Australia are as follows: 3.07% interest in the Darwin Project, a 5.0% interest in the Pluto Project, a 1.0% interest in the Gorgon Project, and a 1.25% interest in the Queensland Curtis LNG Project.

66 Osaka Gas Group, *Osaka Gas Group Annual Report 2011*, 14, 34–35.

67 Ibid., 34.

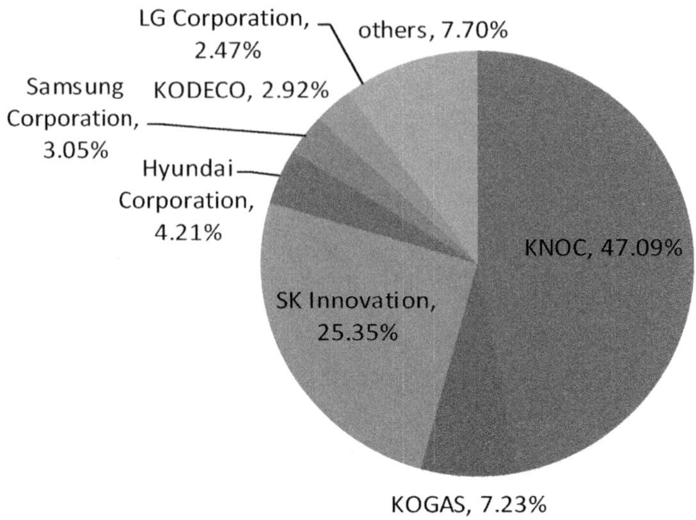

Fig. A.3 Korea's Overseas Oil and Gas Production in 2011 (*Source* Material of Energy & Mineral Resources Development Association of Korea [EMRD])

7.5% of equity for Cordova Gas Resources Ltd., a subsidiary of Mitsubishi.[68] Kansai Electric has equity in the LNG Pluto project in Australia.[69]

Korea's Oil Companies

In Korea, state-owned companies' initiatives in overseas oil development are in contrast to Japan's patterns. The SOEs' contribution in the business is significantly larger than private companies in terms of production level and investment. As of 2011, state-owned companies' oil and gas production was 251 thousand BOE/day, larger than the private sector's production of 214 thousand BOE/day.[70] Decision-making in state-owned companies is affected by greater consideration of energy security than in private companies (Fig. A.3).

It is worthwhile to note that *chaebols* are not so active in oil and gas ventures except in the SK Group. While Korean GTCs, subsidiaries of *chaebols*, had been eager to get involved in oil ventures in the 1980s, unsuccessful

[68] Chubu Electric Power Co., Inc., *Annual Report 2011* (Nagoya: Chubu Electric Power Co. Inc., 2011), 14.

[69] Kansai Electric Power Co. Inc., *Annual Report 2011* (Osaka: Kansai Electric Power Co. Inc., 2011), 20.

[70] Ministry of Knowledge Economy, "Performance of Resource Development and Future Plan," 114th Emergency Economic Response Meeting (Kyunggido, 2012), 5.

performances and an unfavorable business environment following the Asian Economic Crisis discouraged their efforts.

Korea's State-Owned Oil and Gas Companies

Korea has two state-owned companies related to overseas oil and natural gas development. One is KNOC, originally established for oil stockpiling and oil development. The other is KOGAS, which was originally founded as a natural gas wholesaler, but was expanded in the early twenty-first century to conduct overseas oil and gas development. Both companies have been actively involved in E&P in the early twenty-first century, which possibly resulted in a competitive rather than complementary relationship with Korea's private companies. In 2009, the total investment for overseas oil and gas development by the KNOC and the KOGAS was $5.2 billion, compared to the private sector's $0.57 billion.[71]

(a) **The KNOC**

At the end of 2011, the KNOC operated 92 projects, and its daily production reached 219 thousand b/d, accounting for 47.1% of Korea's oil production.[72] It plans to be a major player in the international energy industry with a daily production of 300 thousand b/d and reserves of 2 billion barrels by 2012, as stated in the "Great KNOC 3020" action plan[73] The KNOC has also been aggressive in M&A of foreign oil companies, such as Canada's Harvest Energy, Kazakhstan's Sumbe, and the UK's DANA Petroleum.[74]

Since its inception as a wholly owned SOE in 1979, the KNOC has played an important role in securing Korea's oil supply. Its tasks, which are stipulated in KNOC Law, range from oil stockpiling, to administering subsidies for overseas oil development on behalf of the Korean government and oil development in the domestic continental shelf and foreign countries.[75] In

[71] Ministry of Knowledge Economy, "Historical Investment for Overseas Resource Development," *Press Release*, Jan. 19, 2010.

[72] Ministry of Knowledge Economy, "Performance of Resource Development and Future Plan," 5; Korea National Oil Corporation, "Overview of Oil Development Business," http://www.knoc.co.kr/sub03/sub03_2_1.jsp.

[73] Korea National Oil Corporation, "Overview of Oil Development Business," http://www.knoc.co.kr/ENG/sub01/sub01_3.jsp.

[74] "300 Thoudand Barrel of Daily Production by 2012," *The Korea Energy News*, Jan. 11, 2010.

[75] Article 10 in *PEDCO Law* (Dec. 5, 1978).

addition, the KNOC had been an information source for government officials due to its accumulated knowledge and data from its field operations.[76]

As an SOE, the KNOC operates under a unique set of management principles. Unlike Japan's INPEX, which borrows money from the financial market, KNOC's investment funds are provided by the government based on the KNOC Law.[77] In the case of the expansion of the KNOC in 2008, the Korean government plans to invest U$3.9 billion until 2012.[78] Also, the government can provide loan guarantees for the KNOC and give subsidies for KNOC's projects.[79] KNOC's characteristic as an SOE also provides favorable financing conditions because the applied credit rate is the sovereign rate.[80]

On the other hand, its operations must face scrutiny from three ministries and the National Assembly. While the right to monitor the KNOC had originally been given only to the Board of Audit and Inspection (BAI),[81] the Law on Management of Public Institutions, enacted in 2007, extends that right to the MOSF and the MKE.[82] The finance ministry also has a *de-facto* power to manage SOEs by associating the result of the annual performance evaluation with actions related to personnel, budgets, and bonuses.[83] Also, the Standing Committee on Knowledge Economy of the National Assembly

---

[76] Interview with MKE official, conducted on June 19, 2011.

[77] KNOC'S capital stock stipulated in KNOC Law has changed as follows: US$103 million (enacted in December 1978) -> US$3,735 million (revised in March 1994) -> US$4,208 million (revised in January 1999) -> US$10,764 million (revised in August 2007) -> US$11,818 million (planned as of 2012). This calculation is based on the annual average exchange rate in the WDI. In the case of 2012, US$ = 1,100 won.

[78] Ministry of Knowldge Economy, "KNOC Production Will Be Six Times of Current Level," *Press Release*, June 11, 2008. The dollar amount is calculated based on $1 = 1030 won, as of June 11, 2008.

[79] Articles 14 and 15 in *KNOC Law*.

[80] In 2012, Korea's sovereign rate evaluated by Moody's is A1.

[81] In the case of the BAI's inspection of overseas resources development during 2011–2012, the BAI requested KNOC to suspend a team leader due to his inappropriate conduct during KNOC's M&A of Harvest Trust Energy, a Canadian oil company, and asked the MKE to punish another high-ranking official of KNOC for his indulgence in the projects of Harvest and Iraqi Kurdistan. See Board of Audit and Inspection, "Inspection Result Report: Situation of Overseas Resources Development and Introduction," ed. Board of Audit and Inspection (Seoul, 2012).

[82] Hoon You, Yong-su Bae, and Won-hui Lee, *Konggiŏmnon [Public Enterprise]* (Kyunggido: Pŏmmunsa, 2010), 81–82.

[83] Article 48 in *Law on Management of Public Corporation*, (Jan 19, 2007); Article 27 in *Enforcement Ordinance on Management of Public Corporation* (Mar 27, 2007). For the annual evaluation, the evaluation unit is composed mainly of university professors, and evaluation criteria are as follows: leadership and strategy, management system, the management performance. These criteria require the SOE to follow the government's policy recommendation, such as recruitment of the disabled. See Management Evaluation Unit, "2009 Nyŏndo Konggiŏp, Chunchŏngbukikwan Kyŏngyŏngsilchŏkpyŏngga Pogosŏ: SOC Yuhyŏng [State-Owned Company, Quasi-Public Organization Management Performance Evaluation Report of 2009: Category of Social Overhead Capital (SOC)]," ed. Ministry of Strategy and Finance (Kyunggido: Ministry of Strategy and Finance, 2010).

reviews the management of the KNOC at the annual state affairs inspection in the fall. Furthermore, following President Roh Moo-hyun's "balanced development policy" (*kyunhyŏng palchŏn chŏngch'aek* in Korean), the KNOC is required to move its headquarters from the Seoul metropolitan area to Ulsan, a city in the south eastern part of the Korean Peninsula in October 2014.[84] Associated with this relocation of public institutions, the distances between *Chŏngwadae* in Seoul, the MKE in Sejong Administrative City, and the KNOC in Ulsan has generated concern about how to maintain efficient communication among them.

Unlike Japanese oil companies, the CEO of the KNOC is appointed by the president. Another difference is that CEOs of the KNOC tend to be confidants of the president or retired military generals rather than retired bureaucrats. This practice can be attributed to the custom of respect for the military under Presidents Chun Doo-hwan and Roh Tae-woo. This custom has changed since President Lee Myung-bak's appointing of Mr. Kang Young-won in August 2008, who used to be president of Daewoo International, as CEO of the KNOC.[85]

(b) **The KOGAS**

At the end of 2011, the KOGAS was producing 32 thousand BOE/day, which amounts to 6.8% of Korea's overseas oil and gas production. While the KOGAS was founded in August 1983 as a state-owned natural gas wholesaler, it has expanded its business to E&P since January 1997 and has participated in 17 oil and gas E&P projects, as of October 2011.[86] Since 2006 when energy prices soared, the KOGAS has made rigorous efforts in its upstream business, and in March 2011 the KOGAS Law was revised to expand this part of the company's operations to include oil and natural gas exploration. The momentum for this decisive change came from KNOC's difficult situation in Iraq, where it was on the Iraqi central government's blacklist following a July 2008 oil development deal with Iraqi Kurdistan regional government that was

---

[84] "Construction for KNOC's New Headquarter Building in Ulsan Innovation City Began," *Yonhap News*, Nov. 22, 2011. Korea Energy Economics Institute (KEEI) also moved to Ulsan City for a similar cause.

[85] Hiring outside experts is a practical dimension in the management of Korea's SOEs. Another example is KNOC's hiring of two foreigners to be head of human resource management and lead their own research institute. This is different from INPEX, where no foreigners have been employed in the headquarters. See Korea National Oil Corporation, *Annual Report 2010* (Kyunggido, 2011), 15.

[86] Korea Gas Corporation, "Presentation to Investors," ed. Korea Gas Corporation (Kyunggido, 2011), 9.

not endorsed by the Iraqi central government.[87] The KOGAS stepped in to bid in KNOC's place, and won all three contracts for Iraqi oil and gas fields, which were held by the Iraqi Ministry of Oil in 2009 and 2010. The KOGAS succeeded in winning the Zubair contract in 2009 with its partnership with ENI and Occidental. In addition, the KOGAS was awarded a contract for the Badra oil field with other foreign oil companies such as Gazpromneft and Petronas in 2010. Also, in October 2010 the KOGAS was awarded the Akkas gas field as operator with Kazmunaigas, Kazakhstan's state-owned gas company, as well as the Mansuriyah gas field with TPAO, a Turkish national oil company, and Kuwait Energy Co. in October 2010.[88]

Considering the KOGAS has its special law and public ownership is 60.91%, as of December 31, 2010, the KOGAS is also a state-owned gas company.[89] Like the KNOC, the Korean government and the National Assembly maintain their influence over the management of the KOGAS through the appointment of the CEO, supervision, inspection, and annual performance evaluation. The current CEO of the KOGAS is Choo Kang-soo, who used to work as Executive Director with President Lee Myung-bak, the then-CEO, in Hyundai Energy and Resources (HER).[90] One example of government intervention in KOGAS's operation is the use of the profits from overseas oil and gas equity investment. Despite members of National Assembly continuously pointing out the need for reinvestment into E&P in the annual state affairs inspections since 2005,[91] the MKE ordered the KOGAS to use 344.5 billion won (about $300 million) of its dividend

87 "Korea and Iraq Agreed to Co-Develop Basra," *Yonhap News*, Feb. 24, 2009.

88 Petroleum and Contracts Licensing Directorate, "Iraq's First Petroleum Licensing Round: Zubair Contract Area–Bidding Results," ed. Republic of Iraq Ministry of Oil (Baghdad, 2009); Petroleum and Contracts Licensing Directorate, "Iraq's Second Petroleum Licensing Round: Badra Contract Area–Bidding Results," ed. Republic of Iraq Ministry of Oil (Baghdad, 2009); Petroleum and Contracts Licensing Directorate, "Iraq's Third Petroleum Licensing Round: Akkas Contract Area–Bidding Results," ed. Republic of Iraq Ministry of Oil (Baghdad, 2010).

89 Korea Gas Corporation, "Current Shareholder," Korea Gas Corporation, http://www.kogas.or.kr/kogas_kr/html/investor/investor_12.jsp. Public shareholders, as of December 31, 2010, are: the central government (26.86%), Korea Electric Power Corporation (24.46%), and local governments (9.59%).

90 "Hyundai Energy and Resources Established with Appointing Lee Myung-Bak as CEO," *Yonhap News*, Sept. 1, 1990.

91 Arguing the need to reinvest profits for E&P were not only ruling party members, such as Oh Young-sik, Kim Dŏk-gyu and Choi Chŏl-gyu, but also opposition party members, such as Kwon Sŏn-taek, a member from the People Center Party.

income to lower the domestic retail price of natural gas.[92] The MKE maintains this policy direction as of 2011.[93] Furthermore, like the KNOC, the KOGAS was required to move its headquarters out of the Seoul metropolitan area by 2014. However, it moved not to Ulsan with the KNOC, but to Daegu.

Private Companies in Overseas Oil Development

While Japanese private companies have been aggressively involved in overseas oil development, Korean private companies are less active in this area. With the noticeable exceptions of SK Innovation and Daewoo International, Korean *chaebols* have invested a very small portion of the equity in oil development. While Korea's *chaebol* had engaged in overseas oil development aggressively in the 1980s, as the focus of their business became diversified, their interest in oil ventures diminished. However, it is worthwhile to note that in the first phase of Korea's oil venture in the 1980s, it was not SOE but private companies that had initiated the business as seen in KODECO's successful oil discovery in Indonesia.

(a) **Vertically Integrated Oil Companies**

(1) **SK Innovation**

SK Innovation is a private holding company that evolved from *Yukong*, Korea's former state-owned oil refiner that was sold to SK Group in 1980. Since its attempt to acquire the Karimun oil field in Indonesia in 1983, the company has been active in overseas oil development, and it achieved a few successes including the Yemen Marib oil exploration in 1984.[94] In 2011, its oil and gas daily production was 64 thousand BOE/day, which accounts for 13.7% of Korea's equity oil and gas,[95] and it is Korea's biggest private oil

[92] Chŏl Sohn, "Controversy on KOGAS's Using the Benefit from Equity Investment to Lower Gas Retail Prices," *Seoul Economic Daily*, Sept. 22, 2005; Ahn Kyung-ju, "344.5 Billion Won of Dividend from Overseas Equity Investment Will Be Refelcted in Gas Retail Pricing," *E-Today*, Oct 22, 2008; Jae-eun Kim, "KOGAS Used Money Reserved for Reinvestment to Price Lowering," *Korea Economic Daily*, Oct. 21, 2008.

[93] Interview with MKE official, conducted on April 19, 2011.

[94] The Marib oil field discovery in 1984 was driven by the Hunt Oil, the US independent oil company. Interview with Keun-Wook Paik, senior research fellow at Oxford Institute for Energy studies (OIES) at the University of Oxford, on Oct. 7, 2012.

[95] SK Innovation, "Presentation to Investors" (Seoul, 2012), 22.

company. As of 2010, oil and gas E&P is one of the company's main business areas, and E&P accounts for approximately 15% of operating profit and 8% of sales.[96] Its proactive operation in oil E&P can be attributed partly to its independence from Western oil majors.

The characteristics of the company as a vertically integrated private company are likely to make the company's management less affected by the government's concerns about the oil and gas self-sufficiency ratio than that of the KNOC, an upstream-only state-owned oil company. As a private company, economic factors such as profit and risk are SK Innovation's main concerns in investment decision-making. Contributing to the nation's self-sufficiency ratio is not its main concern, despite a recommendation by the government to take this into account. This explains its decision to sell its stake in the Brazilian offshore field to Maersk Oil, a Danish oil company, despite a successful repayable loan from the government and an expectation that the field would be retained in order to enhance Korea's oil self-sufficiency ratio.[97]

(2) **GS Caltex**

GS Caltex, formerly LG Caltex, got involved in oil E&P later than SK Innovation. Since it partnered with Caltex in 1967, most of its crude oil has been supplied by Chevron, and Chevron has maintained its influence over decision-making related to business, based on equal ownership between GS Holdings and Chevron.[98] Therefore, E&P participation by GS Caltex did not begin until 2003, and even then, the investment level has been limited to four projects with Chevron in Cambodia, Bangladesh, and Vietnam, and two Thailand projects with Caltex. It is worth noting that, in addition to GS Caltex, GS Holdings has also participated in seven oil and gas E&P projects based on its independent status from Chevron.[99]

---

[96] SK Innovation, *2010 Annual Report: Innovate Change for Tomorrow's Energy* (Seoul: SK Innovation, 2011), 10–11; Interview with industry insider, conducted on April 18, 2011. Most of SK Innovation's sales come from its refinery business.

[97] Ministry of Commerce Industry and Energy, "SK Energy Will Launch Oil Production at Brazil Bmc-8 Field from the End of July"; SK Innovation, "SK Energy Makes Strategic Oil Development Model through Restructuring Portfolio of Oil Field," *Press Release*, Dec 24, 2010. While the KNOC once sold its equity to other companies, the buyers in this case were Korean domestic financiers. Therefore, KNOC's action did not affect the national self-sufficiency ratio.

[98] Samjong KPMG, "GS Caltex and Its Subsidiary: Audit Report on Consolidated Financial Statements" (Seoul, 2011), 12.

[99] GS Caltex, "Exploration and Production," GS Caltex, http://www.gscaltex.com/eng/business/production.aspx.

(b) General Trading Companies (GTCs or *Chonghapsangsa*)

Unlike Japan, investment by Korean GTCs in overseas oil development has been implemented mostly without the cooperation of other subsidiaries in the group. In addition, the main motivation of Korean GTC's has been to promote exports, which afforded a structure less favorable to oil and gas E&P than did in Japan's GTCs, which are more import-oriented.

(1) Daewoo International

Daewoo International, which used to be a subsidiary of Daewoo Group, has been aggressive in oil and gas E&P and maintains nine projects, as of August 2010. These projects include Daewoo International's operatorship in the Myanmar A-1 and A-3 offshore natural gas development projects, from which natural gas would be exported to China beginning in May 2013.[100] The company is similar to Japan's MOECO, which tries to accumulate the capacity to operate oil and gas fields independently. Pohang Steel Corporation (POSCO) merged with Daewoo International in August 2010 in order to expand POSCO's business and stabilize the supply of materials.[101]

(2) Samsung C&T Corporation

Samsung C&T Corporation, a subsidiary of Samsung Group, has considered oil exploration, such as E&P in Peru, since 1982.[102] The company began its E&P projects in 1987 with Malaysia's SK-7 oil field exploration. The company has nine production fields, as of 2011. After Lee Kun-hee, Chairman of Samsung Group, emphasized overseas resource development in June 2011, the company became aggressive in its foreign oil ventures.[103] As part of this revised agenda, in November 2011 Samsung merged with Parallel Petroleum, a US oil company, a first for Korean private oil developers.[104]

---

[100] Daewoo International, "Daewoo International IR Material" (Seoul, 2010), 22–27.

[101] Dong-wook Kwon, "Daewoo International Embraced to Posco in Ten Years," *YTN News*, Aug. 30, 2010.

[102] Ministry of Energy and Resources, *Tongnyŏkjawonhaengchŏng Simnyŏnsa [10-Year History of Ministry of Energy and Resources]* (Kyunggido: Ministry of Energy and Resources, 1988), 389.

[103] Lee Kun-hee, Chairman of Samsung Group has mentioned that if Samsung Corporation had made efforts in resource development, the company would have reached to the level of 70% of Mitsubishi Corporation. See Mingki Choi, "Samsung Corporation Tries Resource-Rich Country through Lee Kun-Hee," *EBN Industry News*, Dec. 1, 2011.

[104] Chang-min Jang, "Samsung Corporation Took over U.S. Oil and Gas Company," *Korea Economic Daily*, Nov. 30, 2011.

Samsung's investments are made in relatively stable areas like the United States.

(3) **Hyundai Corporation**

Hyundai Corporation was the pioneer in Korea's overseas resource development, as seen in the Drayton coal mine development since the late 1970s. The company's target in the oil business is profit-maximization and it depends on equity investments in projects such as Yemen's Marib, Oman LNG, and Qatar LNG.[105] The company became a subsidiary of Hyundai Heavy Industry (HHI) in 2009, and it is expected that Hyundai Corporation will become more proactive, owing to HHI's healthy financial situation. Hyundai Corporation established Hyundai Energies and Resources (HER) together with HHI and Hyundai Oil Bank in April 2011.[106]

(4) **LG International**

Along with coal and non-ferrous metals, oil, and natural gas development is a main business priority of LG International, a subsidiary of LG Group. Since its investment in the Adang oil field in 1984, which is the first attempt at oil E&P among Korean GTCs, the company has been involved in foreign oil ventures. While focusing on Kazakhstan, the company has expanded investment areas to other continents, such as South America. At the end of 2010, LG invested in eight projects including three production fields.[107]

(c) **Utility Companies**

Traditional energy companies of the 1960s and the 1970s evolved into utility companies in the 1980s and oil and gas developers mainly in the twenty-first century. The companies producing briquettes, which used to be a main fuel source in Korea, shifted their business to urban natural gas supply in the 1980s. Following the saturation of the natural gas market and the competition with district heating, they became interested in oil and gas

[105] Hyundai Corporation, "Global Business Organizer Hyundai Corporation" (Seoul, 2012).

[106] Hyo-jin Jang, "GTCs Get Fund for Resource Development in a Various Way," *E-Today*, Nov. 12, 2012. The establishment of HER is regarded as a reincarnation of the old HER, which was established by Chung Ju-yung, the founder of Hyundai Group. HER was led by Yang, Bong-jin as of 2011.

[107] E-mail interview with industry person, conducted on April 29, 2011.

development.[108] These companies include Daesung, Seoul City Gas, and Samch'olli, and their investment is mainly equity investment in low-risk countries like the United States.

(d) **Shipbuilding and Shipping Companies**

The involvement of shipping company groups in overseas oil development is another feature that distinguishes Korea from Japan.[109] Shipbuilding business groups expanded their business area to oil and gas E&P in the early twenty-first century, when oil prices were rising rapidly. STX Group, a shipping and shipbuilding-centered business group established STX Energy in 2002 to diversify its future profit-earning sources. In order to strengthen its energy business, STX Group also hired a retired high official from the MKE.[110] Daewoo Shipbuilding and Marine Engineering (DSME) founded DSME ENR in 2007, and the company's oil E&P is expected to provide opportunities to export maritime plants, such as LNG careers and FPSO (floating production, storage, and offloading) units.[111] In order to lower investment risks, both companies prefer oil and gas fields in the production stage rather than exploration.[112]

(e) **KODECO Energy**

While its current E&P function is weak, it is worth mentioning KODECO as the first Korean oil developer to succeed in developing oil in a foreign country. KODECO Energy, a subsidiary of Korea Development Company (which had maintained a lumbering business since 1968 in Indonesia), explored a West Madura offshore field based on a production sharing contract (PSC) with PERTAMINA, the Indonesian state-owned oil company, and

---

108 Interview with a Korean industry person, conducted on April 13, 2011.

109 It seems that A.P. Moller-Maersk Group (Maersk), a Danish business conglomerate, is the model of this evolution of shipping companies. Maersk Group started as a shipping company in 1904, and Odense Steel Shipyard was established in 1919. In 1962, Maersk received a concession for oil and gas exploration and production in the Danish sector of the North Sea, and consequently, Maersk Oil was established. In 2010, Maersk Oil provided the group with US$1.7 billion of its total US$5.0 billion of profit. See A.P. Moller-Maersk Group, "Annual Report 2010, A.P. Moller-Maersk Group," *Press Release*, Feb. 23, 2011.

110 Jong-wook Park, "Lee Byung-Ho, CEO for Trading and Business in STX," *Maeil Economic Daily*, July 6, 2009. As of 2012, STX Energy is the only oil developer that has hired a former MKE high official as CEO.

111 Sang-yong Kim, "Global Korea, We Make It Ealier: Daewoo Shipbuilding & Marine Engineering," *Seoul Economic Daily*, May 25, 2011.

112 Jung-hyung Jo, "Lee Byung-ho, CEO of STX Energy," *ET News*, Mar. 19, 2012; Daewoo Shipbuilding & Marine Engineering, "Future Value Generator DSME ENR" (Seoul, 2012), 7.

Korea's first equity oil was imported in 1984. However, mismanagement of the oil field and financial difficulties after the Asian Economic Crisis diminished KODECO's E&P business. In May 2011, the contract of KODECO with PERTAMINA for West Madura offshore oil field was extended.[113]

Institutional Setting Between State and Business: Industrial Associations

Japan Petroleum Development Association (JPDA)

The JPDA, which was established on November 20, 1961, has made efforts to reflect the opinions of the private sector in government policymaking. In particular, it has provided opinions for the management of JOGMEC, the state funder for the private sector in overseas oil and gas development. The association also disseminates information and advertises the importance of oil development on behalf of 19 member companies.[114]

Energy and Mineral Resources Development Association of Korea (EMRD)

Korea's EMRD, which was established in 2008, represents the opinions of private sector actors related to overseas resource development. While the EMRD is roughly equivalent to Japan's JPDA as an industry association, it also has some similarities with JOGMEC in terms of its function and structure. The EMRD deals with both metal and oil and gas resources and has a Petroleum Team and a Mineral Resources Team.[115] The EMRD also reviews member companies' applications for government loans.[116] However, there are also a few distinct differences, namely that the EMRD is not a public agency but a membership association composed of oil and mineral resources developers. Moreover, member companies include not only resource developers,

113 PERTAMINA, "PERTAMINA Operator Blok WMU."

114 Japan Petroleum Development Association, "Contents of Work," www.sekkoren.jp/summary/sum3.htm; Japan Petroleum Development Association, "Member Companies," www.sekkoren.jp/summary/sum5.htm.

115 Energy and Mineral Resources Development Association of Korea, "Organization," www.emrd.or.kr/english/intro/organization.jsp; Japan Oil Gas and Metals National Corporation, "Organization," www.jogmec.go.jp/english/aboutus/organization.html.

116 Ministry of Knowldge Economy, "Improving Operation System to Enhance Transparency of Loan Delegation Committee on Overseas Resources Development."

but also state-owned banks, resource-related research institutes, construction companies, and accounting firms.[117]

Conclusion

This chapter illustrates how the state and businesses interact in overseas oil and gas development in Japan and Korea. As illustrated above, these two countries do not show a substantial difference in the structure at the state level. However, the substantial contribution of the private sector in Japan's overseas oil development is quite distinct from Korea's situation where state-owned energy companies account for a majority of the direct production and investment.[118]

In Japan, private companies produce all oil and gas from foreign countries, and JOGMEC, the state institution for overseas oil development, is involved only in financing, R&D, and information gathering, not direct production. Among private companies, upstream-only companies such as INPEX and JAPEX produce more than half of the self-developed oil and gas, with only a small role in GTCs.

On the other hand, in Korea, state-owned energy companies produce more than half the country's oil and gas. State involvement in this area has further expanded since 2003 in response to an oil price spike. The KNOC, Korea's state-owned oil company, was reinvigorated through massive financial investment from the state, as well as through increased M&A activities. Also, during this era, the KOGAS, Korea's natural gas wholesaler, expanded its function to oil and gas production, especially in Iraq. Among private companies, only SK Innovation produces and invests in a significant portion of Korea's overseas oil and gas development.

This distinction between Japan and Korea in overseas oil and gas development boils down to the question of why the Japanese state could not create or strengthen an SOE, while the Korean state could, despite the fact that

[117] Lee, "Government and Business Speed up Overseas Resources Development." While government officials favored this centralized system, some private sector people were skeptical about the establishment of a new association. One criticism concerned the appointment of Jung, Gyu-chang, former Chief of Gas Bureau at MOTIE, as the Vice Chairman and head of standing employees. See Park, "EMRD Became Common Eyesore. In This Week Launched. Without New Functions, It Would Be Redundant."

[118] This fact is parallel to Calder's finding that the Japanese private sector is the principal agent of economic success. See Calder, *Strategic Capitalism: Private Business and Public Purpose in Japanese Industrial Finance.*

both the Japanese and Korean governments attempted to found SOEs. This research explained the significant reasons by examining Japan in the 1960s and the early twenty-first century and Korea in the 1970s and the early twenty-first century through the lens of veto player theory.

Appendix B: Chronology of Postwar Overseas Energy Investment of Korea and Japan

See Table B.1.

S.-i. Oh, *Overseas Energy Investment of Korea and Japan*,
https://doi.org/10.1007/978-981-99-0285-9

Table B.1 Chronology of overseas energy investment

Japan	Time line (year)	Korea
(Jan) AOC discovers oil in the Neutral Zone between Saudi Arabia and Kuwait **(June)** North Sumatra Oil Development Cooperation (NOSODECO) was established for the oil production at Northern Sumatra oil field	1960	
(April) First oil developed by AOC arrives in Japan **(Nov)** JPDA (Japan Petroleum Development Association) is established	1961	
	1962	**(Oct)** *Yukong*, Korea's state-owned refiner, is established with the partnership of Gulf Oil
(Feb) North Sumatra Offshore Petroleum Exploration (a former entity of INPEX) is established	1966	
(June) Six-Day War breaks out **(Oct)** JPDC (Japan Petroleum Development Corporation) is established	1967	
(July) Mitsui Group establishes MOECO (Mitsui Oil Exploration Corporation)	1969	
(Feb) Mitsubishi Group establishes its oil developer	1972	
(Jan) Sumitomo establishes its oil developer **(Feb)** JODCO is established and acquired ADMA concession from BP **(July)** MITI Minister Nakasone establishes ANRE (Agency for Natural Resources and Energy) **(Oct)** Yom Kippur War breaks out and the first oil shock takes place	1973	**(Oct)** Yom Kippur War breaks out and the first oil shock takes place

(continued)

Table B.1 (continued)

Japan	Time line (year)	Korea
(Jan) Agreement on Joint Development of continental shelf is signed between Korea and Japan	1974	**(Dec)** Korea's National Assembly ratifies the joint development agreement between Korea and Japan
	1976	**(Jan)** President Park announced the oil discovery in Pohang
(June) Japan's Diet ratifies the joint development agreement between Korea and Japan	1977	
	1978	**(Jan)** Ministry of Energy and Resources is established **(Oct)** Hyundai Yanghaeng negotiates on oil development with PERTAMINA but failed
(Jan) Political unrest in Iran affects international oil market, which is called the second oil shock	1979	**(Jan)** Political unrest in Iran affects international oil market, which is called the second oil shock **(Mar)** PEDCO (Korea Petroleum Development Corporation) is established
(June) Nizu Sekiyu Kaihatsu begins oil exploration in Gulf of Bohai, China	1980	**(Nov)** Sunkyung becomes the owner of *Yukong*, the former entity of SK Energy
	1981	**(May)** KODECO signs PSA with PERTAMINA for West Madura offshore oil exploration
(Nov) Sumitomo joins oil exploration in the North Sea	1984	**(Apr)** KODECO begins exploratory oil production in West Madura **(July)** Oil is discovered in Yemen Marib, where Korean consortium joins with Hunt Oil **(Aug)** The first oil developed by KODECO arrives in Korea
	1991	**(Sept)** Overseas Petroleum Development Association is established
	1993	**(Feb)** President-elect Kim Young-Sam dissolves the Ministry of Energy and Resources

(continued)

Table B.1 (continued)

Japan	Time line (year)	Korea
(June) Oil production in Sakhalin-II, where Mitsui and Mitsubishi are major shareholders, begins	1999	**(June)** OPDA is merged to KPA (Korea Petroleum Association)
(Feb) AOC's concession with Saudi Arabia is expired **(Aug)** Petroleum Committee under MITI recommends to abolish oil self-sufficiency ratio target **(Aug)** INPEX discovers natural gas in Indonesia's Abadi offshore field	2000	
	2001	**(Aug)** KNOC consortium, including Conoco Philips, finds a giant-size oil reserve in Vietnam's offshore 15–1 field
(July) JNOC Dissolution Law passes the Diet	2002	
(Jan) AOC's concession is expired and technical service agreement is agreed between AOC and Kuwait	2003	
(Feb) JOGMEC is established	2004	**(Sept)** President Roh Moo-hyun's energy diplomacy begins
	2005	**(Nov)** Oil concession on Yemen Marib is not extended
(Apr) INPEX and Teikoku are merged to form Inpex Holdings **(Sept)** Prime Minister Koizumi visits Central Asia for energy diplomacy	2006	
(July) JOGMEC increases limit of equity capital investment from 50 to 75%	2007	
(Mar) The guideline for resources acquisition is approved in the Cabinet	2008	**(Feb)** EMRD is established **(June)** MKE announces the expansion plan for KNOC

(continued)

Table B.1 (continued)

Japan	Time line (year)	Korea
(Dec) JAPEX is awarded Iraqi Garaff oil field with PETRONAS	2009	**(July)** KOGAS is awarded Iraqi Zubair oil field with ENI consortium **(Dec)** KOGAS is awarded Iraqi Badra oil field with Gazpromneft consortium
(Oct) INPEX decides to withdraw from Azadegan oil project due to the US sanction to Iran	2010	**(Sept)** KNOC merges DANA Petroleum
(Mar) The Fukushima nuclear power plant is shut down due to Tohoku earthquake, which caused spike in LNG imports	2011	**(Mar)** The MOU and HOT for the oil development in Abu Dhabi offshore is signed **(May)** KODECO's concession of West Madura with PERTAMINA is renewed
(Sept) First LNG Producer–Consumer Conference is held	2012	
(May) The U.S. DOE approved the export of US LNG to Japan	2013	**(Mar)** KNOC discovers oil in the field of Hawler in the Iraqi Kurdistan **(Oct)** MOTIE announces restructuring plan for KNOC and KOGAS
	2014	**(Nov)** KNOC sells NARL **(Dec)** KNOC discovers 100 million barrel of oil in the UAE. The 14th Energy Committee decides no more new investment of KNOC
	2015	
(Nov) JOGMEC Law is revised to make JOGMEC enable to support development project and M&A	2016	**(Jan)** Government stops providing companies with success repayable loan
	2017	**(Jan)** Government resumes providing companies with success repayable loan
(July) Cabinet approves 5th Strategic Energy Plan. INPEX starts natural gas production in Ichthys offshore field	2018	**(May)** MOTIE asks Prosecutor's Office to investigate KNOC's Harvest project and KOGAS's West Cut Bank project
(July) INPEX got approval from Indonesian government on the Abadi LNG **(Sept)** Mitsui & Co. and JOGMEC decide to invest in Arctic LNG-2 Project	2019	**(July)** KNOC and GS Energy start oil production in the Haliba field in the UAE **(Oct)** DANA discovers natural gas in the Tolmount field in North Sea

(continued)

Table B.1 (continued)

Japan	Time line (year)	Korea
(Mar) Japan's New International Resource Strategy is established **(July)** First Yamal LNG arrived in Japan	2020	**(Jan)** DANA sells 25% of share of Tolmount field to Premier Oil **(Feb)** KNOC receives right to explore in the East Sea

Bibliography

"1,405-B.-Yen Losses Feared for JNOC - MITI Horiuchi." *Jiji Press*, July 28, 1998.

"300 Thoudand Barrel of Daily Production by 2012." *The Korea Energy News*, Jan. 11, 2010.

"4 Western Oil Companies Inactive in Continental Shelf Development." *Dong-A Ilbo*, Jan. 17, 1976.

"90th National Assembly Regular Session Ended."*Maeil Economic Daily*, Dec. 18, 1974.

A.P. Moller-Maersk Group. "Annual Report 2010, A.P. Moller-Maersk Group." *Press Release*, Feb. 23, 2011.

"Actively Implement Overseas Resources Development and Import." *Maeil Economic Daily*, Dec. 21, 1977.

"Administrative Reform Execution Conference Meeting Suggeted Abolition of JNOC and Idea to Integrate Oil Developers." *Asahi Shinbun*, Aug. 23, 2001.

Ahn, Kyung-ju. "344.5 Billion Won of Dividend from Overseas Equity Investment Will Be Refelcted in Gas Retail Pricing." *E-Today*, Oct. 22, 2008.

Amsden, Alice H. *Asia's Next Giant: South Korea and Late Industrialization*. New York: Oxford University Press, 1989.

Amyx, Jennifer Ann. *Japan's Financial Crisis: Institutional Rigidity and Reluctant Change*. Princeton: Princeton University Press, 2004.

Amyx, Jennifer Ann, and Peter Drysdale. *Japanese Governance: Beyond Japan Inc*, Politics in Asia Series. London; New York: RoutledgeCurzon, 2003.

AOC Holdings. *Annual Report 2011*. Tokyo: AOC Holdings, 2011.

"Approximately 5-Year Temporary Law, No Power to Licence Facility, Keidanren's Attitude on Petroleum Industry Law." *Asahi Shinbun*, Feb. 17, 1962, 1.

S.-i. Oh, *Overseas Energy Investment of Korea and Japan*,
https://doi.org/10.1007/978-981-99-0285-9

"Appointment of 24 Members for Energy & Resources Response Committee." *Dong-A Ilbo*, Apr. 18, 1978.

Arabian Oil Company. *The 35-Year History of the Arabian Oil Company 1958–1993: A Bridge between the Arabian Gulf and Japan in War and Peace*. Tokyo: Arabian Oil Company, 1995.

Arabian Oil Company. "Agreements with Kuwait." Arabian Oil Company. http://www.aoc.co.jp/e/pro/agreement.html.

Arisawa, Hiromi. *Enerugi Seisaku No Shindankai: Ōshū No Enerugi Kakumei [New Stage of Energy Policy: European Energy Revolution]*. Tokyo: Daiyamondosha, 1963.

Armstrong, Margaret, Rafael D'Arrigoet, Carlos Petter and Alain Galli. "How Resource-Poor Countries in Asia are Securing Stable Long-term Reserves: Comparing Japan's and South Korea's approaches." *Resources Policy* 47 (2016): 51–60.

Bae, June-Hoon. "A Study on the Reorganization of the Presidential Office: The Case of President Myung-Bak Lee & Quot's Early Presidential Office." *Journal of Secretarial Studies* 18, no. 2 (2009): 47–67.

"Big Demanders Show Disapproval to the Color of Control in Petroleum Industry Law Bill." *Asahi Shinbun*, Feb. 2, 1962.

"Big General Trading House is Eager for Oil Business." *Asahi Shinbun*, June 12, 1966.

"Blind Investment Produced Increase of Debt by 18 Trillion Won." *Newsis*, Oct. 12 2012.

Bloomberg. "Inpex Inks $20 Billion deal on Abadi LNG Project, the Largest Investment Ever in Indonesia." *Japan Times*, June 17, 2019.

Board of Audit and Inspection. "Inspection Result Report: Situation of Overseas Resources Development and Introduction." Seoul, Board of Audit and Inspection, 2012.

Board of Audit and Investigation. "Investigation Result Report: Status of Financial and Business Management of Public Corporation." Seoul: Board of Audit and Investigation, 2013.

Board of Audit of Japan. "Coverage of Audit." http://www.jbaudit.go.jp/english/effort/coverage.html#header.

Bok, Moon-Soo. "Analysis on the Fiscal Relationship between Government and National Assembly with Respect to Fiscal Democracy in Korea." *Korean Governance Review* 14, no. 3 (2007): 55–90.

Bremner, Brian. "Sorry, We're Not Paying." *Business Week*, June 3, 2002.

Brown, Gilbert T. *Korean Pricing Policies and Economic Development in the 1960s*. Baltimore: John Hopkins University Press, 1973.

Buchanan, James M., and Gordon Tullock. *The Calculus of Consent, Logical Foundations of Constitutional Democracy*. Ann Arbor: University of Michigan Press, 1962.

"Budget for New Year is 4,533.8 Billion Won." *Maeil Economic Daily*, Nov. 14, 1978.

"Budget, Unskillful Finish." *Asahi Shinbun*, Feb. 28, 1967.

"Buoyant Expectation, Oil Development." *Maeil Economic Daily*, Jan. 16, 1976.

"Bureaucracy Overhaul Key to Reforming Public Firms." *Kyodo News*, July 9, 2001.

Cabinet Secretariat. "Junichiro Koizumi Administration." http://www.kantei.go.jp/foreign/rekidaisouri/koizumi_e.html.

Cabinet Secretariat. "Statement by Prime Minister Junichiro Koizumi at the First Meeting of the New Cabinet." http://www.kantei.go.jp/foreign/koizumispeech/2001/0426setuji_e.html.

Calder, Kent E. *Crisis and Compensation: Public Policy and Political Stability in Japan, 1949–1986*. Princeton, NJ: Princeton University Press, 1988.

Calder, Kent E. "Elites in an Equalizing Role: Ex-Bureaucrats as Coordinators and Intermediaries in the Japanese Government-Business Relationship." *Comparative Politics* 21, no. 4 (1989): 379–403.

Calder, Kent E. *Embattled Garrisons: Comparative Base Politics and American Globalism*. Princeton, NJ: Princeton University Press, 2007.

Calder, Kent E. "Fueling the Rising Sun." *Harvard International Review* 19, no. 3 (1997): 24.

Calder, Kent E. "Japan's Energy Angst and the Caspian Great Game." *NBR Analysis* 12, no. 1 (2001).

Calder, Kent E. *Japan's Stealth Reform: The Key Role of Political Process*, Asia-Pacific Policy Paper Series. Washington, DC: The Edwin O. Reischauer Center for East Asian Studies, 2005.

Calder, Kent E. *Korea's Energy Insecurities: Comparative and Regional Perspectives*. Washington, DC: Korea Economic Institute of America, 2005.

Calder, Kent E. *Strategic Capitalism: Private Business and Public Purpose in Japanese Industrial Finance*. Princeton, NJ: Princeton University Press, 1993.

Caldwell, Martha Ann. "Petroleum Politics in Japan: State and Industry in a Changing Policy Context." The University of Wisconsin—Madison, 1981.

Campbell, John Creighton. *Contemporary Japanese Budget Politics*. Berkeley: University of California Press, 1977.

"Cautious on Abolition of Housing Finance Corporation." *Asahi Shimbun*, Sept. 15, 2001.

Chang, Soon-won. "Choi Choong-Gyong Said Chung Woon-Chan Not to Talk More About Benefit-Sharing." *E-Daily*, Mar. 16, 2011.

Cho, Dong-sung. *Han'gukŭi Chonghapmuyŏksangsa: Chaetowa Kwalli [Korea's General Trading Company: Institution and Management]* 2nd ed. 2 vols. Vol. 1. Seoul: Bŏmmunsa, 1984.

Cho, Dong-sung. *Han'gukŭi Chonghapmuyŏksangsa: Ponjilgwa Chŏllyak [Korea's General Trading Company: Nature and Strategy]* 2nd ed. 2 vols. Vol. 2. Seoul: Bŏmmunsa, 1984.

Cho, Dong-sung. *Kukchaechawollon: Sŏgyumunchaerŭl Chungsimŭro [International Natural Resources: Mainly on Petroleum]*. 2nd ed. Seoul: Pagyŏngsa, 1983.

Cho, Kap-che. "1976nyŏn P'ohangsŏgyupalgyŏnŭn Kajjayŏtta [Pohang Oil Discovery in 1976 Was Bogus]." *Monthly Chosun*, Mar. 2000.

Cho, Ki-suk. "The Ideological Orientation of 2008 Candlelight Vigil Participants: Anti-American, Pro-North Korean Left or Anti-Neoliberalism?" *Korean Political Science Review* 43, no. 3 (2009): 125–48.

Cho, Lee-Jay, and Yoon Hyung Kim. *Economic Development in the Republic of Korea: A Policy Perspective*. Honolulu, Hawaii: East-West Center: Distributed by the University of Hawaii Press, 1991.

Cho, Mahn-gyu. "KNOC's Success in Exploration is Only One since 2008." *Energy & Environment News*, Sept. 26, 2011.

Cho, Mee-young and Michael Erman. "South Korea Wins Biggest Energy Deal with Canada's Harvest." *Reuters*, Oct. 22, 2009.

Cho, Yoon-je. *Hankugŭi Kwŏllyŏkkuchowa Kyŏngjechongch'aek: Saeroun Chŏngch'i, Kyŏngjeŭi T'ŭrŭl Ch'ajasŏ [State Governance Structure and Economic Policies in Korea: Pursuing for New Politics and Economy]*. Paju: Hanul, 2009.

Choi, Haangsoon. "Collaboration between the President and the Ruling Party: Focused on Analysis of Affecting Factors in Korean Regimes." *Korean Public Management Review* 22, no. 4 (2008): 183–204.

Choi, Mingki. "Samsung Corporation Tries Resource-Rich Country through Lee Kun-Hee." *EBN Industry News*, Dec. 1, 2011.

Choi, Youn-sik. "Personalization of Power and Yushin Constitution." *Journal of Korean Political and Diplomatic History* 33, no. 1 (2011): 69–98.

Chu, Chong-kuk, Youn, Kŭn-young and Shin, Ho-kyŏng. "Mosf Will Be Dinosaur Ministry?" *Chosun Ilbo*, Jan. 16, 2008.

Chu, Yong-joong. "The Right of Personnel Management Returns to Minister?" *Chosun Ilbo*, June 24, 2009.

Chubu Electric Power Co. Inc. *Annual Report 2011*. Nagoya: Chubu Electric Power Co. Inc., 2011.

Chung, Se-jin. "40 Trillion Won of Debt, Repressed Charge Hike of State-Owned Energy Companies for 3 Years." *Dong-A Ilbo*, Aug. 29, 2011.

"Committee on Knowledge Economy Reproved Insolvent Overseas Resource Development." *Yonhap News*, Sept. 26, 2011.

"Comprehensive Energy Plan Finalized, Nuclear Power Generation Corproation Establshed by 1980." *Maeil Economic Daily*, June 1, 1974.

"Comprehensive Energy Plan Including Wholesaler Will Be Examined." *Asahi Shinbun*, Mar. 7, 1962.

"Confrontation on the JNOC Reform Reappear, LDP Horiuchi Argues to Abolish Loan Gurantee." *Asahi Shinbun*, Dec. 13, 2001.

"'Consider Abolition of JNOC' Former METI Minister Drive Old Enemy into Corner." *Asahi Shinbun*, June 22, 2001.

"Consolidation through the Private-Oriented for Making a Core Company is General Tendency." *Asahi Shinbun*, Nov. 29, 2002.

"Construction for KNOC's New Headquarter Building in Ulsan Innovation City Began." *Yonhap News*, Nov. 22, 2011.

"Consultations to Be Held before Dissolution of Oil Corporation." *BBC Monitoring Asia Pacific—Economic*, July 25, 2001.

"Contract Dispute Halts Marib Operations in Yemen." *Oil & Gas Journal*, Nov. 28, 2005.

Cox, Gary W., Frances McCall Rosenbluth, and Michael F. Thies. "Electoral Reform and the Fate of Factions: The Case of Japan's Liberal Democratic Party." *British Journal of Political Science* 29, no. 1 (1999): 33–56.

Cumings, Bruce. *Korea's Place in the Sun: A Modern History*. 1st ed. New York: Norton, 1997.

Cunningham, David E. "Veto Players and Civil War Duration." *American Journal of Political Science* 50, no. 4 (2006): 875–92.

Curtis, Gerald L. *The Japanese Way of Politics*, Studies of the East Asian Institute. New York: Columbia University Press, 1988.

Daehan News. "Pakchŏnghŭi Taetongnyŏng Sinnyŏn Tŭkpo [President Park Chung-hee's New-Year Special News]." Korea, 1976.

Daewoo International. "Daewoo International Ir Material." Seoul, 2010.

Daewoo Shipbuilding & Marine Engineering. "Future Value Generator DSME ENR." Seoul, 2012.

"Debt Ratio." *Han'gukkyŏngje Magazine*, April 1998.

Democratic Party. "Indulgence for Grand National Party's Illegal Rushing through Does Not Make Sense." *Press Release*, Dec. 15, 2010.

Development division under Petroelum Advisory Committee. "Interim Report of Basic Policy Subcommittee." Tokyo, 2000.

"Diminishing Role Lowers Political Power, from MITI to METI." *Asahi Shinbun*, May 31, 2002.

"Dissolution of JNOC Delayed One More Year, Special Company Not Included for Legislation." *Asahi Shinbun*, May 2, 2002.

"Division of Commerce and Industry Approved Petroleum Industry Law." *Asahi Shinbun*, Mar. 2, 1962.

Duran, Mirza. "Novatek Delivers 1st Yamal LNG cargo to Japan via NSR." *LNG Prime*, July 24, 2020.

E-Daily. "President Asked Three Tasks." *Chosun Ilbo*, Mar. 17, 2008.

"Economic Ministerial Meeting Agreed to Establish the Ministry Taking Full Charge of Resource and Submit Bill to Assembly in Coming September." *Maeil Economic Daily*, Aug. 13, 1977.

Economic Planning Board. *Kaepalnyŏndaeŭi Kyŏngjech'ŏngch'aek: Kyŏngjekihwoegwŏn 20nyŏnsa [Economic Policy in the Era of Development: 20-Year History of the Economic Planning Board]*. Seoul: Economic Planning Board, 1982.

"Editorial: Abolition of JNOC is Model of Special Corporation Reform." *Asahi Shinbun*, July 8, 2001.

"Efficient Solution in Entering the 1980s: Backgroud and Meaning of New Year's Administrative Policy Blueprint." *Kyung-hyang Shinmun*, Oct. 4, 1978.

"Eligibility of Oil Import License Expanded to the Private Companies." *Maeil Economic Daily*, May 29, 1980.

Energy and Mineral Resources Development Association of Korea. "Organization." http://www.emrd.or.kr/intro/organ.jsp.

Energy and Mineral Resources Development Association of Korea. "Organization." www.emrd.or.kr/english/intro/organization.jsp.
Energy Intelligence. "The Energy Intelligence Top 100: Ranking the World's Oil Companies." *Petroleum Intelligence Weekly*, 2003.
Energy Intelligence. "The Energy Intelligence Top 100: Ranking the World's Oil Companies." *Energy Intelligence*, 2010.
Energy Law. Mar. 3, 2006.
"Energy Minister Jang's Official Trip to Strengthen Resource Diplomacy in American Continent." *Kyung-hyang Shinmun*, Apr. 12, 1978.
"Energy Ministry Concluded No Oil in Pohang." *Maeil Economic Daily*, Mar. 20, 1982.
"Energy Ministry Disappointed by Budget Cutting." *Kyung-hyang Shinmun*, Sept. 20, 1978.
Enforcement Ordinance on Management of Public Corporation. Mar. 27, 2007.
Evans, Peter B. *Embedded Autonomy: States and Industrial Transformation*. Princeton Paperbacks. Princeton, NJ: Princeton University Press, 1995.
Evans, Peter B., Dietrich Rueschemeyer, Theda Skocpol, Social Science Research Council (U.S.). "Committee on States and Social Structures, Joint Committee on Latin American Studies, and Joint Committee on Western Europe." *Bringing the State Back In*. Cambridge Cambridgeshire; New York: Cambridge University Press, 1985.
"Ex-Foreign Minister Tanaka to Join DPJ." *Kyodo News International*, Aug. 15, 2009.
"Examine Reformation of State-Owned Company: Don't Make 'Bright Beginning, Dull Finish', Fix It." *E-Daily*, July 16, 2010.
Executive Order for Presidential Committee of Future and Vision. Executive Order 20652. Dec. 29, 2008.
Federation of Korean Industries. "Chŏn'gyŏngnyŏn 50 Nyŏnsa: Kichŏgŭi 50nyŏnŭl Nŏmŏ Hŭimang 100nyŏnŭro [50-Year History of the Federation of Korean Industries: Beyong 50-Year of Miracle to 100-Year of Hope]." Seoul: Federation of Korean Industries, 2011.
First Oil. "Kraken 9/02b-5z Tests at 4,550 Barrels of Oil Per Day." *Press Release*, Sept. 2011.
First Oil. "Signed SPA for Celtic Oil Acquisition." *Press Release*, Apr. 2011.
"'For Oil Development, Support Private Business Rather Than Tax Input,' President of Idemitsu Criticized METI." *Asahi Shinbun*, Oct. 4, 2001.
"Fukuda's Proposal Did Not Make an Agreement." *Asahi Shinbun*, Apr. 26, 2002.
Fukui, Haruhiro, and University of California Berkeley. Center for Japanese and Korean Studies. *Party in Power: The Japanese Liberal-Democrats and Policy-Making*, Publications of the Center for Japanese and Korean Studies. Berkeley: University of California Press, 1970.
"Full Text of President Park's Administrative Policy Speech." *Maeil Economic Daily*, Oct. 4, 1978.

Ganghof, Steffen. "Promises and Pitfalls of Veto Player Analysis." *Swiss Political Science Review* 9, no. 2 (2003): 1–25.

"Government and Ruling Party Review Administrative Reform from Zero-Base." *Asahi Shinbun*, Aug. 30, 1966.

"Government Considers Establishment of Ministry of Resource." *Kyung-hyang Shinmun*, Sept. 3, 1975.

Government Structure Law. July 17, 1948.

"Gov't and Ruling Party Deceided to Establish PEDCO." *Maeil Economic Daily*, Aug. 12, 1977.

"Gov't, Coalition Finalize Plans to Disband State-Run Oil Firm." *Kyodo News*, Nov. 22, 2001.

"Gov't Noticed Japan Its Decision to Develop Zone-VII Independently." *Kyung-hyang Shinmun*, Jan. 17, 1976.

Grand National Party. "History." http://www.hannara.or.kr/ohannara/hannara/hannara_05.jsp.

Grand National Party. *Illyukukka, Hŭimanggongdongch'ae Taehanminkuk [No. 1 Nation, Community of Hope, Republic of Korea]*. Edited by Grand National Party. Seoul: Bookmark, 2007.

Grand National Party. "Unprecedented Violence in Plenary Session by Opposition Party." *Press Release*, Dec. 8, 2010.

Grant, Wyn. "Pressure Groups, Politics and Democracy in Britain." New York: Philip Allan, 1989.

GS Caltex. "Exploration and Production." GS Caltex. http://www.gscaltex.com/eng/business/production.aspx.

Ha, Soo-jung. "Trade Industry Requested Ruling Party to Support Overseas Resource Development." *E-Daily*, Aug. 13, 2004.

Haggard, Stephan, and Mathew D. McCubbins. *Presidents, Parliaments, and Policy*, Political Economy of Institutions and Decisions. Cambridge; New York: Cambridge University Press, 2001.

Han, Pyŏng-do, Kim, Tae-nyŏn, Lee, Kwang-jae and Seo, Kab-won. *Enŏchi Munje, Saeroun Daeannŭl Ch'ajasŏ [Energy Problem, Pursuing New Alternatives]*, Policy Material for the Regular Session in 2004. Seoul: Legislature Research Center, 2004.

Han, Sang-jin. "Issue: Overcharged for M&A? Low Quality Mine? With Low Prices, Many Assets Would Not Be Profitable." *Shin Dong-A*, August 2011.

"Happy News of Dragon Dream That Excited the Beginning of the Year. Oil Discovered!" *Kyung-hyang Shinmun*, Jan. 15, 1976.

Harrison, Selig S., and Carnegie Endowment for International Peace. *China, Oil, and Asia, Conflict Ahead?* New York: Columbia University Press, 1977.

Hatley, Allen G. *The Oil Finders: A Collection of Stories About Exploration*. Utopia, Tex: Centex Press, 1995.

Hayao, Kenji. *The Japanese Prime Minister and Public Policy*, Pitt Series in Policy and Institutional Studies. Pittsburgh: University of Pittsburgh Press, 1993.

"He is Good at Speaking English, So He Touched Other's Business." *Kyung-hyang Shinmun*, Feb. 6, 1978.

Hein, Laura Elizabeth. *Fueling Growth: The Energy Revolution and Economic Policy in Postwar Japan*, Harvard East Asian Monographs. Cambridge, MA: Council on East Asian Studies Distributed by Harvard University Press, 1990.

Henisz, Witold J., and Bennet A. Zelner. "Interest Groups, Veto Points, and Electricity Infrastructure Deployment." *International Organization* 60, no. 01 (2006): 263–86.

Henisz, Witold J., and Edward D. Mansfield. "Votes and Vetoes: The Political Determinants of Commercial Openness." *International Studies Quarterly* 50, no. 1 (2006): 189–211.

Heo, Nam-hun. "Yuchŏn'gaepal Chiwondaech'aek [Report to the President: Measures to Support Oil Development]." Edited by Ministry of Energy and Resources. Kyunggido: Presidential Archive, 1982.

Heo, Young. *Hankuk Hŏnbŏmnon [Korean Constitution]*. Seoul: Pakkyŏngsa, 2011.

Hong, Duk-ryul. "Study on Korea's Big Business Organization: Mainly on FKI." *Society and History* 36 (1992): 56–98.

Horiuchi, Mitsuo. "Tsushōsangyōshō No Chibu Sekiyukōdan Wo Kokuhatsusuru [Denounce the JNOC, MITI's Disgrace]." *Bungeishunzū*, November 1998.

Huber, Evelyne, Charles Ragin, and John D. Stephens. "Social Democracy, Christian Democracy, Constitutional Structure, and the Welfare State." *American Journal of Sociology* 99, no. 3 (1993): 711–49.

Hughes, Llewelyn. "The Limits of Liberalism?: Long-Run Petroleum Prices and Government Intervention in Petroleum Markets in Japan, France, and the United States." Thesis (Ph.D.), Massachusetts Institute of Technology, 2009.

Huntington, Samuel P., and Harvard University. Center for International Affairs. *Political Order in Changing Societies*. New Haven: Yale University Press, 1968.

Hyundai Corporation. "Global Business Organizer Hyundai Corporation." Seoul, 2012.

"Hyundai Energy and Resources Established with Appointing Lee Myung-Bak as CEO." *Yonhap News*, Sept. 1, 1990.

Idemitsu Kosan Co. Ltd. *Annual Report 2011*. Tokyo: Idemitsu Kosan Co. Ltd., 2011.

Idemitsu Kosan Co. Ltd. "Resource Development: Oil and Gas." http://www.idemitsu.com/products/resource/oil.html.

Idemitsu Oil & Gas Co. Ltd. "Message from the President." Idemitsu Oil & Gas Co., Ltd. http://www.iog-idemitsu.co.jp/eng/company/greeting.html.

Idemitsu Oil & Gas Co. Ltd. "Project: Thailand." Idemitsu Oil & Gas Co. Ltd. http://www.iog-idemitsu.co.jp/eng/project/thailand/index.html.

IEA. "Country Profile: Japan." https://www.iea.org/countries/japan (Accessed Dec. 4, 2022).

IEA, Japan Natural Gas Security Policy. https://www.iea.org/articles/japan-natural-gas-security-policy (Accessed Dec. 4, 2022).

IEA. "LNG importing prices in selected countries, 2010–2019." https://www.iea.org/data-and-statistics/charts/lng-import-prices-in-selected-countries-2010-2019 (Accessed Dec. 5, 2022).

Ikuo, Nagazima. *Shigengaikō Rensenrenpai: Azadeganyuden No Satetsu [Consecutive Defeats in Resource Diplomacy: Set-Back of Azadegan Oil Field]*. Tokyo: Yosensha, 2009.

Immergut, Ellen M. *Health Politics: Interests and Institutions in Western Europe*, Cambridge Studies in Comparative Politics. Cambridge England; New York, NY, USA: Cambridge University Press, 1992.

"Improving Demand and Supply of Oil." *Maeil Economic Daily*, May 24, 1975, 5.

"In the Draft of Petroelum Industry Law, the Power to Enter for Inspection Removed." *Asahi Shinbun*, Mar 7, 1962.

INPEX Corporation. "Annual Report 2011: Firm Commitment to Sustainable Energy." Tokyo, INPEX, 2011.

INPEX Corporation. "Annual Report 2021: A New Wind for Energy." Tokyo: INPEX, 2021.

INPEX Corporation. "Corporate Governance." http://www.inpex.co.jp/english/company/governance.html#gov04.

INPEX Corporation. "Ichthys Project: Outline." https://www.inpex.co.jp/english/ichthys/index.html?__utma=47397631.138530629.1655342165.1669535386.1670394784.4&__utmb=47397631.1.10.1670394784&__utmc=47397631&__utmx=-&__utmz=47397631.1670394784.4.3.utmcsr=inpex.co.jp|utmccn=(referral)|utmcmd=referral|utmcct=/&__utmv=-&__utmk=88167134.

INPEX Corporation. "INPEX-operated Ichthys LNG Project Commences LNG Shipment." *Press Release*, Oct. 23, 2018.

INPEX Corporation. "INPEX Receives Approval of Revised Plan of Development for Abadi LNG Project, Masela Block, Indonesia." *Press Release*, July 16, 2019.

INPEX Corporation. "INPEX to Commence Exploratory Offshore Drilling Offshore Shimane, Yamaguchi Prefectures in Japan." *Press Release*, Jan. 17, 2022.

INPEX Corporation. "Southeast Asia: Main Projects: Abadi LNG Project." https://www.inpex.co.jp/english/business/asia.html.

Investopedia U.L.C. "Dictionary: Golden Share." http://www.investopedia.com/terms/g/goldenshare.asp#axzz1jpNkr6bf.

"Iran: Japan Looks to Iran to Replace Assets Lost in Neutral Zone." *Middle East Economic Digest*, Nov. 10, 2000.

"Ishihara Takes Issue with METI's JNOC Dissolution Plans." *Kyodo News*, Apr. 23, 2002.

Itochu Oil Exploration Co. Ltd. "Itochu's Activities: Azerbaijan (Btc Pipeline)." Itochu Oil Exploration Co. Ltd. http://www.itochuoil.co.jp/e/world_index-e.html.

Itochu Oil Exploration Co. Ltd. *Itochu Corporation Annual Report 2010*. Tokyo: Itochu Oil Exploration Co. Ltd. 2010.

Jang, Chang-min. "Samsung Corporation Took over U.S. Oil and Gas Company." *Korea Economic Daily*, Nov. 30, 2011.

Jang, Hyo-jin. "GTCs Get Fund for Resource Development in a Various Way." *E-Today*, Nov. 12, 2012.

Japan Business Federation. "Policy Proposals Environment and Energy." http://www.keidanren.or.jp/en/policy/index07.html.

"Japan Communist Party's Proposal on Special Corporation Reform." *Asahi Shinbun*, July 19, 2001.

Japan House of Councillors, Standing Committee on Commerce and Industry. *28th Committee Meeting*, Interpellation, Apr. 27, 1962.

Japan House of Councillors, Standing Committee on Diet Management. *16th Committee Meeting*, Interpellation, June 8, 1967.

Japan House of Councillors, Standing Committee on Commerce and Industry. *19th Committee Meeting*, Interpellation, July 20, 1967.

Japan House of Representative, Standing Committee on Commerce and Industry. *21st Committee Meeting*, Interpellation, Mar. 23, 1962.

Japan House of Councillors, Standing Committee on Economy, Trade and Industry. *24th Committee Meeting*, Interpellation, July 16, 2002.

Japan House of Councillors, Standing Committee on Economy, Trade and Industry. *25th Committee Meeting*, Interpellation, July 18, 2002.

Japan House of Representative, Standing Committee on Commerce and Industry. *23rd Committee Meeting*, Interpellation, Mar. 28, 1962.

Japan House of Representative, Standing Committee on Commerce and Industry. *24th Committee Meeting*, Interpellation, Mar. 29, 1962.

Japan House of Representative, Standing Committee on Commerce and Industry. *27th Committee Meeting*, Interpellation, Apr. 10, 1962.

Japan House of Representative, Standing Committee on Commerce and Industry. *28th Committee Meeting*, Interpellation, Apr. 11, 1962.

Japan House of Representative, Standing Committee on Commerce and Industry. *2nd Committee Meeting*, Interpellation, Dec. 17, 1963.

Japan House of Representative, Standing Committee on Commerce and Industry. *37th Committee Meeting*, Interpellation, May 18, 1965.

Japan House of Representative, Plenary Session. *14th Plenary Meeting*, Interpellation, May 18, 1967.

Japan House of Representative, Standing Committee on Commerce and Industry. *18th Committee Meeting*, Interpellation, June 9, 1967.

Japan House of Representative, Standing Committee on Commerce and Industry. *26th Committee Meeting*, Interpellation, July 4, 1967.

Japan House of Representative, Standing Committee on Commerce and Industry. *27th Committee Meeting*, Interpellation, July 5, 1967.

Japan House of Representative, Standing Committee on Commerce and Industry. *28th Committee Meeting*, Interpellation, July 11, 1967.

Japan House of Representative, Standing Committee on Economy, Trade and Industry. *22nd Committee Meeting*, Interpellation, June 12, 2002.

Japan House of Representative, Standing Committee on Economy, Trade and Industry. *23rd Committee Meeting*, Interpellation, June 26, 2002.

Japan House of Representatives, Standing Committee on Economy, Trade and Industry. *24th Committee Meeting*, Interpellation, June 28, 2002.

Japan House of Representative, Standing Committee on Economy, Trade and Industry. *25th Committee Meeting*, Interpellation, July 2, 2002.

Japan House of Representative, Standing Committee on Economy, Trade and Industry. *26th Committee Meeting*, Interpellation, July 3, 2002.

Japan House of Representatives, Standing Committee on Economy, Trade and Industry. *27th Committee Meeting*, Interpellation, July 5, 2002.

Japan House of Representatives, Standing Committee on Economy, Trade and Industry. *10th Committee Meeting*, Interpellation, Apr. 2, 2003.

Japan National Oil Corporation. *Sekiyukōdan Nijūnensi [20-Year History of Japan National Oil Corporation]*. Tokyo: Japan National Oil Corporation, 1987.

Japan National Oil Corporation. *Sekiyukōdanshi (1997<Heisei9> Nendo-2004<Heisei16>Nendo) [The History of JNOC (1997-2004)]*. Tokyo: Japan National Oil Corporation, 2005.

Japan Oil Gas and Metals National Corporation. "About Us." http://www.jogmec.go.jp/english/aboutus/index.html.

Japan Oil Gas and Metals National Corporation. "Organization." www.jogmec.go.jp/english/aboutus/organization.html.

Japan Oil, Gas and Metals National Corporation Law. July 26, 2002.

Japan Petroleum Consultants. Ltd. *Japan Petroleum & Energy Yearbook 1978*. Tokyo: Japan Petroleum Consultants. Ltd., 1978.

Japan Petroleum Development Association. "Contents of Work." www.sekkoren.jp/summary/sum3.htm.

Japan Petroleum Development Association. "Member Companies." www.sekkoren.jp/summary/sum5.htm.

Japan Petroleum Development Association. *Sekiyu Kogyō Renmei Nijunen No Ayumi [20 Year-Footprints of Japan Petroleum Development Association]*. Tokyo: Japan Petroleum Development Association, 1982.

Japan Petroleum Development Association. *Wagakuni Sekiyu Tennengasukaihatsu No Genjō to Kadai [The Situation and Issues of Our Country's Oil and Natural Gas Development]*. Tokyo: Japan Petroleum Development Association, 2009.

Japan Petroleum Development Association. *Wagakuni Sekiyu Tennengasukaihatsu No Genjō to Kadai [The Situation and Issues of Our Country's Oil and Natural Gas Development]*. Tokyo: Japan Petroleum Development Association, 2001.

"Japan Petroleum Development Corporation Will Be Established." *Asahi Shinbun*, July 29, 1966.

Japan Petroleum Exploration Co. Ltd. *Annual Report 2011*. Tokyo, 2011.

Japan Petroleum Exploration Co. Ltd. *Sekiyushigenkaihatsu Kabushikikaisha Junensi [The 10-Year History of Japex]*. Tokyo: Japan Petroleum Exploration Co. Ltd., 1977.

Japan Petroleum Exploration Co. Ltd. *Uchiyu Hyakugojuokunenni Sakuita Hana: Sekiyushigenkaihatsu Kabushikikaisha Gojunensi [Flower in the 15 Billion Years of Space: 50-Year History of Japex]*. Tokyo: Japan Petroleum Exploration Co. Ltd., 2006.

"Japan Reforms - Only to Remain Much the Same." *Petroleum Intelligence Weekly*, Dec. 3, 2001.

"Japan to Abolish National Oil Firm." *Jiji Press English News Service*, Nov. 22, 2001.

"Japan to keep interest in Russia's Sakhalin-2 gas project." *Kyodo News*, Aug. 20, 2022.

"Japan/JNOC/Oil Development-2: Equity Crude Target Intact." *Dow Jones Newswire*, Aug. 21, 1998.

"Japanese News Papers Reported with Interest in Oil Discovery. Stimulate Continental Shelf Development." *Dong-A Ilbo*, Jan. 16, 1976.

"Japanese Premier Calls for "Full Abolishment" of JNOC." *Platts Commodity News*, Nov. 21, 2001.

Jeon, Jeong-hong. "President Lee Met Barzani to Discuss Oil Development in Kurdistan." *Maeil Economic Daily*, Feb. 13, 2008.

Jeon, Min-Hui. "300 Thousand B/D of Daily Production by 2012, 'World Top-50s'." *Korea Energy News*, Jan. 11, 2010.

Jeon, Joon-bum. "Acquisition and Profit... Australian Gas Field Lee Administration Invested Became Good Asset." *Chosun Biz*, Aug. 24, 2022.

Jeon, Jin-Young. "Party Cohesion of the 18th National Assembly." *Korean Party Studies Review* 9, no. 2 (2010): 119–39.

Jeon, Soo-yong. "Heavily-indebted KNOC Will Sell Overseas Assets and Restructure. Managers of 3rd Rank Will be Reduced by 10%." *Chosun Biz*, Mar. 12, 2019.

Jing, Hong, and Lee Hyeon-Woo. "Different Types of Violent Conflict in Korean National Assembly: Consensual Conflict, Discordant Conflict, and Resistant Conflict." *Korean Party Studies Review* 10, no. 2 (2011): 221–50.

"JNOC's Consolidated Financial Statement Shows That JNOC's Asset is More Than 4 Trillion Yen, Quasi-Major Level." *Asahi Shinbun*, Aug. 31, 2001.

"JNOC, Difficult Mediation between Dissolution and Privatization, Burden of Jewel Beetle Color Decision." *Asahi Shinbun*, Apr. 17, 2002.

"JNOC Says Failed Exploration Cost 372 Billion Yen." *Reuters*, Dec. 3, 1997.

Jo, Jung-hyung. "Lee Byung-Ho, CEO of STX Energy." *ET News*, Mar. 19, 2012.

JOGMEC. "Financial Assistance to Japanese Companies." https://www.jogmec.go.jp/english/oil/oilgas_10_000007.html (Accessed Sept. 11, 2022).

Johnson, Chalmers A. *Japan's Public Policy Companies*, AEI-Hoover Policy Studies. Washington: American Enterprise Institute for Public Policy Research, 1978.

Johnson, Chalmers A. *MITI and the Japanese Miracle: The Growth of Industrial Policy, 1925–1975*. Stanford, CA: Stanford University Press, 1982.

Jones, Leroy P. *Public Enterprise and Economic Development: The Korean Case*. Seoul: Korea Development Institute, 1975.

Jones, Leroy P., and Il SaKong. *Government, Business, and Entrepreneurship in Economic Development: The Korean Case*, Studies in the Modernization of the Republic of Korea, 1945–1975. Cambridge, MA: Council on East Asian Studies, Harvard University: Distributed by Harvard University Press, 1980.

Jung, Eui-jin. "Government Is Busy in Selling Overseas Mines. It Sold 4 Mines with Loss of 80 Billion Won." *Korea Economic Daily*, Jan. 17, 2022.

Jung, Heon-joo. "Reassessing Reform Outcomes in South Korea and Japan a Decade after Crisis: The Paradox of Political Leadership." Thesis (Ph.D.), University of Pennsylvania, 2008.

Jung, Hyŏng-su. "Kishi Said Delayment of Ratification Ridiculous. Agreement Proposed by Japan." *Dong-A Ilbo*, May 2, 1977.

Jung, Sang-pil. "KNOC Was Weakened Due to the Administration's Greed. Resource Development Ignored." *Energy Platform News*, Aug. 2, 2021.

Junichiro, Yamaoka. *Tanaka Kakuei Fujirareta Shigen Senryaku: Sekiyu, Uran, Sosite Americatono Toui [Tanaka Kakuei's Sealed Resource Strategy: Oil, Uranium and Battle with the U.S.]*. Tokyo: Soshisha, 2009.

JX Holdings. Inc. *Annual Report 2011: Bold Strategies To...* Tokyo: JX Holdings. Inc., 2011.

JX Holdings. Inc. *Jx: From Integration to Sustaining Power*. Tokyo: JX Holdings. Inc., 2010.

Kaihara, Hiroshi. "The Advent of a New Japanese Politics: Effects of the 1994 Revision of the Electoral Law." *Asian Survey* 47, no. 5 (2007): 749–65.

Kang, Chan-gu. "Yoon Administration Sell Resource Public Corporation, Either? Lost Decade in Overseas Resource Development." *Weekly Chosun*, Oct. 12, 2022.

Kansai Electric Power Co. Inc. *Annual Report 2011*. Osaka: Kansai Electric Power Co. Inc., 2011.

Kastner, Scott L., and Chad Rector. "International Regimes, Domestic Veto-Players, and Capital Controls Policy Stability." *International Studies Quarterly* 47, no. 1 (2003): 1–22.

"Keidanren Decided It Initiate JPDC." *Asahi Shinbun*, Oct. 18, 1966.

Kerbo, Harold R., and John A. McKinstry. *Who Rules Japan?: The Inner Circles of Economic and Political Power*. Westport, CT: Praeger, 1995.

Kim, Chan-jin. "Teindonesiachawonhyŏmnyŏksaŏpch'uchinhyŏnhwang: Tongnyŏkchawonpupoko [Korea-Indonesia Resource Cooperation Projects: Ministry of Energy and Resources Reported]." Edited by The Office of First Senior Economy Secretary of Chŏngwadae. Seoul: Chŏngwadae, 1979.

Kim, Chul-hoon. "KNOC Sold All Shares of Peruvian Oil Company, Which Is a Good Signal for Improving Financial Structure." *Global Economic*, Jan. 21, 2021.

Kim, Eun Mee. *Big Business, Strong State: Collusion and Conflict in South Korean Development, 1960–1990*, Suny Series in Korean Studies. Albany: State University of New York Press, 1997.

Kim, Ho-Seob. "Presidential Advisory Committees on National Policies: Actual Conditions, Unknown Performance and Future Tasks." *Korean Review of Organizational Studies* 1, no. 2 (2004).

Kim, In-hwan. "KOGAS Has Never Implemented Exploration Since 2013." *The Korea Industry Daily*, Oct. 22, 2020.

Kim Jae-eun. "Kogas Used Money Reserved for Reinvestment to Price Lowering." *Hankook Kyungje*, Oct. 21, 2008.

Kim, Ji-yoon. "Energy Resources Drive Cost Korea Billions." *Korea JoongAng Daily*, Nov. 12, 2014.

Kim, Kwang-Soo. "On Presidential Transition in Korea." *Public Law Journal* 10, no. 3 (2009): 315–42.

Kim, Mi-young. "Kim Je-nam, member of National Assembly, said Park Administration Executed Massive Restructuring." *e-Today*, Apr. 9, 2015.

Kim, Pyong-guk, and Ezra F. Vogel. *The Park Chung Hee Era: The Transformation of South Korea*. Cambridge, MA: Harvard University Press, 2011.

Kim, Sae-hyung. "There Is No Principle That Bureaucrat Has to Retire." *E-Daily*, Apr. 22, 2008.

Kim, Sang-gon and Yun-Ja Kim et al. *Kukka Aenŏji Chŏngch'aekkwa Hankugŭi Ch'ŏnyŏn'gasŭsanŭp Yŏn'gu [Research on National Energy Policy and Korea's Natural Gas Industry]*. Seoul: Nohkiyŏn, 2008.

Kim, Sang-yong. "Global Korea, We Make It Ealier: Daewoo Shipbuilding & Marine Engineering." *Seoul Economic Daily*, May 25, 2011.

Kim Shin. "New Year's Interview: Open Uri Party Member Kim Tae-Nyon." *G&E TIMES*, Jan. 11, 2006.

Kim, Timothy. "Park Geun-hye, What Does She Want Through Prosecutors' Investigation On Resource Diplomacy." *Business Post*, Apr. 9, 2015

Kim, Woo-bo. "No Budget Had East Sea Natural Gas Development Stopped." *Seoul Economic Daily*, Feb. 4, 2022.

Kim, Young-taek. "2010 Overseas Resource Development Symposium." *Tomato News*, Dec. 2, 2010.

King, Gary, Robert O. Keohane, and Sidney Verba. *Designing Social Inquiry: Scientific Inference in Qualitative Research*. Princeton, NJ: Princeton University Press, 1994.

Kingdon, John W. *Agendas, Alternatives, and Public Policies*. Boston: Little, Brown, 1984.

"Kishi, Sato and Ikeda Decided to Included NOSODECO in Japan EXIM Bank Loan." *Asahi Shinbun*, Mar. 11, 1960.

Kishima, Shinichi. "Japan PM's Reform Battle Heats up as Bills Stall." *Reuters News*, Apr. 25, 2002.

Koh, Byung Chul. *Japan's Administrative Elite*. Berkeley: University of California Press, 1989.

Koh, Eun-gyul. "Moon Administration Invested Half. New Administration's Financing Is Important." *Newsis*, Apr. 7, 2022.

Kohli, Atul. *State-Directed Development: Political Power and Industrialization in the Global Periphery*. Cambridge, UK; New York: Cambridge University Press, 2004.

Kohli, Atul. "Where Do High Growth Political Economies Come From? The Japanese Lineage of Korea's 'Developmental State'." *World Development* 22, no. 9 (1994): 1269–69.

Koike, Masanari. "Japan's Overseas Oil Development and a Role of Technology." In *GraSPP Working Paper Series*. Tokyo: The University of Tokyo, 2008.

"Koizumi Branded 'Mr. NATO'—No Action, Talk Only." *Kyodo News International*, Dec. 5, 2001.

"Koizumi Reforms; Old Guard Champions Vested Interests." *Daily Yomiuri*, Dec. 21, 2001.

"Koizumi Seeks Quick Dissolution of State Oil Firm." *Kyodo News*, Apr. 16, 2002.

"Koizumi Seeks to Dissolve State Oil Firm." *Agence France-Presse*, July 10, 2001.

"Korea and Iraq Agreed to Co-Develop Basra." *Yonhap News*, Feb. 24, 2009.

Korea Confederations of Trade Unions. "KCTU's Demonstration to Stop Marketization and Privatization." *Press Release*, Mar. 24, 2008.

Korea Confederations of Trade Unions. "The Launching of Struggle Headquater to Stop Marketization and Privatization, and to Strengthen Social Publicness." *Press Release*, Mar. 10, 2008.

Korea Energy Economic Institute. *Enŏchi Chŏngch'aek Pyŏnchŏnsa [The History of Energy Policy]*. Kyunggi-do: Korea Energy Economics Institute, 2006.

Korea Gas Corporation. "Current Shareholder." Korea Gas Corporation. http://www.kogas.or.kr/kogas_kr/html/investor/investor_12.jsp.

Korea Gas Corporation. "Presentation to Investors." Edited by Korea Gas Corporation. Kyunggido, 2011.

"Korea Might Develop Joint Development Zone Based on the Result of UN Law of the Sea Conference." *Dong-A Ilbo*, Jan. 27, 1976.

Korea National Assembly. "Ich'ŏnp'allyŏndo Cheirhoe Ch'ukakyŏngjŏngyesan [The First Supplementary Budget in 2008]." Seoul: Korea's National Assembly, 2008.

Korea National Assembly, Special Committee on Budget Settlement. *The First Supplementary Budget Draft of 2008*, Interpellation, Sept. 9, 2008.

Korea National Assembly, Standing Committee on Foreign Affairs. *11th Committee Meeting*, Interpellation, Dec. 13, 1974.

Korea National Assembly, Standing Committee on Foreign Affairs. *12th Committee Meeting*, Interpellation, Dec. 16, 1974.

Korea National Assembly, Special Committee on Budget. *7th Meeting: Examination Reference Material*, Interpellation, Nov. 2, 1975.

Korea National Assembly, Standing Committee on Internal Affairs. *20th Committee Meeting*, Interpellation, Dec. 5, 1977.

Korea National Assembly, Standing Committee on Commerce and Industry. *14th Committee Meeting*, Interpellation, Oct. 23, 1978.

Korea National Assembly, Standing Committee on Commerce and Industry. *15th Committee Meeting*, Interpellation, Oct. 26, 1978.

Korea National Assembly, Standing Committee on Construction and Transportation. *Inspection of Korea Railway Corporation*, Interpellation, Oct. 5, 2005.

Korea National Assembly, Standing Committee on Knowledge Economy. *5th Committee Meeting*, Interpellation, Sept. 10, 2008.

Korea National Assembly, Standing Committee on Knowledge Economy. *6th Committee Meeting*, Interpellation, Sept. 17, 2008.

Korea's National Assembly Budget Office. "Ichŏnp'allyŏndo Cheirhwae Ch'ukakyŏngjŏngyesananmit Kikŭmunyongkyehwaekpyŏnkyŏngan Punsŏk [The Analysis on the First Supplementary Budget and the Change of the Management of Fund in 2008]." Seoul: Korea's National Assembly Budget Office, 2008.

Korea National Assembly Secretariat. *Taehanmin'gukkukhoe Yuksimnyŏnsa [60-Year History of the National Assembly of the Republic of Korea]*. Seoul: Korea National Assembly Secretariat, 2008.

Korea National Oil Corporation. *Annual Report 2010*. Kyunggido 2011.

Korea National Oil Corporation. *Han'gukhyŏng Sŏgyumaeichŏ Yuksŏngŭl Wihan Han'guksŏgyukongsa Hyŏksinpangan [The Innovation Plan of Korea National Oil Corporation to Promote Korean-Style Oil Major]*. Kyunggido: Korea National Oil Corporation, 2006.

Korea National Oil Corporation. *KNOC 25: Han'guksŏgyukongsa Isibonyŏnŭi Yŏksa [KNOC 25: 25-Year History of Korea National Oil Corporation]*. Kyunggido: Korea National Oil Corporation, 2005.

Korea National Oil Corporation. "KNOC Agreed on Implementation Plan of Oil-SOC Deal with Iraqi Kurdistan Regional Government." *Press Release*, Sept. 25, 2008.

Korea National Oil Corporation. "KNOC Discovered Oil Again in the Field of Hawler." *Press Release*, Nov. 11, 2013.

Korea National Oil Corporation. "KNOC Made PSC Contract for Exploration of Iraqi Bazian Field." *Press Release*, Nov. 12, 2007.

Korea National Oil Corporation. "KNOC Signed Contracts with Iraqi Kurdistan Regional Government for 8 Oil Fileds." *Press Release*, June 25, 2008.

Korea National Oil Corporation. "Korean Consortium Agreed on Package Deal-Style Resource Development with Iraqi Kurdistan Regional Government." *Press Release*, Feb. 14, 2008.

Korea National Oil Corporation. "More than 100 Million barrel of Oil was Discovered in UAE, Premier League of Oil Industry." *Press Release*, Dec. 26, 2014.

Korea National Oil Corporation. "Oil Discovery in the field of Hawler in Iraqi Kurdistan." *Press Release*, Mar. 27, 2013

Korea National Oil Corporation. "Opinet." 2011.

Korea National Oil Corporation. "Overview of Oil Development Business." http://www.knoc.co.kr/ENG/sub01/sub01_3.jsp.

Korea National Oil Corporation. "Overview of Oil Development Business." http://www.knoc.co.kr/sub03/sub03_2_1.jsp.

Korea National Oil Corporation. "260 Million barrel of Commercial Discovery in the Field of Hawler." *Press Release*, Apr. 2, 2014.

"Korea Petroleum Association Launched in This Month." *Maeil Economic Daily*, July 5, 1980.

Korea Petroleum Development Corporation. *Han'guksŏgyukaepalgongsa Sibonyŏnsa, 1979–1994 [15-Year History of Korea Petroleum Development Corporation, 1979–1994]* Seoul: Korea Petroleum Development Corporation, 1994.

Krauss, Ellis S., and Robert Pekkanen. *The Rise and Fall of Japan's LDP: Political Party Organizations as Historical Institutions*. Ithaca, NY: Cornell University Press, 2011.

Ku, Cha-ryong. "China's Hunting for Company is Our Concern." *Dong-A Ilbo*, July 6, 2009.

Kubo, Iwao and Kato Shigeru. *Zukai Shōsha Gyōkai Hendobuku [An Illustrative Guide to the Trading Companies]*. Tokyo: Tōyōkeizaishinbunsha, 2008.

"Kukhoepŏp [The National Assembly Law]." Edited by National Assembly, 1972.

Kwak, Sung-yeung. *The Korean Economy at a Crossroad: Development Prospects, Liberalization, and South-North Economic Integration*. Westport, Connecticut: Praeger, 1994.

Kwon, Dong-wook. "Daewoo International Embraced to Posco in Ten Years." *YTN News*, Aug. 30, 2010.

Kwon, Tae-soo. "Cheirhwae Ch'ukakyŏngjŏngyesananmit Chŏllyŏksanŏpkipankikŭm Unyongkyehwaekpyŏnkyŏngan [The Inspection Report on the First Supplementary Budget and the Changed Proposal for the Management of Utility Industry Foundation Fund]." Edited by The Committee on Knowledge Economy. Seoul: Korea's National Assembly, 2008.

Kyodo News. "Utilities Got 68 Ex-Bureaucrats Via 'Amakudari'." *Japan Times*, May 4, 2011.

Lardner, Peter. "Azeri Leader's Japan Visit May Unlock Caspian Oil." *Reuters*, Jan. 23, 1998.

Laver, Michael, and Kenneth A. Shepsle. "Coalitions and Cabinet Government." *The American Political Science Review* 84, no. 3 (1990): 873–90.

"Law on Government Organization." Edited by National Assembly. National Assembly, 1948.

Law on Management of Public Corporation. Jan. 19, 2007.

"LDP Executive Council Chair Horiuch Shows Intention to Abolish JNOC Possibly in This Year." *Asahi Shinbun*, July 12, 2001.

"LDP Executive Council Chair Proposed to Establish 3-Year Term Liquidation Organization for JNOC Reform." *Asahi Shinbun*, Nov. 10, 2001.

"LDP Executive Council Chairman Horiuchi Accepted Support for Oil Development Conditionally." *Asahi Shinbun*, Aug. 3, 2001.

"Lead: Ministers Agree on State Involvement in Oil Projects." *Kyodo News*, July 17, 2001.

"Lead—State-Run Oil Firm to Be 'Virtually' Dissolved in FY 2003." *Kyodo News*, May 2, 2002.

Lee, Byung-chul. "Flare of East Sea gas field disappeared. Exploration is Halted in Korea." *Dong-A Ilbo*, Apr. 29, 2022.

Lee, Hak-ryul. "Government Studies Various Options for Expansion of KNOC." *Money Today*, May 2, 2008.

Lee, Ho. *Nanŭn Asŭpaltŭ Kkallin Kirŭn Kachi Anŭnda [I Don't Go on the Asphalted Road.]*. Seoul: Ollim, 2007.

Lee, Jae-hang. "Kim Kyung Man, Member of National Assembly said we should make more efforts for exploration of oil near Korean Peninsula." *Straight News*, Oct. 20, 2022

Lee, Je-ho. "Clash on Direct Import of Natural Gas between Opposition Party and Government." *Naeil Shinmun*, Feb. 9, 2011.

Lee, Jin-ho. "Government and Business Speed up Overseas Resources Development." *etnews*, Feb. 14, 2008.

Lee, Jin-sok. "Black hole China, 'Adverse Effect on Resource Supply,' HRI Report." *Korea Economic Daily*, July 22, 2009.

Lee, Jung-hwan. "Overseas Resource Development, Which Has Been Neglected for a Decade, Should Fall Into Place." *Gas News*, May 6, 2022.

Lee, Joon-hyung. "KNOC Sold Anchor Oil Field, for Which It Had Invested 1.2 Trillion won, at a Clearance Price." *Asia Economy*, Sept. 6, 2022.

Lee, Na-gyung. "Due to Covid-19, Eleven Public Corporations, Including Inchon International Airport Corporation and Korea Racing Authority, Went Into Red." *Energy Economic News*, May 2, 2021.

Lee, Sŭng-gu. "The Oil Men, the Guard of Petroleum War Frontier." *Kyung-hyang Shinmun*, Feb. 23, 1981.

Lee, Sung-ho. "Bright Beginning, Dull Finish in 1975: The Establishment of PEDCO." *Kyung-hyang Shinmun*, Dec. 16, 1975.

Lee Youn-Ho. "Part of Utility Will Be Privatized." *Yonhap News*, May 9, 2008.

"Liberal Democratic Party Approved the Draft Bill Conditionally." *Asahi Shinbun*, Mar. 6, 1962.

Liberal Democratic Party of Japan. "Liberal Democratic Party - Terms in Office of Top Party Leaders." http://www.jimin.jp/english/the_president/leaders/104392.html.

Lijphart, Arend. *Parliamentary Versus Presidential Government*, Oxford Readings in Politics and Government. Oxford; New York: Oxford University Press, 1992.

Linz, Juan J., and Arturo Valenzuela. *The Failure of Presidential Democracy*. Baltimore: Johns Hopkins University Press, 1994.

"Listen to Horiuchi, LDP Executive Council Chairman, Arguing the Dissolution of JNOC." *Asahi Shinbun*, Apr. 20, 2002.

"Listen to the Three New High Officials of the LDP." *Asahi Shinbun*, Apr. 29, 2001.

Macalister, Terry. "DANA Board Relents in KNOC Battle." *Guardian*, Sept. 24, 2010.

MacIntyre, Andrew J. *The Power of Institutions: Political Architecture and Governance*, Cornell Studies in Political Economy. Ithaca: Cornell University Press, 2003.

Management Evaluation Unit. "2009 Nyŏndo Konggiŏp, Chunchŏngbukikwan Kyŏngyŏngsilchŏkpyŏngga Pogosŏ: SOC Yuhyŏng [State-Owned Company, Quasi-Public Organization Management Performance Evaluation Report of

2009: Category of Social Overhead Capital (SOC)]." Edited by Ministry of Strategy and Finance. Kyunggido: Ministry of Strategy and Finance, 2010.

Marcel, Valérie, and John V. Mitchell. *Oil Titans: National Oil Companies in the Middle East*. London, Baltimore: Chatham House; Brookings Institution Press, 2006.

Market Watch. "Biden's Climate Envoy Kerry Says Natural Gas Could Be a 'Bridge Fuel,' but Emissions Capture Is Key." *Spectra*, Mar. 14, 2022.

Marubeni Corporation. "Our Business: Energy Division-2." www.marubeni.com/division/energy_2.html.

Masumi, Junnosuke. *Contemporary Politics in Japan*. Berkeley: University of California Press, 1995.

McCubbins, Mathew D., and Michael F. Thies. "As a Matter of Factions: The Budgetary Implications of Shifting Factional Control in Japan's LDP." *Legislative Studies Quarterly* 22, no. 3 (1997): 293–328.

"Meet Aspiration of People General, New Era of Constitutional Politics." *Kyunghyang Shinmun*, Aug. 31, 1987.

"METI's Proposal: JNOC Will Be Integrated with MMAJ, Financing Will Be Subsidy." *Asahi Shinbun*, Oct. 30, 2001.

"METI and Advisory Council Untied to Support That Jnco is Needed for the Nation." *Asahi Shinbun*, June 21, 2001.

"METI Did Not Agree with Kantei's Proposal. The Final Conclusion after Holidays." *Asahi Shinbun*, May 1, 2002.

METI Establishment Law. 108. July 18, 1999.

"METI Has No Plan to Revise Bills for JNOC Abolition." *Kyodo News*, Apr. 17, 2002.

"METI Hiranuma: Ok to Abolish Japan Natl Oil Corp-Kyodo." *Reuters*, July 12, 2001.

Migdal, Joel S. *Strong Societies and Weak States: State-Society Relations and State Capabilities in the Third World*. Princeton, NJ: Princeton University Press, 1988.

Mill, John Stuart. *A System of Logic, Ratiocinative and Inductive: Being a Connected View of the Principles of Evidence and the Methods of Scientific Investigation*. 8th ed. New York: Harper & Brothers, 1874.

Ministry of Commerce, Industry and Energy. "The Biggest Organizational Restructuring since the Consolidation between the Ministry of Trade and Industry and the Ministry of Energy and Resources." *Press Release*, June 9, 2006.

Ministry of Commerce, Industry and Energy, ed. *Namkikosipŭn Iyakidŭl: Yŏktae Sanggong Tongjapuchanggwan Aesaeichip [Story That I Want to Hand Down: Essay Collection of All Successive Ministers of the Ministry of Trade and Industry and the Ministry of Energy and Resources]*. Kyunggido: Ministry of Commerce, Industry and Energy, 1999.

Ministry of Commerce, Industry and Energy. "Oil and Gas Self-Sufficiency Ratio Will Be 28% in 10 Years: The Third Overseas Resources Development Plan Established." *Press Release*, Aug. 7, 2007.

Ministry of Commerce, Industry and Energy. "SK Energy Will Launch Oil Production at Brazil BMC-8 Field from the End of July." *Press Release*, July 30, 2007.

Ministry of Economy, Trade and Industry. "Information on Advisory Committee." http://www.enecho.meti.go.jp/info/committee/index.htm.

Ministry of Economy, Trade and Industry. "History of METI." http://www.meti.go.jp/english/aboutmeti/data/ahistory2009.html.

Ministry of Economy, Trade and Industry. *Enerugi Hakusho 2010 [Energy White Paper 2010]*. Tokyo: Ministry of Economy Trade and Industry, 2011.

Ministry of Economy, Trade and Industry. "Member List of Petroleum-Subcommittee under Comprehensive Resources and Energy Investigation Committee." Tokyo, 2002.

Ministry of Economy, Trade and Industry. "METI Published Our Country's Oil and Natural Gas Self-Sufficiency Ratio as of 2020." *Press Release*, Nov. 24, 2021.

Ministry of Economy, Trade and Industry. "New International Resource Strategy." Ministry of Economy, Trade and Industry, March 2020

Ministry of Economy, Trade and Industry. "Sekiyukōdanhō Oyobi Kinzokukōkyōzigyōdanhō No Haishi Tōnikansuru Hōritsu [Law Related to Abolition of Law for Japan National Oil Corporation, Law for Metal Mining Agency of Japan, Etc.]." Edited by Japanese Diet, 2002.

Ministry of Economy, Trade and Industry. "The Road to Securing LNG." *METI Journal*, May Issue, 2014. https://www.meti.go.jp/english/publications/pdf/journal2014_05b.pdf.

Ministry of Energy and Resources. *Tongnyŏkjawonhaengchŏng Simnyŏnsa [10-Year History of Ministry of Energy and Resources]*. Kyunggi-do: Ministry of Energy and Resources, 1988.

"Ministry of Finance Decided Not to Permit Any Public Companies." *Asahi Shinbun*, Dec. 3, 1963.

"Ministry of Finance Will Repress Establishment of New Public Corporation." *Asahi Shinbun*, Aug. 30, 1966.

Ministry of Foreign Affairs of Japan, *Diplomatic Bluebook 2018*. Tokyo: Ministry of Foreign Affairs of Japan, 2018.

Ministry of Foreign Affairs of Japan, *Japan's Enernrgy Diplomacy*. Tokyo, Ministry of Foreign Affairs of Japan, 2019.

Ministry of International Trade and Industry. *Tsūshōsangyō Seisakushi [The History of International Trade and Industry Policy]*. 17 vols. Vol. 15. Tokyo: Tsūshōsangyōchōsakai, 1991.

Ministry of International Trade and Industry. *Tsūshōsangyōsesakushi [The History of International Trade and Industry Policy]*. 17 vols. Vol. 17. Tokyo: Tsūshōsangyōchōsakai, 1991.

Ministry of International Trade and Industry. *Tsūshōsangyōsesakusi [The History of International Trade and Industry Policy]*. 17 vols. Vol. 13. Tokyo: Ministry of International Trade and Industry, 1991.

Ministry of Knowledge Economy. "Establish National Energy Basic Plan, Foundation of Green Growth." *Press Release*, Aug. 27, 2008.

Ministry of Knowldge Economy. "Energy Committee Launched, Showing Energy Vision." *Press Release*, Oct. 5, 2010.

Ministry of Knowldge Economy. "History: Ministry of Knowledge Economy." http://www.mke.go.kr/intro/intro/history07.jsp.

Ministry of Knowldge Economy. "History: Ministry of Trade and Industry." http://www.mke.go.kr/intro/intro/history01.jsp.

Ministry of Knowldge Economy. "Ich'ŏnpallyŏndo Chisikkyŏngjepu Ŏmmupokocharyo: Chisik Hyŏksinjutohyŏng Sanŏpkangguk Kŏnsŏl [The Ministry of Knowledge Economy Annual Work-Plan of 2008: The Construction of Innovation-Leading Industrial Power]." Edited by Ministry of Knowldge Economy. Kyunggido: Ministry of Knowldge Economy, 2008.

Ministry of Knowldge Economy. "Improving Operation System to Enhance Transparency of Loan Delegation Committee on Overseas Resources Development." *Press Release*, Sept. 8, 2008.

Ministry of Knowldge Economy. "KNOC Production Will Be Six Times of Current Level." *Press Release*, June 11, 2008.

Ministry of Knowledge Economy. "Historical Investment for Overseas Resource Development." *Press Release*, Jan. 19, 2010.

Ministry of Knowledge Economy. "KNOC Succeeded in Overtaking Harvest." *Press Release*, Oct. 22, 2009.

Ministry of Knowledge Economy. "Performance of Resource Development and Future Plan." 26. Kyunggido, 2012.

Ministry of Knowledge Economy. "President Held National Energy Advisory Meeting." *Press Release*, Nov. 3, 2004.

Ministry of Knowledge Economy. "SK Innovation's Selling Shares of 3 Fields in Brazil Approved." *Press Release*, July 13, 2011.

Ministry of Knowledge Economy. "This Year Investment for Overseas Oil and Gas Development Amounts to $7.8 billion." *Press Release*, Jan. 24, 2011.

Ministry of Strategy and Finance. "2009 Evaluation Report for Management Performance of State-Owned Companies (SOC)." Kyunggido: Ministry of Strategy and Finance, 2010.

Ministry of Strategy and Finance. "Advancement of State-Owned Company Launched Including Consolidation of Housing Corporation and Land Corproation." *Press Release*, Aug 11, 2008.

Ministry of Strategy and Finance. "Ministry of Strategy and Finance." Edited by Ministry of Strategy and Finance. Kyunggido: Ministry of Strategy and Finance, 2011.

"Ministry of Trade and Industry Considers Establishing PEDCO in 1974." *Dong-A Ilbo*, July 24, 1974.

Ministry of Trade, Industry and Energy. "Government is Supporting Discovery and Development of Oil and Natural Gas in 3 Areas Including the Continental Shelf in the East Sea by Investing about US$10 mil in KNOC in 2022.(Explanation on the Seoul Economic Daily's report that No Budget Had East Sea Natural Gas Development Stopped." *Press Release*, Feb. 4, 2022

Ministry of Trade, Industry and Energy. "Remove Distressed Asset and Upgrade Capability." *Press Release*, June 29, 2016.

"MITI Asked Industrial Structure Delegation Committee to Review Crude Oil Policy Company." *Asahi Shinbun*, June 5, 1962.

"MITI Gets LDP Nod for Reconstruction Plan for JNOC." *Dow Jones Newswires*, Sept. 28, 1998.

"MITI/Nat'l Oil -2: Submits Restructuring Plan to Diet." *Dow Jones Newswires*, Sept. 28, 1998.

Mitsubishi Corporation. "MC Signs MOU with Korea Gas Corporation in Natural Gas Business " *News Release*, Feb. 25, 2011.

Mitsubishi Corporation. *Mitsubishi Corporation 2011*. Tokyo, 2011.

Mitsui Oil Exploration Co. Ltd. "Settlement of Claims Arising from Gulf Oil Spill." *News*, May 20, 2011.

Mitsui Oil Exploration Co. Ltd. "Shareholders." Mitsui Oil Exploration Co. Ltd. http://www.moeco.co.jp/en/about/stockholder.html.

Mitsui & Co. Thailand. Ltd. "Company Profile." http://www.mitsui.co.th/en/company_profile/company_overview.php.

Mitsui & Co. and JOGMEC. "Final Investment Decision for Arctic LNG 2 Project in Russia." *Press Release*, Sept. 5, 2019.

Mochizuki, Mike. "Managing and Influencing the Japanese Legislative Process: The Role of Parties and the National Diet." Thesis (Ph.D.), Harvard University, 1982., 1982.

"MOF Will Not Permit Public Corporation." *Asahi Shinbun*, Jan 10, 1966.

Moon, Angela. "Korea National Oil Acquires Half of Peru Petro-Tech." *Reuters*, Feb. 6, 2009.

"Mr. Horiuchi Asked Reform Headquarter to Stop the Expenditure of Special Account to Oil Stockpiling and Development." *Asahi Shinbun*, Aug. 4, 2001.

"Mr. Tahara Criticized METI Minister Japan Does Not Have Energy Strategy and JNOC Should Be under Review." *Asahi Shinbun*, Aug. 1, 2001.

Mughan, Anthony. *Media and the Presidentialization of Parliamentary Elections*. Houndmills, Basingstoke, Hampshire; New York: Palgrave, 2000.

Mulgan, Aurelia George. "Japan's 'Un-Westminster' System: Impediments to Reform in a Crisis Economy." *Government and Opposition* 38, no. 1 (2003): 73–91.

Mulgan, Aurelia George, and Australian National University. Asia Pacific School of Economics and Management. *Japan's Failed Revolution: Koizumi and the Politics of Economic Reform*. Canberra: Asia Pacific Press, Asia Pacific School of Economics and Management, 2002.

Naito, Yosuke. "Bureaucrats Victors as Diet Draws to Close." *Japan Times*, July 27, 2002

Nakano, Jane. "Japan to Invest in the Latest Russian LNG Project." Center for Strategic and International Studies, July 12, 2019. https://www.csis.org/analysis/japan-invest-latest-russian-lng-project (Accessed Nov. 22, 2022).

National Assembly of the Republic of Korea. "History." http://www.assembly.go.kr/renew10/asm/ifn/yat_04.jsp.

"Necessary Minimum Regulation, Keidanren Response Committee Finalized Opinion." *Asahi Shinbun*, Feb. 22, 1962.

"New Right Confederation Announced the Support for Lee Myung-Bak." *Yonhap News*, Nov. 28, 2007.

"Newsbrief: Japan." *Platt's Oilgram News*, Nov. 18, 1997.

"Nippon Mitsubishi Oil Head Urges Gov't to Keep JNOC Functions." *Kyodo News*, July 18, 2001.

Nippon Oil. *Nihonsekiyu Hyakunenshi [100-Year History of Nippon Oil]*. Tokyo: Nippon Oil, 1988.

"No Need of Tax Input in Oil Development, President of Idemitsu Showed Real Intention Publicly (Osaka)." *Asahi Shinbun*, Sept. 22, 2001.

NS Energy. "Abadi project." https://www.nsenergybusiness.com/projects/abadi-lng-project/ (Accessed Nov. 13, 2022).

O, Won-chol. "Wonnyukonggŭpkyosŏp Ch'ulchangboko [Debrief on the Result of Official Trip for Crude Oil Negotiation]." Edited by Office of the President of Korea Secretariat. Seoul, 1973.

Obayashi, Yuka. "Japan's 2021 LNG Imports Slip, Gives Up World's Top Buyer Spot to China." *Reuters*, Jan. 20, 2022.

O'Connor, Kaori, and Mitsubishi Sh*oji Kabushiki Kaisha. *Japanese Business Language: An Essential Dictionary*. London KPI: Distributed by Routledge & Kegan Paul, 1987.

O, Won-chol. *Enŏchi Chŏngch'aekkwa Chungdongjinch'ul [Korean Style Economic Development: Energy Policy and Going Middle East]*. Vol. 6, Hankukhyŏng Kyŏngjekŏnsŏl [Korean Style Economic Development]. Seoul: Kia Economic Institute, 1997.

O, Won-chol. *The Korea Story: President Park Jung-Hee's Leadership and the Korean Industrial Revolution*. 1st. ed. Seoul, Korea: Wisdom Tree, 2009.

O'Reilley, Robert F. "Veto Points, Veto Players, and International Trade Policy." *Comparative Political Studies* 38, no. 6 (2005): 652–75.

Oak, Dong-suk, and Young-jin Ham. "Institutional Framework of Ministries for Fiscal Policy in Korea." *Korean Review of Orgarnizational Studies* 6, no. 1 (2009): 195–231.

Office of Legacy Management, Department of Energy. "Timeline of Events: 2013." https://www.energy.gov/lm/timeline-events-2013 (Accessed Nov. 22, 2022).

Office of the President. "Korea Opens the Door to UAE Abu Dhabi Oil Field in 37 Years." *Press Release*, Mar. 13, 2011.

Office of the President. "President Lee Wins International Prize, Oil Deal." *Press Release*, Mar 15, 2011.

Oh, Se-young. "Moon Administration had neglected Overseas Resource Development, but now Changed Stance on Resource Security." *Energy Economy*, Dec. 14, 2021.

Oh, Young-hwan. "Seosomun Forum: Korean Politics Is Really Following Japan?" *Korea JoongAng Daily*, June 11, 2012.

"Oil Fund Will Be Established." *Maeil Economic Daily*, Feb. 11, 1976.

"Oil Shingikai Supported Crude Oil Corporation for Overseas Resources Development." *Asahi Shinbun*, Aug. 21, 1965.

Okimoto, Daniel I. *Between MITI and the Market: Japanese Industrial Policy for High Technology*. Stanford, Calif.: Stanford University Press, 1989.

"Organizational Inflation Accumulating." *Kyung-hyang Shinmun*, Dec. 14, 1977.

Osaka Gas Group. *Osaka Gas Group Annual Report 2011*. Osaka: Osaka Gas Group, 2011.

Paik, Keun Wook. *Sino-Russian Oil and Gas Cooperation: The Reality and Implications*. Oxford: Oxford University Press, 2012.

"PAJ Did Not Support Miti's Crude Oil Corporation at the Meeting." *Asahi Shinbun*, July 28, 1965.

Park, Chung-hee. *Korea Reborn: A Model for Development*. Englewood Cliffs, NJ: Prentice-Hall, 1979.

Park, Chung-hee. *Our Nation's Path; Ideology of Social Reconstruction*. [2nd ed. Seoul: Dong-A Pub. Co., 1964.

Park, Dong-suk. "Thank You, Minister Youn Jin-sik!" *E-Daily*, Jan. 16, 2008.

Park, Jong-wook. "Lee Byung-Ho, CEO for Trading and Business in STX." *Maeil Economic Daily*, July 6, 2009.

Park, Kyung-mee. "Party Nomination and Organization in the 18th General Election: The Grand National Party and the Democratic Party." *Korean Party Studies Review* 7, no. 2 (2008): 41–63.

Park, Soon-wook, Kim, Hŭng-soo. "Skyrocketing Oil Price Hurts Korean Economy?" *Chosun Ilbo*, Jan. 4, 2008.

Park, Sung-gyu. "MKE Announced without Knowing Details of Oil Deal." *Newsis*, Mar. 13, 2011.

Park, Yang-soo. "EMRD Became Common Eyesore. In This Week Launched. Without New Functions, It Would Be Redundant." *Munhwa Ilbo*, Mar. 10, 2008.

Park, Yong-bum. "In Order to Avoid the Second Failure of 'Addax'" *Maeil Economic Daily*, July 15, 2009.

PEDCO. *Han'guksŏkyukaepalgongsa Sibonyŏnsa 1979–1994 [15 Year-History of PEDCO, 1979–1994]*. Seoul: PEDCO, 1994.

PEDCO Law. Dec. 5, 1978.

"PEDCO Will Be Established in 1975." *Maeil Economic Daily*, Jan. 31, 1975.

People's Solidarity for Participatory Democracy (PSPD). *PSPD Issue Report: First 2 years of Moon Jae-in Administration, How Much is Achieved in Removing Deep-rooted Corruption*. https://www.peoplepower21.org/pspd_press?mod=document&uid=1642054 (Accessed Oct. 2, 2022).

Persily, Larry. "Japan Eyes Support for US LNG Project Expansions to Shore Up Its Energy Needs." *Alaska Journal of Commerce*, May 17, 2022.

"Personals." *Oil & Gas Journal*, Oct. 4, 1982.

PERTAMINA. "PERTAMINA Operator Blok WMU." *Press Release*, May 9, 2011.

Petroelum Association of Japan. *Sengo Sekiyu Sangyōsi [Postwar History of Oil Industry]*. Tokyo: Japan Petroleum Association, 1985.

Petroleum and Contracts Licensing Directorate. "Iraq's First Petroleum Licensing Round: Zubair Contract Area–Bidding Results." Edited by Republic of Iraq Ministry of Oil. Baghdad, 2009.

Petroleum and Contracts Licensing Directorate. "Iraq's Second Petroleum Licensing Round: Badra Contract Area–Bidding Results." Edited by Republic of Iraq Ministry of Oil. Baghdad, 2009.

Petroleum and Contracts Licensing Directorate. "Iraq's Second Petroleum Licensing Round: Garraf Contract Area-Bidding Results." Edited by Republic of Iraq Ministry of Oil. Baghdad, 2009.

Petroleum and Contracts Licensing Directorate. "Iraq's Third Petroleum Licensing Round: Akkas Contract Area–Bidding Results." Edited by Republic of Iraq Ministry of Oil. Baghdad, 2010.

"Petroleum Development Corporation Launched Earlier." *Maeil Economic Daily*, Jan. 30, 1978.

"Petroleum Industry Law and Others Are Informally Decided to Be Submitted to Coming Diet Session." *Asahi Shinbun*, Mar. 6, 1962.

"Petroleum Industry Law is against Liberalization. Steel Industry Decided to Oppose." *Asahi Shinbun*, Feb. 16, 1962.

"Petroleum Industry Law Needed after Liberalization." *Asahi Shinbun*, Dec. 15, 1961.

"Petroleum Industry Law Will Be Submitted to Regular Diet Session." *Asahi Shinbun*, Nov. 25, 1961.

"Petroleum Institution Will Be Established." *Maeil Economic Daily*, Jan. 17, 1976.

"Placid Full-Scale Exploration in the Dream of Oil Rich Nation." *Dong-A Ilbo*, Jan. 16, 1976.

Policy Planning Committee. *Ch'amyŏjŏngbu Chŏngch'aek Pokosŏ 2–16: Haeoejawŏn Hwakpo Kyŏngjaeng [Roh Moo-Hyun Administration White Paper 2–16: The Competition to Secure Foreign Resources Supply]*. Seoul: Policy Planning Committee, 2008.

"Pressure to Kantei, LDP PARC's Hearing of CEFP." *Asahi Shinbun*, June 6, 2001.

"Privatization in Korea: Against the Tide." *The Economist*, Oct. 2, 2008.

"Proposal of Businesses for Petroleum Industry Law, Identical among Oil, Coal and Electricity." *Asahi Shinbun*, Feb. 15, 1962.

"Quick Survey 2: Names of DPJ Members Not Resisting Koizumi's Reform." *Weekly Asahi*, June 22, 2001.

"Quick Survey: Names of LDP Members Resisting Koizumi's Reform." *Weekly Asahi*, June 15, 2001.

"Reason to Pass Late in the Night." *Kyung-hyang Shinmun*, Dec. 17, 1974.

"Representative of Keidanren Proposed to MITI Minister on Petroleum Industry Law." *Asahi Shinbun*, Feb. 28, 1962.

"Resolved Political Storm by Pursuing for Desire of People General." *Kyung-hyang Shinmun*, Oct. 13, 1987.

"Resource Agency is Considered to Sever Energy Issues from MTI." *Kyung-hyang Shinmun*, Aug. 26, 1977.

"Resources Response Committee Established." *Maeil Economic Daily*, July 12, 1973.

Reuters Staff. "Japan's JBIC to Provide loan to Novatek-led Yamal LNG Project." *Reuters*, Nov. 24, 2016

Reuters Staff. "Japan Passes Law to Allow JOGMEC to Invest in Foreign Oil, Gas Firms." *Reuters*, Nov. 11, 2016. https://www.reuters.com/article/us-japan-jogmec-idUSKBN1360UC (Accessed Aug. 15, 2022).

"Role of Japan National Oil Necessary: METI's Hirose." *Kyodo News*, June 14, 2001.

Ryu, Si-hoon. "Approach to North American Company That Has Negative Feeling Towards China Works." *Korea Economic Daily*, Oct. 22, 2009.

Sachs, Jeffrey. *Developing Country Debt and the World Economy*, A National Bureau of Economic Research Project Report. Chicago: University of Chicago Press, 1989.

"Sacked Oil Agency Chief Returns Quarter of Allowance." *Japan Energy Scan*, Sept. 6, 1999.

SaKong, Il. *Korea in the World Economy*. Washington, DC: Institute for International Economics, 1993.

Samjong KPMG. "GS Caltex and Its Subsidiary: Audit Report on Consolidated Financial Statements." Seoul, 2011.

Samuels, Richard J. *The Business of the Japanese State: Energy Markets in Comparative and Historical Perspective*, Cornell Studies in Political Economy. Ithaca: Cornell University Press, 1987.

Sartori, Giovanni. *Comparative Constitutional Engineering: An Inquiry into Structures, Incentives, and Outcomes*. New York: New York University Press, 1994.

Sartori, Giovanni. *Parties and Party Systems: A Framework for Analysis*. Cambridge [Eng.]: Cambridge University Press, 1976.

Scalapino, Robert A., and Social Science Research Council (U.S.). *The Foreign Policy of Modern Japan*. Berkeley: University of California Press, 1977.

Schlesinger, Arthur M. *The Imperial Presidency*. Boston: Houghton Mifflin, 1973.

Schumpeter, Joseph Alois. *The Theory of Economic Development: An Inquiry into Profits, Capital, Credit, Interest, and the Business Cycle*, Social Science Classics Series. New Brunswick, NJ: Transaction Books, 1983.

Schumpeter, Joseph Alois, Vittorangelo Orati, and S. Bhagwan Dahiya. *Economic Theory in the Light of Schumpeter's Scientific Heritage*. 3 vols. Rothak, India: Spellbound Publications, 2001.

Schwartz, Frank J. *Advice and Consent: The Politics of Consultation in Japan*. Cambridge, UK; New York: Cambridge University Press, 1998.

"Sekiyu Renmei Criticized the Reply of Energy Committee." *Asahi Shinbun*, Feb. 25, 1967.

Seo, Kyung-ho. "Dominant Power in Economic Policy-Making Transferred from Government to Legislature." *Korea JoongAng Daily*, May 28, 2012.

Shepsle, Kenneth A. "Institutional Arrangements and Equilibrium in Multidimensional Voting Models." *American Journal of Political Science* 23, no. 1 (1979): 27–59.

Shin, Hyun-don. "National Energy Security Has Been Abandoned for a Decade." *Gas News*, July 15, 2022.

Shin, Sang-gil. "Enŏjipumun Chŏnmunillyŏgŭi Hyoyulchŏgin Hwakpopangan Yŏnku [Study on Efficient Securing of Experts in Energy Sector]." Kyunggido: Korea Energy Economics Institute, 1998.

Shin, Yong-oak. "Sovereignty Principle and Economy Orders of the Constitution of Korea Made in 1948." *The Journal for the Studies of Korean History* 17 (2004): 209–39.

Shinoda, Tomohito. *Koizumi Diplomacy: Japan's Kantei Approach to Foreign and Defense Affairs*. Seattle: University of Washington Press, 2007.

Shugart, Matthew Soberg. "Leaders, Rank and File, and Constituents: Electoral Reform in Colombia and Venezuela." *Electoral Studies* 11, no. 1 (1992): 21–45.

Shugart, Matthew Soberg, and Martin P. Wattenberg. *Mixed-Member Electoral Systems: The Best of Both Worlds?*, Comparative European Politics. Oxford, UK New York: Oxford University Press, 2001.

Sim, Ji-yŏn. "Park Chung-Hee Chŏngbuhaŭi Chŏngdangkutopunsŏk(2): 1972–1979 [The Analysis of Party Formation During Park Chung-Hee Administration (2): 1972–1979]." *Korean Party Studies Review* 1, no. 1 (2002).

SK corporate culture office. *SK 50nyŏn: P'aekiwa Chisŏngŭi Yŏjŏng [SK Group's 50 Year: The Journey of Aspiration and Intelligence]*. Seoul: SK corporate culture office, 2006.

SK Innovation. *2010 Annual Report: Innovate Change for Tomorrow's Energy*. Seoul: SK Innovation, 2011.

SK Innovation. "Presentation to Investors." Seoul, 2012.

SK Innovation. "SK Energy Makes Strategic Oil Development Model through Restructuring Portfolio of Oil Field." *Press Release*, Dec. 24, 2010.

"Sŏgyukaepalkongsabŏp [Korea Petroleum Development Corporation Law]." Edited by Ministry of Energy and Resources, 1978.

Sohn, Chŏl. "Controversy on Kogas's Using the Benefit from Equity Investment to Lower Gas Retail Prices." *Seoul Economic Daily*, Sept. 22, 2005.

Sohn, Hyun-duk. "Prepare for Era of $100 of Oil Price." *Maeil Economic Daily*, July 18, 2004.

Special Corporation Reform Headquarters. "Tokushu Hojin Nado Seiri Gōrika Keikaku [Plan for Disposal and Rationalization of Special Corporations]." Tokyo: Special Corporation Reform Headquarter, 2001.

Statistics Korea. "Status of Overseas Resources Development." https://www.index.go.kr/potal/main/EachDtlPageDetail.do?idx_cd=1167 (Accessed Nov. 11, 2022).

"Stitched Confrontation among 3 Coalition Partners, Discuss the Structure." *Asahi Shinbun*, Apr. 16, 2001.

Stockwin, J. A. A. *Governing Japan: Divided Politics in a Resurgent Economy*. 4th ed. Malden, Mass.: Blackwell Pub., 2008.

Stockwin, J. A. A. "The Japanese Socialist Party under New Leadership." *Asian Survey* 6, no. 4 (1966): 187–200.

Stockwin, J. A. A. "Reshaping of Japanese Politics and the Question of Democracy." *Asia-Pacific Review* 9, no. 1 (2002): 45–59.

Subcommittee on JNOC Asset Evaluation & Liquidation Review. "Sekiyukōdanga Hoyūsuru Kaihatsukanrenshisan No Shorini Kansuru Hōshin [Plan Related to Disposal of Development-Related Asset the JNOC Possesses]." Edited by Comprehensive Resource & Energy Examination Committee. Tokyo, 2003.

Suh, Je-jin. "Formation and Characteristics of Class Organization of Korean Capitalist Class: Mainly on the FKI." *Society and History* 16 (1989): 169–97.

Suleiman, Ezra N. *Bureaucrats and Policy Making: A Comparative Overview*. New York: Holmes & Meier, 1984.

Sumitomo Corporation. *Annual Report 2011*. Tokyo: Sumitomo Corporation, 2011.

Sumitomo Corporation. "Our Business: Mineral Resources, Energy, Chemical & Electronics Business Unit." http://www.sumitomocorp.co.jp/english/business_overview/resource_chemical/outline15b.html.

Sumitomo Corporation. "Share Sale of Petro Summit Investment UK Ltd." *News release*, Nov. 13, 2009.

Sung, Dong-won. "A Comparative Study on the Participation of Policy Financial Institutions in Project Finance for Overseas Resource Development: The Case of Korea and Japan." Thesis (Ph.D.), Korea University, 2021.

Sung, Han-yong. "President's Power to Nominate Party Candidate is the Retreat of Party Democracy." *The Hankyoreh*, Dec. 23, 2007.

Suzuki, Shigeru. *Nihon No Enerugi Kaihatsu Seisaku [Japan's Energy Development Policy]*. Tokyo: Mineruba shobō, 1985.

Swedberg, Richard, and Ola Agevall. *The Max Weber Dictionary: Key Words and Central Concepts*. Stanford, Calif.: Stanford Social Sciences, 2005.

"Taehanmin'guk Hŏnbŏp [The Constitution of Republic of Korea]." Edited by The Constitution Delegation Committee, 1962.

"Taehanmin'guk Hŏnbŏp [The Constitution of the Republic of Korea]." Edited by The Committee for Drafting Constitution and Government Organization Law, 1948.

Takenaka, Kiyoshi. "Talks on Long-Delayed Siberia Pipeline Begin." *Reuters*, Jan. 14, 1998.

Tarō, Shōji. *Yamashita Tarō to Hinomaru Genyu [Mr. Yamashita Taro and Hinomaru Crude Oil]*. Tokyo: Enerugi Horamu, 2007.

"The Bulldozer: A President-Elect Starts to Roll." *The Economist*, Jan. 17, 2008.

The Bureau of Government Legislation. "Organization of Ministry of Energy and Resources." Seoul, 1977.

The Platform of Democratic Party. Oct. 3, 2010.

The Platform of Grand National Party. Jan. 9, 2006.

"The Term of Party Chair Extended. Party Constitution Changed." *Kyung-hyang Shinmun*, Oct. 4, 1967.

Thayer, Nathaniel B. *How the Conservatives Rule Japan*, Studies of the East Asian Institute, Columbia University. Princeton, NJ: Princeton University Press, 1969.

"The Import of Developed Coal Will Be Focused, Energy Minister Jang Reported. Spr Will Also Be Expanded." *Maeil Economic Daily*, Feb 5, 1979.

The 17th Presidential Transition Committee. *Che Sipch'iltae Taet'ongnyŏngjik Insuwiwŏnhoe Paeksŏ: Sŏnggong Kŭriko Nanum [The White Paper of the 17th Presidential Transition Committee: Success and Sharing]*. 2 vols. Vol. 2. Seoul: The 17th President Transition Committee, 2008.

The 17th Presidential Transition Committee. *Che Sipch'iltae Taet'ongnyŏngjik Insuwiwŏnhoe Paeksŏ: Sŏnggong Kŭriko Nanum [The White Paper of the 17th Presidential Transition Committee: Success and Sharing]*. 2 vols. Vol. 1. Seoul: The 17th President Transition Committee, 2008.

The Constitution of Japan. 1st.

The Democratic Party of Japan. "Members Profile." http://www.dpj.or.jp/english/member/?detail_4032=1.

The National Assembly. "The Current Status of the National Assembly Members." http://www.assembly.go.kr/renew10/mem/mem/mem_search.jsp.

The National Assembly of the Republic of Korea. "Assembly Introduce: The Seats and Votes of Respective Parties." http://www.assembly.go.kr/renew10/asm/ifn/yat_04.jsp.

The National Assembly Secretariat. "Budget Review Report for the Fiscal Year of 1976." Seoul: the National Assembly Secretariat, 1975.

Treaty Bureau of Ministry of Foreign Affairs and Trade, ed. *Tongbuga Haeyangbŏmnyŏnggwa Yuenhaeyangbŏphyŏbyakchip [Maritime Law of Northeast Asia and United Nations Convention on Law of the Sea]*. 1 vols. Vol. 1. Seoul: Ilchogak, 2006.

Tsebelis, George. "Decision Making in Political Systems: Veto Players in Presidentialism, Parliamentarism, Multicameralism and Multipartyism." *British Journal of Political Science* 25, no. 03 (1995): 289–325.

Tsebelis, George. *Veto Players: How Political Institutions Work*. Princeton, NJ: Princeton University Press, 2002.

Uchiyama, Yu. *Koizumi and Japanese Politics Reform Strategies and Leadership Style*. Hoboken: Taylor & Francis, 2010.

Uhm, Nam-sok. "FKI Asked the Support for Oversea Resources Development." *Yonhap News*, June 23, 2004.

Uriu, Robert M. *Troubled Industries: Confronting Economic Change in Japan*, Cornell Studies in Political Economy. Ithaca, N.Y.: Cornell University Press, 1996.

"US Oil Capitals Watch Petroleum Industry Law Bill with Strong Interest." *Asahi Shinbun*, Feb. 17, 1962, 4.

Watanabe, Mika. "Japan's MITI Submits National Oil Restructuring Plan to Diet." *Dow Jones Newswire*, July 28, 1998.

Watanabe, Mika. "Japan Oil Chief: Indus Still Needs Japan Natl Oil Corp. Aid." *Dow Jones Newswires*, July 15, 1998.

Watanabe, Mika. "JNOC/Debt/Japan Oil Investments -3: Govt Role Needed." *Dow Jones Newswires*, June 10, 1998.

Waterbury, John. *Exposed to Innumerable Delusions: Public Enterprise and State Power in Egypt, India, Mexico, and Turkey*, Political Economy of Institutions and Decisions. Cambridge England; New York, NY, USA: Cambridge University Press, 1993.

Weaver, R. Kent, and Bert A. Rockman. *Do Institutions Matter?: Government Capabilities in the United States and Abroad*. Washington, DC: The Brookings Institution, 1993.

"Widening the Gap between Kantei and LDP, Mr. Aoki's Nail." *Asahi Shinbun*, May 30, 2001.

"Will Ministry of Finance under Pressure Accept New Public Corporation." *Asahi Shinbun*, Dec. 19, 1966.

"Withering Koizumi Cabinet Having Historical Mission Takes Risk of Life by Dissolving Diet in July." *Asahi Shinbun*, Apr. 29, 2002.

World Bank Open Data. https://data.worldbank.org/indicator/BN.GSR.MRCH.CD?end=2021&locations=JP&start=1999&view=chart (Accessed Dec 4, 2022).

Wolferen, Karel van. "The Japan Problem." *Foreign Affairs* 65, no. 2 (1986): 288–303.

Wolferen, Karel van. *The Enigma of Japanese Power: People and Politics in a Stateless Nation*. 1st Vintage Books ed. New York: Vintage Books, 1990.

Wolferen, Karel van. *The Enigma of Japanese Power: People and Politics in a Stateless Nation*. 1st ed. New York: A.A. Knopf, 1989.

Woo-Cumings, Meredith. *Race to the Swift: State and Finance in Korean Industrialization*, Studies of the East Asian Institute. New York: Columbia University Press, 1991.

Wybrew-Bond, Ian, Jonathan P. Stern, and Oxford Institute for Energy Studies. *Natural Gas in Asia: The Challenges of Growth in China, India, Japan and Korea*. Oxford; New York: Oxford University Press, 2002.

Yanaga, Chitoshi. *Big Business in Japanese Politics*, Yale Studies in Political Science 22. New Haven: Yale University Press, 1968.

Yonhap News. "Democratic Party Saved Hong Joon-Pyo?" *Chosun Ilbo*, Sept. 17, 2008.

Yonhap News. "Parties Agreed on Supplementary Budget Rapidly." *Chosun Ilbo*, Sept. 17, 2008.

Yoon, Jung-ho. "Revised Bill for Sejong City Rejected in the National Assembly: 105 vs. 164." *Chosun Ilbo*, June 30, 2010.

Yoshitake, Kiyohiko. *An Introduction to Public Enterprise in Japan*. Beverly Hills Calif.: Sage Publications, 1973.

You, Hoon, Yong-su Bae, and Won-hui Lee. *Konggiŏmnon [Public Enterprise]*. Kyunggido: Pŏmmunsa, 2010.

You, Jae-joon. "Overseas Resource Development as Mid and Long-term Policy Should Be Implemented Consistently: Budget for Success Repayable Loan was Removed in the Next Year's Budget." *Gas News*, Nov. 21, 2015.

You, Youn-jong. "Today's Bookmark: Kim Dae-Jung and Kim Jong-P'il Agreed on Single Presidential Candidate in 1997." *Dong-A Ilbo*, Nov. 2, 2004.

"Yujŏnghoe Asked Gov't to Establish PEDCO." *Kyung-hyang Shinmun*, Jan. 17, 1976.

Index

S.-i. Oh, *Overseas Energy Investment of Korea and Japan*, https://doi.org/10.1007/978-981-99-0285-9

C

D

E

Printed by Printforce, the Netherlands